YACHTMASTER: OFFSHORE & OCEAN

CW01639200

YACHTMASTER
OFFSHORE & OCEAN

A Guide to the Navigational and Meteorological Requirements of the R.Y.A./D.T.I. Yachtmasters' Certificates

BY

PETER CLISSOLD
MASTER MARINER
FELLOW OF THE ROYAL INSTITUTE OF NAVIGATION

Yachtmaster: Offshore & Ocean

ERRATA

PAGE 44 Question 9. *bears* 037° *not* 057°.
Question 10. From the position in Question 9 *not* 8.
Question 12. last words. *Omit* in 8.

GLASGOW
BROWN, SON & FERGUSON, LTD., NAUTICAL PUBLISHERS
52 DARNLEY STREET, G41 2SG

Copyright in all Countries signatories to the Berne Convention
All rights reserved

First Edition – 1968
Second Edition – 1977

ISBN 0 85174 261 0

© 1977 BROWN, SON & FERGUSON, LTD., GLASGOW, G41 2SG

Printed and Made in Great Britain

Dedicated to

'They that go down to the sea in ships
And . . . see . . . His wonders in the deep.'

PREFACE

'YACHTMASTER: OFFSHORE & OCEAN' is designed especially to cover the navigational and meteorological requirements of the Royal Yachting Association/Department of Trade & Industry Yachtmaster examinations. Mathematics have been kept to a minimum, but sufficient theory is included to explain the methods used. For the seamanship and signals requirements, the student is referred to books on Seamanship: *Basic Seamanship*, by the same author and publisher as *Yachtmaster*: *Offshore & Ocean*, covers the syllabus.

The student is recommended to work at least some of the many exercises given throughout the book, comparing his answers with those given at the end, to consolidate his grasp of the theory and method. Sufficient extracts from the *Nautical Almanac*, *Admiralty Tide Tables*, etc., are given to allow this to be done, but a book of Nautical Tables is also required for many of the problems. Any of the standard Nautical Tables: *Bowditch's* (U.S.), *Burton's*, *Inman's*, *Norie's* are suitable, and *Reed's Nautical Almanac* (of any year) also contains sufficient tables. Rich's small *Traverse and Other Tables* is suitable for the Offshore syllabus only. Burton's *Four-Figure Navigation Tables* are recommended as convenient and sufficient (and are referred to throughout this book as *Burton's*). As they do not include the Table of Amplitudes, that table is reproduced here by permission from Burton's larger *Nautical Tables*. It can be cut out and pasted in the Four-Figure Tables.

Two practice charts are included with the book. They cannot be corrected up to date and, therefore, should not be used for practical navigation. For Chart Work the following instruments are required: a soft pencil and soft rubber. Parallel rulers (the transparent type, marked from 0° to 360°, are best) are used by most professional navigators, but some prefer a pair of set squares, or for use in small craft, a Douglas (square) protractor and straight edge. A pair of compasses. Useful extras: a pair of dividers; a protractor.

Further Reading

The Complete Coastal Navigator. C. H. Cotter, F.R.I.N.
Hollis & Carter.

The Complete Nautical Astronomer. C. H. Cotter, F.R.I.N.
Hollis & Carter.

The Art of Astronomical Navigation. S. M. Burton, F.R.I.N.
Brown, Son & Ferguson.

Compasses in Small Craft. Cmdr. C. A. Luard, R.N.
Brown, Son & Ferguson.

Radar in Small Craft. P. Clissold, F.R.I.N.
Thomas Reed Publications.

Meteorology. C. W. Roberts, F.R.Met.S. M.R.I.N.
Thomas Reed Publications.

The Haven-Finding Art. A History of Navigation from Odysseus to Captain Cook. E. R. G. Taylor, Hon.M.R.I.N.
Hollis & Carter.

ACKNOWLEDGEMENTS

FOR the photography of Station Pointer, Sextant, Compass and Azimuth Mirrors I am most grateful to the makers, Messrs. Kelvin Hughes Division, Henry Browne & Son, Ltd., and East Berks Boat Co.

For the Map of Paths of Tropical Cyclones I am grateful to Mr. David Niddrie and the editor of *The Geographical Magazine* in which it first appeared. The Form for taking down Weather Bulletins is reproduced from *The Nautical Magazine* by kind permission of Lt.-Commander L. B. Philpott, D.S.C., R.N.R., and the Editor. The Amplitude Table is reproduced from *Burton's Nautical Tables* by kind permission of the author and publisher.

I am indebted to the Controller of H.M. Stationery Office and the Hydrographer of the Navy for permission to reproduce the Chart Symbols and Abbreviations No. 5011 and Areas used in Weather Forecasts; to the Controller of H.M.S.O. and the Liverpool Observatory and Tidal Institute for extracts from the *Admiralty Tide Tables*; to the Controller of H.M.S.O. for extracts from *List of Radio Signals*, Vol. V; and for the extracts from *The Nautical Almanac*, and I am grateful to the Royal Yachting Association for permitting me to reproduce the syllabus for Yachtmaster Offshore and Yachtmaster Ocean.

Antony Butler has kindly allowed me to draw on his expertise in Preparing for a Cruise, and Geoffrey Fanner briefed me about Navigating High-Speed Craft. I have, too, benefitted from the comments of many students whose combined experiences greatly exceeds my own. Charles Roberts collaborated with me in writing our earlier *Yachtmaster Coastal*, and much of his work is gratefully included in its successor.

P.C.

CONTENTS

OUTLINE OF THE YACHTMASTERS' CERTIFICATE SCHEME

THE scheme of training consists of shore-based and practical courses which, together with periods gaining practical experience at sea, enable the newcomer to offshore sailing to achieve the standard of proficiency required for the award of a Yachtmasters' Certificate.

The Yachtmasters' (Offshore) and Yachtmasters' (Ocean) Certificates are awarded by the Royal Yachting Association with the sanction of the Department of Trade and Industry to candidates who have proved their ability during shore-based and practical courses of instruction and satisfy the requirements as to seatime.

The R.Y.A.—D.T.I. Yachtmasters' (Offshore) Certificate.

This Certificate provides for the cruising yachtsman who wishes to take charge of any offshore yacht. The syllabus contains a comprehensive, practical nautical education.

The course leading to this award includes shore-based instruction in navigation and meteorology, practical instruction in yacht handling and seamanship and, before the award of a certificate, candidates must have acquired considerable seatime and practical experience in cruising yachts.

Offshore Certificate Training Scheme.

The Offshore Certificate Training Scheme is divided into three parts:

1. *Shore-Based Course.*

   This is a course of instruction in yacht navigation, meteorology, and general seamanship, designed for use by a recognised Further Education Establishment during a session of evening classes which provide a minimum of 48 hours of formal instruction.

2. *Practical Course.*

   This is a course of instruction in practical seamanship and yacht handling. The course will last six days, but may be completed in three weekends rather than six consecutive days.

3. *Seatime and Final Examination.*

   Before undergoing the Final Examination for a Yachtmasters' (Offshore) Certificate, candidates must have completed a specified amount of seatime and have kept a Log Book to provide a record of their experience.

SEATIME

Before examination for a Yachtmaster's Certificate a candidate must have acquired some experience at sea. This experience must be recorded in the Log Book and must include the following:

1. **20** *days on board a cruising yacht as an active member of the crew.*

   For the purpose of this requirement a day is defined as a period of 24 consecutive hours. Parts of days may not be included in the total, but a day is not invalidated by a candidate leaving the yacht for a few hours during a cruise.

2. **500** *miles logged in a cruising yacht at sea.*

   At sea means outside any harbour, natural or artificial, in which a cruising yacht could secure or anchor for a prolonged period.

3. **12** *hours at sea by night as an active member of the crew of a cruising yacht.*

   Night is defined as the period between sunset and sunrise.

   Each entry in the Log Book must be signed by the skipper of the yacht in which the cruise has been made.

   Candidates claiming experience during which they are themselves acting as skipper may sign the entries themselves.

   Candidates for the full certificate must complete this seatime in sailing yachts.

   Candidates for certificates endorsed 'Power-driven craft only' must complete a majority of the seatime requirements in motor yachts.

Special Provisions for Motor Yachtsmen.

Both certificates may be issued to yachtsmen who go to sea in sailing or motor yachts. Those who intend to go to sea only in motor yachts may omit the sections of the practical course which relate only to sailing craft, and their certificate will be endorsed 'Power-driven craft only'.

Highly-Experienced Yachtsmen.

Experienced yachtsmen who wish to qualify for the award of a certificate may be examined practically at sea. Candidates for this examination must be in possession of the following items, which they will be required to present to the Examiner:

1. Certificate of sea experience, signed by the Flag Officer of an R.Y.A. affilitated club on the appropriate form available from the R.Y.A.

2. Any other documents relating to the candidate's competence or experience (e.g. a yacht club's certificate of competence) should also be presented.

Experienced yachtsmen who feel that they should attend either a shore-based course or a practical course before undertaking the examination, in order to complete their knowledge, may do so.

Examinations will be conducted at R.Y.A. recognized examination centres and may involve the major part of a day at sea in a cruising yacht. Candidates will also be examined orally on their knowledge of all parts of the syllabus for the shore-based and practical course.

Candidates for this examination should apply to the R.Y.A. for a list of examination centres, and dates on which examinations are to be held, and they should then apply direct to the centre for examination.

Syllabus for the Yachtmaster's Offshore Certificate

Depth of Knowledge required: A=Thorough; B=Working; C=Outline

Item	*Subject*	*Broad detail to be covered*	
1.	Definition of Position Course and Speed	1. Latitude and Longitude	B
		2. Nautical mile and cable	B
		3. True bearings and courses	B
		4. The knot	C
2.	Navigational drawing instruments	1. Parallel rulers	B
		2. Dividers and compasses	B
		3. Proprietary plotting instruments	B
3.	Navigational charts and publications	1. Suppliers—Admiralty, Stanford, etc.	B
		2. Information shown on Admiralty charts	B
		3. Chart symbols—Chart 5011	B
		4. Standard Chartwork	B
		5. Projections—Mercator and gnomonic	C
		6. Navigational publications in common use	C
		7. Chart correction	B
4.	Dead reckoning and Estimated Position	1. Definition of D.R. and E.P.	A
		2. Working up D.R. and E.P. by plotting on a chart	A
5.	The position line	1. Sources of position lines	B
6.	Position fixing	1. Techniques of visual fixing	A
		2. Horizontal angle fixing	B
		3. Running fixes	A
		4. Radio fixes	B
		5. Fixes containing a mixture of position lines	B
		6. Deviation of position from a line of soundings	B
		7. Ranges by dipping distances	C
7.	Basic Coastal Navigation	1. Routine for navigating a yacht in coastal waters	A
		2. Strategy of course laying	B
8.	Tides	1. Causes of tides—springs and neaps	C
		2. Tide tables—Admiralty and Reeds	B
		3. Tidal levels and datums	B
		4. Times and heights at standard ports	B
		5. The rule of twelfths	B
		6. Corrections for secondary ports	B
		7. Tidal anomalies (Solent, etc.)	C
9.	Tidal Streams	1. Tidal stream atlas	B
		2. Tidal diamonds	B
		3. Tidal stream information in Sailing directions and Reeds	B
		4. Allowance for tidal streams in computing a course to steer	B
		5. Tide rips, overfalls and races	B
		6. Tidal stream observation by buoys, beacons, etc.	B

Item	*Subject*	*Broad detail to be covered*	
10.	Buoyage	1. Uniform systems—outline differences between lateral and cardinal	C
		2. The lateral system	A
		3. Limitations of buoys as navigational aids	C
11.	Lights	1. Characteristics	B
		2. Ranges—Visual, luminous and nominal ranges	C
		3. Light lists—Admiralty and Reeds	B
12.	Pilotage	1. Harbour regulations and control signals	B
		2. Method of pre-planning	B
		3. Clearing lines	A
13.	The magnetic compass	1. Allowance for variation. Change of variation with time and position	B
		2. Siting of compass and causes of deviation	C
		3. Deviation, Allowance for	B
		4. Steering and hand bearing compasses	B
		5. Swing for deviation (but not correction)	C
14.	Echo-sounders and lead line	1. Principle of operation	C
		2. Types available—Rotating noon, dial and pointer, recording	C
		3. Reduction of soundings	C
		4. 'Second trace' echoes	C
		5. Marking of lead line	C
15.	Radio-direction finding	1. Radio beacons. ALRS, Vol. II and Reeds	B
		2. D/F receivers. Types of aerials. Method of operation. Audio and visual nul points. Reciprocal bearings	B
		3. The consol system and lattice charts	C
16.	Logs (speed and distance measuring)	1. Types and principles of operation, comparison of the following types: (*a*) Towed (*b*) Pressure (*c*) Impeller	C
17.	Deck Log	1. Importance of log as yacht's official document	B
		2. Layout of log, hourly and occasional entries	B
18.	Meteorology	1. Basic met. terms, the Beaufort scale	B
		2. Air masses	B
		3. Cloud types	B
		4. Weather patterns associated with pressure and frontal systems	B
		5. Weather forecasts and simple met. instruments	B
		6. Ability to sketch a weather map from a shipping forecast	B
		7. Land and see breezes	B
		8. Sea-fog—forecasting	C
19.	Anchoring	1. Types of anchors and principles of of operation	B
		2. Selection of an anchor berth	B

Item	*Subject*	*Broad detail to be covered*	
20.	Rule of the Road	1. A sound knowledge of the international regulations for prevention of collision at sea, except seaplane regulations and detailed regulations for the construction and position of lights	A
21.	Safety at Sea	1. Personal safety, use of lifejacket, safety harnesses and lifelines	B
		2. Fire prevention and fire fighting	B
		3. Distress signals	B
		4. Coastguards	C
		5. First aid, contents of first aid kit and treatment of likely accidents and seasickness	B
		6. Action to be taken in heavy weather	B
22.	Signals	1. Ability to send and receive Morse code by light at slow speed, and basic knowledge of signalling procedure	B
		2. Ability to recognise the alphabetical flags and numeral pennants of the International Code of Signals	B
		3. Knowledge of the urgent and important meanings of single letters	B
		4. The law and customs relating to flags	C
23.	Navigation in restricted visibility	1. Precautions to be taken in fog	B
		2. Limitations to safe navigation imposed by fog	B
24.	Passage planning and making	1. Preparation of charts and notebook for route planning and use at sea	B
		2. Customs regulations as they apply to yachts	B
		3. M notices	B
		4. Chartwork exercises as they involve passage making in coastal waters	B

The R.Y.A.—D.T.I. Yachtmaster's (Ocean) Certificate is an additional qualification in Astro Navigation open only to holders of the Yachtmaster's offshore certificate.

SYLLABUS FOR THE SHORE-BASED COURSE FOR THE YACHTMASTER'S (OCEAN) CERTIFICATE

This certificate is an additional qualification which indicates that the holder has sufficient knowledge of Astro Navigation to navigate out of sight of land and outside the coverage of radio fixing aids.

The syllabus suggested for use by Further Education Establishments which intend to provide courses leading to the award of this certificate is based on the use of the Air Navigation Tables, and goes no more deeply into the theory of the subject than is absolutely necessary for candidates to understand how to use the tables.

Candidates for the Yachtmaster's Ocean examination will not be required to attend a course of instruction if they do not wish to do so, but those who do may be awarded a certificate of completion of the course.

Details of establishments which provide courses for the ocean certificate are obtainable from Local Education Authorities or from the R.Y.A.

Candidates who wish to be assessed for this certificate must submit to the R.Y.A. the following navigational records, compiled at sea in a yacht in open waters over a minimum period of 12 hours:

(*a*) A track chart showing an observed position derived from at least four position lines obtained from sights of stars, planets or the moon, and an observed position derived from a sun—run—meridian altitude.

(*b*) All workings associated with the planning and reduction of the above sights.

(*c*) The result of, and working associated with, a compass check by bearing of a heavenly body.

All written work submitted must be accompanied by an application form (available from the R.Y.A.) and by the examination fee, which is required to cover the cost of examination, marking of papers and issue of certificate.

Syllabus for the Yachtmaster's Ocean Certificate

Depth of Knowledge required: A=Thorough; B=Working; C=Outline

Item	*Subject*	*Broad detail to be covered*
1.	The earth and the celestial sphere	1. Definition of observer's zenith and position of a heavenly body in terms of latitude, longitude, GHA and declination 2. Right angle relationships, latitude and co-lat, declination and polar distance. Tabulation of declination in nautical almanac. 3. Relationships between GHA, longitude and LHA. The tabulation of GHA in the almanac. Rate of increase of hour angle with time.
2.	The *PZX* Triangle	1. The tabulated components of the triangle, LHA, Co-lat and polar distance. 2. The calculable components, zenith distance and azimuth. 3. Relationship between zenith distance and altitude. 4. Introduction to the tabular method of solution in the air navigation tables and the basic sight form.
3.	The sextant	1. Practical guide to the use and care of a sextant at sea. 2. Conversion of sextant altitude to true altitude, application of dip, index error and refraction. 3. Correction of side error, perpendicularity, index error and collimation error.
4.	Measurement of time	1. Definition of and relationship between GMT, LMT, standard time and zone time. 2. Rating of chronometers and watches.
5.	Meridian Altitudes	1. Forecasting of time of meridian altitude. 2. Reduction of meridian altitude sights.
6.	Sun sights	1. The reduction and plotting of sun sights using air navigation tables. 2. The plotting of a sun—run—meridian altitude.
7.	Planet sights	1. Identification of planets from LHA and declination. 2. Reduction of planet sights using air navigation tables (twilight sights only).
8.	Straight planning	1. Calculations of times of sunrise, sunset, civil and nautical twilights. 2. Star identification using air navigation tables.
9.	Star sights	1. Reduction of star sights using air navigation tables. (Seven selected stars only.) 2. Plotting of starsights. Corrections to position for precession and nutation.

Item	*Subject*	*Broad detail to be covered*
10.	The pole star	1. Determination of latitude from sights of polaris.
11.	Moon sights	1. Special corrections applicable only to the moon.
12.	Compass checking	1. Use of azimuth tables. (No mention need to be made of bearing amplitude.)
13.	Great circle sailing	1. The comparison of rhumb lines and great circles. 2. Vertices and composite tracks. 3. The computation of a series of rhumb lines approximating to a great circle by use of the gnomonic and Mercator projections.
14.	Passage planning	1. Publications available to assist with planning of long passages (routing charts and ocean passages of the world).
15.	Practical exercises	1. Revision exercises using all types of sights covered in the syllabus.

RYA/DTI
YACHTMASTERS
COMPLETION CERTIFICATE

To be signed by the examiner on completion of examination for Yachtmaster's (Offshore) Certificate.

Name of Candidate ....

Address..

....

has been examined at sea in the yacht.........................and* has demonstrated to my satisfaction that he has sufficient knowledge and practical experience to be fit to command a cruising yacht at sea. The certificate is to be endorsed 'Power-driven craft only'.**
The candidate can distinguish red, green and white.

Signature of Examiner ....

Notes.
This 'form' is not a Yachtmasters' Certificate. A Certificate will be issued by the R.Y.A. upon receipt of this entire booklet, duly completed where necessary as to Certificates and with this page completed and signed by the Examiner, provided that the applicant has satisfied any other requirements that may, from time to time, be required.

*Delete this line if candidate has completed oral examination only.

**Delete this sentence if the candidate has been examined for the full Certificate.

PART I

OFFSHORE NAVIGATION

'The Arte of Navigation demonstrateth how, by the shortest good way, by the aptest direction and in the shortest time, a sufficient ship . . . be conducted.'

John Dee, 1570.

CHAPTER 1

PRELIMINARY

Reference book needed for this chapter: *Burton's Nautical Tables*.

NAVIGATION is the art of taking a vessel safely from one place to another. Most of the problems in navigation consist in finding the answers to the two questions: 'Where are we?' and 'Where do we go from here?'

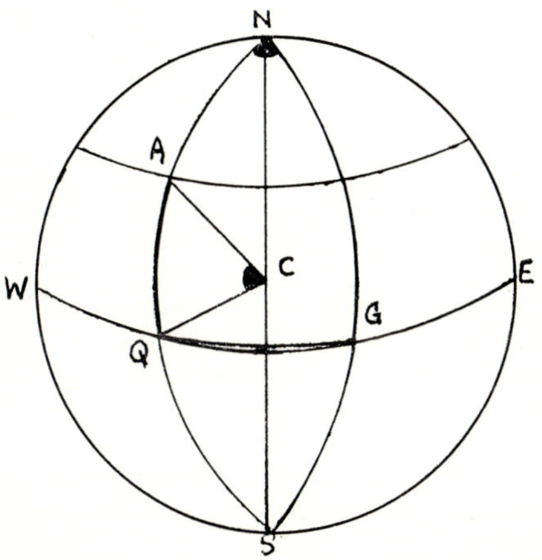

The figure represents the earth, with *C* its centre. *A* is a place on the earth's surface in North Lat. and West Long.

WQGE is the equator.

NWS, *NQS*, *NGS* and *NES* are all meridians, *NGS* is the meridian passing through Greenwich.

The arc *AQ*, or the angle *ACQ* is the Lat. of *A*.

The arc *QG*, or the angle *GNA* is the Long. of *A*.

The small circle through *A* is called a parallel of Lat. because it is parallel to the equator.

FIG. 1.1

To enable us to define our position when far from the land, the earth's surface is covered by a network of imaginary lines called a graticule. This is formed by the intersection of 'small' circles and 'great' circles, each of which is numbered for reference.

(*a*) *A Great Circle* is a circle on the surface of a sphere whose plane passes through the sphere's centre. The shortest distance between two points on the earth's surface is an arc of a great circle.

(*b*) *A Small Circle* is a circle on the surface of a sphere whose plane does not pass through the sphere's centre.

The Equator is a great circle midway between the earth's poles. Its plane is at right angles to the earth's axis.

Meridians are half-great circles joining the poles.

Parallels of Latitude are small circles whose planes are parallel to the plane of the equator.

Position. To describe position we need either two lines of reference or a single point of reference.

On the earth the lines of reference are (*a*) the equator and at right angles to it (*b*) the meridian passing through Greenwich, sometimes called the Prime Meridian.

The Latitude of a place is its distance north or south of the equator; more precisely the arc of a meridian or corresponding angle at the earth's centre, contained between the equator and the parallel of latitude through the place. It is measured in degrees (°), minutes (′) and seconds (″) of arc, from 0° to 90° north or south of the equator.*

The Longitude of a place is an arc of the equator or corresponding angle at the pole, contained between the prime meridian and the meridian of the place. It is measured from 0° to 180° east or west of the Greenwich meridian.

EXAMPLE: Lat. 50° 37′ 12″ N., Long. 0° 25′ 45″ E. or, since seconds are rarely used, Lat. 50° 37′·2 N., Long. 0° 25′·75 E. This position is 10 miles south-east of Beachy Head. Beachy Head could be taken as the point of reference, in which case the position may be given as '10 miles 135° from Beachy Head', or 'Beachy Head, bearing 315°, distance 10 miles.'

Depths. All newly-published charts will show depths (and heights) in metres, or metres and decimetres. Written thus: 5_7, indicates 5·7 metres. Older charts show depths in fathoms (heights in feet) or fathoms and feet. Written thus: 5_3, indicates 5 fathoms 3 feet. The units used on the chart are always shown near the title, and Admiralty charts showing metres have 'Depth in metres' printed in magenta on the bottom border. A conversion table-feet-metres-fathoms is also given. (*See* page 128.)

Depth Sounders or **Echo Sounders.** The depth of water can clearly be of great importance to the navigator; and depth sounders, constructed on the echo principle, are extremely useful. They show the depth either by a needle pointing on a dial, or a neon-light, which flashes at the depth of water, a type which can be more easily read at night. The dials of either sort may be had graduated in fathoms and feet—a switch permitting either to be used; or in metres.

* 60″=1′, 60′=1°

Principle. The speed at which sound-waves travel through water (4900 ft./sec.) is constant. Signals of such high frequency that the ear cannot detect them are transmitted to the sea bottom, and their echoes received, by a transducer. The time between transmission and reception is measured electronically and the result converted to indicate the depth of water. Transmissions follow each other so rapidly that a continuous record of the depth can be obtained, and accuracy to within 1 ft. is achieved.

Distance. Distance is measured *in sea or nautical miles* (*M*). A nautical mile is the length of 1′ of arc of a meridian. This length varies because the earth is not perfectly spherical and the average length, in Latitude 44½°, known as the International Nautical Mile, is 1852 metres, or 6076·2 ft.

Short distances are measured in *cables*. A cable is one-tenth of a sea mile, often taken in practice as 200 yards.

Speed is measured in knots. One *knot* is a speed of 1 *M* per hour. (One knot is nearly equal to ·5 metres per second. One metre per second equals 1·944 knots.) Speed can be estimated by engine revolutions, but experience is needed in judging the amount of speed lost by propeller slip. Slip greatly increases in rough weather.

Logs. Distance and speed are measured by instruments known as logs.

Towing Logs. Towing logs trail a 'spinner' astern at the end of a log-line. This rotates at so many times per mile and the distance is recorded on a 'clock' aboard the vessel. The rotator should not be 'streamed' (put overboard) in very shallow water in case the blades of the rotator are damaged by contact with the bottom, and must be 'handed' (hauled inboard) before there is any risk of the line becoming foul of buoys or moorings. When handing the log, in order to prevent kinks forming as the line is hauled in, the inboard end is first disconnected and paid out again as the rotator is hauled in. When the rotator is on board (be careful not to damage the blades) the line can be hauled in and coiled down.

Bottom Logs. A small impeller projects beneath the hull, whose blades are turned by the passage of the vessel through the water. The revolutions can be made to register distance run, speed, or both.

All logs are liable to get fouled by weed. A towing log must be handed, while the impeller can be withdrawn inside the hull for clearance. To see that the log runs clear, a frequent look at the speed-dial should be taken. The most recent type of bottom log, operating on an electromagnetic principle, consists of a transducer which projects only a few millimetres below the hull and is incapable of becoming fouled by weed.

The Hand Log. A simple log can be employed, based on the principle of the 'English Log' (first described by William Bourne in 1574), which measures the speed through the water. It can be used for speeds up to 10 k, and is particularly effective at the lower speeds. If a 'log' is dropped overboard from a vessel travelling at the speed of 1 k, she will leave it astern at the rate of 8 ft. $5\frac{1}{3}$ in. every 5 sec. If this log is attached to a line and the line is marked at intervals of 8 ft. $5\frac{1}{3}$ in., the number of marks which pass over the stern in 5 sec. indicates the speed.

A ship's log consisted of a fan-shaped piece of board (the log-ship or log-chip), weighted at the curved edge so that it floated vertically, and secured by a three-legged bridle to the log-line. The log-line, some 50 fathoms of substantial line kept rolled on a reel, was marked every 23·6 ft. with a short length of line with 1, 2, 3, etc., knots (hence the use of the word knot as the unit of nautical speed), and used with a 14-sec. sand-glass. This apparatus can be scaled down to a more convenient size for a yacht. The log-line may be of stout fishing-line; 18 fathoms should be enough for a top speed of 8 k. This line can be kept wound on a board with a 'V' cut in each end. There must be enough stray line to allow the log-ship to reach the water and drift clear of the vessel's wake before measurement of the speed is made. About five fathoms should be enough stray line, and a figure-of-eight knot or a piece of bunting is worked into the line at this distance from the log-ship as a starting-point. Then at 8 ft. $5\frac{1}{3}$ in. from this mark put a piece of line with one knot in it, and at a further 8 ft. $5\frac{1}{3}$ in., a second piece of line with 2 knots in it, and so on. Short lengths of tape, instead of knotted lines, may be used if preferred as being less likely to foul the line.

In metric measure, for the speed of one International Nautical Mile (1852 metres) per hour, the distance that the marks should be apart is 2·57 m.

The log should be hove every half hour, and the speed entered in the log-book, and whenever the speed of the yacht is seen to change. Procedure: make the end of the line fast, so that the whole log cannot be lost overboard. Drop the log-ship overboard and pay out the whole line, so that it is seen to tow all clear. Haul in the line until the stray line-mark is held at the gunwale. Now let the line run freely over your fingers for exactly 5 sec. and then stop it. The nearest mark now indicates your speed. If you are holding the line halfway between the third and fourth knots, you are making $3\frac{1}{2}$ k.

At very slow speeds allow the line to run out for 10 sec. and halve the measured speed. A stop-watch is best for measuring the time. With a rough, following sea, the log-ship is likely to move in the water, and your speed will be rather greater than measured. On a passage with a steady breeze it is worth checking your log-speed with the speed actually achieved to see if the log reads accurately or whether it under-estimates the true speed.

Direction. True direction is given with reference to the true meridian (i.e. the direction of the North Pole) and usually measured to the nearest degree, clockwise, from north, from 0° to 360°. It is then written as a three-figure group. Thus North-East, which is 45° clockwise from North, is written 045°; South-East, 135°, and North-West, 315°.

Direction may also be given with reference to the magnetic or compass meridians, that is, the direction in which the magnetic pole lies, or in which the compass needle points, and is then written as 030° M or 030° C., respectively.

The instrument giving direction is the *Compass.* There are two quite distinct kinds. The *gyro compass* depends for its directive force upon the properties of a gyroscope. A gyroscope, if permitted to spin freely, will maintain its direction in space, and gravity can be made to cause the axis of spin to align itself in the true meridian. The gyro needs electrical power to keep it spinning. It points to the true, or geographical, north pole. The *magnetic compass* is described in Chapter IV.

On older compasses, direction is measured from North (0°) towards East or West (90°) and similarly from South towards East or West. This is called *quadrantal notation* and is written thus: N. 30° E., S. 45° W., etc. It is occasionally convenient to use quadrantal notation when making certain calculations.

When such precision is unnecessary, or unattainable, points of the compass, instead of degrees, are used. As there are 32 points of the compass, each point equals $11\frac{1}{4}$°. The conversion of points to degrees is given inside the back cover of *Burton's.*

Bearing is the direction of one place from another and can be given as true, magnetic or compass, according to the meridian of reference.

Bearings can also be obtained by radio direction-finders (RDF), but these, unless combined with a compass, as they sometimes are, give only the 'relative' bearing—that is, the angle measured clockwise from the ship's head at the moment when the bearing is taken. Relative bearings are written, e.g. '045° relative', that is 45° on the starboard bow. 270° relative is abeam to port.

Radio Direction-Finders consist of a receiver and an aerial which can be rotated. Aerials are of two types; the loop, suitable for mounting on the cabin-top of a steel-built yacht, and the ferrite-rod, which may be operated, protected from spray, inside the cabin of a wooden or GRP yacht. In some models the receiver and aerial are combined, and may also include a small compass, which enables compass bearings, instead of relative bearings, to be taken. Many receivers are capable of receiving long and medium waves, and can be used to obtain weather forecasts, time-signals and news. They may be sufficiently powerful to pick up Consol signals also (*see* page 29).

The aerial is directional: that is, the strength of the signal received varies according to the direction in which the aerial is turned. It is not possible to judge when the strongest signal is being received, but at right angles to this bearing is a 'null' position, where the signal fades out completely, and this position can be judged quite accurately.

The aerial may be fixed with the zero of the 360° scale pointing exactly ahead (as is the compass lubber-line), so that when the aerial is rotated the clockwise angle from the ship's head—or relative bearing—is shown. When a compass is combined with the aerial the DF is held and rotated until the 'null' position is located. The *compass bearing* is then read. (*See* page 26.)

The most recent type of DF (Autofix 100) gives a *visual* indication of the bearing. When the instrument is tuned to a station and switched on, a large pointer immediately and continuously homes onto it.

Errors of the Magnetic Compass.

(*a*) *Magnetic Variation.* The magnetic compass, when undisturbed, points to the magnetic poles, which are situated at a considerable distance from the geographical or 'true' poles. Furthermore, the positions of the magnetic poles are slowly and continuously moving. The angle which an undisturbed compass needle makes with the true meridian is called the variation. Its value alters with geographical position and also changes from year to year. In a few places in the world, iron-ore deposits beneath the seabed are sufficiently strong to attract compass needles locally. Where this is so a warning is given in the Sailing Directions and upon the appropriate charts. Variation is named easterly when magnetic North lies to the right of true North. The variation and annual change are shown on charts.

(*b*) *Deviation.* Every vessel built of iron or steel and others which contain, for example, an engine, have magnetic properties which deflect the compass needle from the magnetic meridian. This deflection is called the deviation of the compass. It is named Easterly when the compass needle is deflected to the right of the magnetic meridian. It can be partially removed by placing magnets and 'soft' iron correctors to nullify the ship's magnetic field, but a residue has to be allowed for. Because the deviation changes with the direction of the ship's head, it is tabulated on a Deviation Card (*see* page 8), or drawn as a curve (*see* page 83), for different directions of the ship's head. As deviation may also change with time and latitude, further observations for finding the deviation have to be made (*see* Chapter 4).

(*c*) The Variation and Deviation combined are known as the *Compass Error*, which is defined as the angle between the true meridian and the compass needle. It is named Easterly when the compass North lies to the right of True North.

Deviation Cards. Deviation may be tabulated against Ship's Head by Compass or Ship's Head Magnetic. It cannot, of course, be tabulated for true headings.

DEVIATION CARD

Ship's Head by Compass	Deviation
N. N.N.E. N.E. etc.	5° W. 6° W. 2° W.

As we shall require to enter the deviation table sometimes with a compass course and sometimes with a magnetic course, it is best to expand a given table into three or four columns. If the headings are given in points they should be converted to equivalent degrees, thus:

Ship's Head Compass Points	Ship's Head Compass Degrees (0° to 360°)	Deviation	Ship's Head Magnetic (0° to 360°)
N.	000°	5° W.	355°
N.N.E.	022½°	6° W.	016½°
N.E.	045°	2° W.	043°
E.N.E.	067½°	3° E.	070½°
E.	090°	4° E.	094°
etc.			

We can now find the deviation for either ship's head by compass or ship's head magnetic, interpolating as necessary. It is important to use the correct column when the deviations are large.

CONVERTING 'TRUE' TO 'COMPASS' AND VICE VERSA

To find the compass course to steer, given a true course, we must apply the compass error. First, the variation which will give us the course relative to the magnetic meridian, then the deviation, which will give us the course by compass.

The variation will be found upon the chart; apply the annual change to bring it up to date.

DEVIATION CARD

Ship's Head Compass	Deviation	Ship's Head Magnetic
000°	**0½°** E.	**000½°**
020°	**4°** E.	**024°**
040°	7° E.	**047°**
060°	6° E.	**066°**
080°	**2°** E.	**082°**
100°	**1½°** W.	**098½°**
120°	**5°** W.	**115°**
140°	6° W.	**134°**
160°	**3½°** W.	**156½°**
180°	**0°**	**180°**
200°	**3°** E.	**203°**
220°	**6½°** E.	**226½°**
240°	**5°** E.	**245°**
260°	**2°** E.	**262°**
280°	**1½°** W.	**278½°**
300°	**5°** W.	**295°**
320°	**7°** W.	**313°**
340°	**3½°** W.	**336½°**
360°	**0½°** E.	**000½°**

The deviation can be found from the Deviation Card or Book. When courses or bearings are expressed in 0°—360° notation, use the following rule in applying variation, deviation or compass error:

'Error East, Compass least;' (i.e. the compass or magnetic course is less than the true course).

'Error West, Compass best;' (i.e. the compass or magnetic course is greater than the true course).

EXAMPLE 1:
'*True to Compass.*'

The Course from *A* to *B* is 107° T. The variation found on the chart and corrected for the current year is 7° W. The deviation is found from the Deviation Card on page 8. Required the compass course.

In the figure, *TS* is the true meridian (always drawn vertically as upon a chart).

Course	107°	*T*
Var.	7	*W*
Course	114	*M*
Dev.	5	*W*
Course	119	*C*

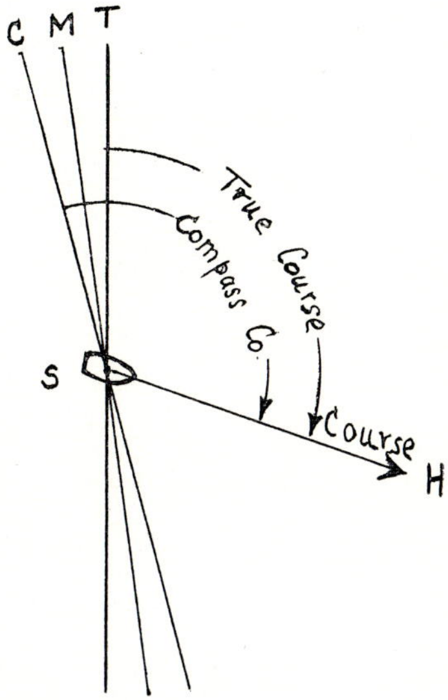

FIG. 1.2

Angle *TSM* is the variation (7° W.), so that *MS* is the magnetic meridian.

Angle *MSC* is the deviation (5° W.), so that *CS* is the direction of the compass needle.

Angle *TSC* is the compass error (13·3° W.).

The (clockwise) angle *TSH* is the true course (107), *CSH* is the compass course, and *MSH* is the magnetic course.

The rule 'Error East; Compass least: Error West, Compass best' also applies when converting bearings taken by compass to true before plotting them on the chart. In this case, since we start with the compass bearing, we first apply the deviation *for the ship's head* (*not* the bearing) and get the magnetic bearing, then the variation to get the true bearing.

EXAMPLE 2:

'Compass to True.'

A compass bearing of a lighthouse is 303°. The ship is steering 211° C. Variation is 10° W. What is the true bearing to plot on the chart?

The true meridian *TS* is drawn first, as always, then *MS* 10° W. to the West (Var. 10° W.). Next *CS* is drawn to the East of *MS*

(Dev. 5° E.). The compass bearing is then drawn 303° clockwise from *CS* and it can then be seen whether one should add or subtract in order to find the magnetic and true bearings.

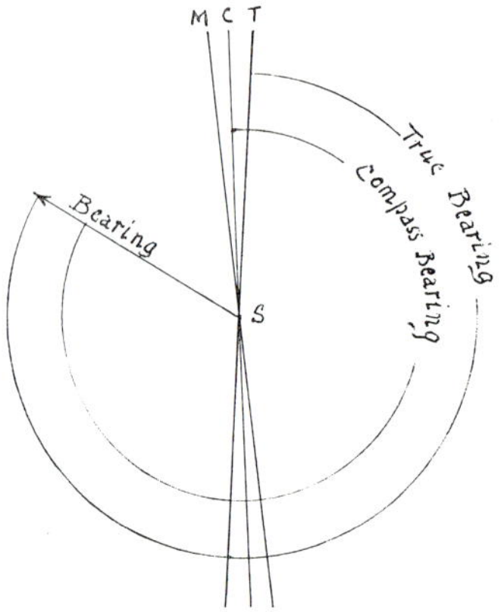

FIG. 1.3

Bearing	303° *C*
Dev.	5° E. for ship's head 211° *C*
Bearing	308° *M*
Var.	10° W.
Bearing	298° *T*

Once the variation and deviation have been found they can be conveniently applied together as Compass Error, until a change of course, or time, produces a change in the deviation.

EXAMPLE 3:

Course 075° C., Variation 12° W., Deviation 3° E. Compass Error is therefore 9° W. Compass bearing of a lighthouse is 072° *C*, what is the true bearing to plot on the chart?

In this case, after drawing *TS*, the true meridian, compass north can be drawn at once 9° W. to the west of true, and there is no need to show the magnetic meridian.

Figures need not be to scale, but they should always be drawn until the principle is fully understood. Draw the true meridian first and then whatever else is known.

Remember: never draw a *compass* course or a *compass* bearing on the chart. It must first be converted to either a *magnetic* or *true* course or bearing.

Bearing	072° *C*
Error	9° W.
Bearing	063° *T*

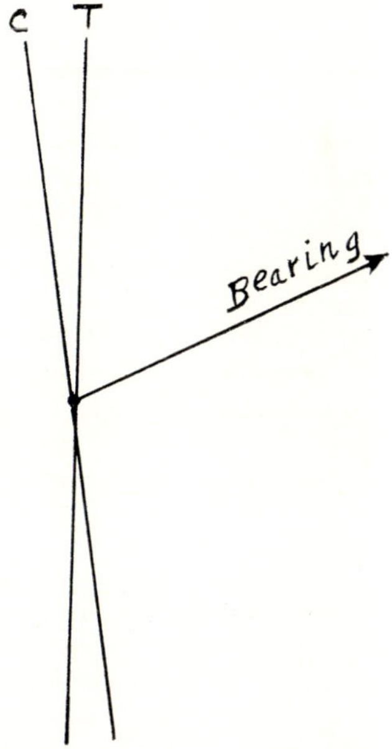

FIG. 1.4

Charts. The navigator needs a chart upon which he can plot positions and measure courses, bearings and distances. Most navigational charts are constructed on the *Mercator principle* because it is very suitable and convenient for these purposes.

(i) *Mercator Charts.* On the Earth's surface meridians converge towards the nearer pole, but on a Mercator chart they are drawn as parallel straight lines with the result that:

(*a*) Any other straight line drawn on the chart cuts all meridians at the same angle, thus courses and bearings can be measured against any meridian.

(*b*) The scale of longitude is constant for all latitudes.

(*c*) The east-west scale of *distance* expands with departure from the equator.

In order that angles and distances should be correctly represented and to keep coast lines the right shape, the north-south scale of distance is also expanded in the same proportion as the east-west scale. *Distances are measured on the scale of latitude abreast the latitude of the place.*

In very high latitudes the scale of distance becomes excessively large and, for this reason, Mercator's projection is not used for latitudes greater than 70°. A further disadvantage is that, unless two places are on the same meridian or both are on the equator, the straight line joining them (called a rhumb line) is not an arc of a great circle and does not, therefore, represent the shortest distance between them. On short distances, however, the difference between rhumb line and great circle is negligible and can be disregarded. The Practice Charts included with this volume are Mercator charts. Great circles (except meridians and the equator) appear as curves when drawn upon Mercator charts.

For large-scale charts, and charts of high latitudes, a gnomonic projection is often used. For very small areas in latitudes below 60° the convergency and curvature is inappreciable.

(ii) *Gnomonic Charts.* Gnomonic charts are useful for plotting Great Circle tracks for long ocean passages. Great circles appear as straight lines on gnomonic charts (*see* Figure), but courses and distances cannot be measured. (*See* Chapter 3, page 67, Figs. 3·7, and 3·9.)

(iii) *Lattice Charts.* Some charts are over-printed with a lattice to enable the ship's position to be found by electronic navigational aids (e.g. Consol Chart, page 31). Decca Lattice Charts (*see* Decca Chart, page 12) have a coloured lattice: red, or green or purple. A special indicator unit called a Decometer contains red, green and purple dials. The pointers on two of the dials are read and, as the ship is on each of the lines indicated, where the lines cross on the chart is the ship's position. A fourth dial, the Lane Identification meter enables the other dials to be set correctly before the start of a voyage or on entering the Decca network.

Admiralty charts cover the greater part of the world's coasts (which, indeed, have been mostly charted by the Royal Navy) and are usually the best obtainable.

In some places, for example in the intricate coastal waters of Scandinavia, a foreign Government chart may be better, though the corresponding Admiralty chart would be perfectly satisfactory for deep-drafted ships which had to keep to seaward of the numerous islands, islets and skerries which abound, and could not attempt the narrow and tortuous channels which a yacht might use.

Special charts for yachtsmen are produced for favourite cruising areas, based, of course, on Government surveys. They comprise Passage charts, showing long stretches of coastline, and are suitable for planning passages and cruises. Coast and Harbour charts are suitable for a coastal voyage. They often include small plans of ports which may be entered *en route.* River and Estuary charts are published of the more popular sailing areas. Often these charts have information about the coast they refer to printed on the back;

pilotage and tidal notes for the harbours, and details of berthing and mooring.

The large size of Admiralty charts requires a corresponding chart table, and chart-plotting boards have been produced to facilitate chart work in a small craft.

Stanford's Chart-Plotting Board consists of a waterproof board with a transparent overlay, marked with magenta meridians, which can be drawn upon with a pencil and cleaned with an eraser. The chart is inserted between board and overlay. The top of the chart is horizontally fixed either:

(*a*) along a True parallel at the top of the board, in which case the magenta meridians marked on the overlay represent *True* meridians; or:

(*b*) laid along a line equal to the local variation. The magenta meridians then represent *Magnetic* meridians; or:

(*c*) laid along a line equal to the Compass Error. The magenta meridians then represent *Compass* meridians. Courses and bearings can thus be directly plotted either True, Magnetic or Compass, as the navigator prefers. A plotting rule made of the same material as the overlay eliminates the need for parallel rulers and dividers.

'Navigaids' produce a series of yachting charts printed on waterproof and untearable plastic of a uniform size. Information about tidal streams, depths, variation, etc., are printed on the left, and on the periphery of the chart is the True Compass Rose, enabling very accurate angles to be measured. The Magnetic Rose is also large and bold. The board is 25 in. by 18 in. and up to 10 charts can be neatly clamped to it in the desired order. A large, separate transparent grid-sheet replaces parallel rulers and dividers. The whole is carried in a wallet with a transparent front, so that charts can be used in the cockpit even in bad weather.

Charts should be correct when purchased, but all charts have to be kept up to date, for navigational marks may at any time have to be altered in character; wrecks obstruct a fairway; shoals extend, etc. The information necessary for chart correction is contained in Admiralty *Notices to Mariners* (*see* next section). It is clearly important that the charts used should be free of error. Chart agents provide a Chart Correction Service, and 'Navigaids' issue monthly bulletins listing the corrections necessary for all 'Navigaid' charts.

Publications.

Admiralty Sailing Directions (*Pilots*) consist of 77 volumes covering the whole world and contain abundant detailed information on each area's geography, government, trade, currency, communications, ports' facilities and regulations, currents, buoyage systems, climate, weather, rivers, tides, etc., together with many illustrations, diagrams and tables.

They are maintained by the issue of Supplements and periodical new editions. A useful addition to them is *The Mariner's Handbook*.

Sailing Directions are compiled from a big-ship point of view, and they can be supplemented by many books written especially for yachtsmen. These may be termed sea-guides, and include volumes on *The English Channel* (Delmar-Morgan); *North Brittany* (Hasler and Coles); *North Biscay* (Coles and Black); *South Biscay* (Brandon); five volumes on the *Central and Eastern Mediterranean* (Denham); *South Baltic* (Townsend); the *Shell Book of South Coast Harbours*, etc.

Notices to Mariners are issued (free) daily, weekly and quarterly, from Admiralty Chart Agents, Shipping Offices, Customs Houses and Consulates, to enable the navigator to correct his charts, Admiralty Pilots, Light Lists, etc., when changes become necessary.

Merchant Shipping Notices ('M Notices'). These are issued in the same way as Notices to Mariners. They contain information of a more general nature, e.g. M554 deals with Radio Distress Procedure.

NEMEDRI.—North European and Mediterranean Routeing Instructions, contains information about minefields, dangerous areas and safe routes.

The Admiralty Lists of Lights and Fog Signals contain descriptions of lights, etc., in greater detail than can be shown on charts.

Admiralty List of Radio Signals. Vol. I, Coast Radio Stations, subsidiary services, including Medical Aid by radio.

Vol. II, Direction-Finding Stations and Radio Beacons.

Vol. III, Meteorological Services and Codes.

Vol. IV, List of Meteorological Observation Stations.

Vol. V, Radio Time Signals; Navigation Warning Systems.

Tidal Atlas showing the *horizontal* movement, i.e. the directions and strength of tidal streams round the British Isles for different times of High Water at Dover.

Admiralty Tide Tables showing the *vertical* movement of the tides, i.e. times and heights of High and Low Water at different places. Vol. I deals with European waters.

(Tidal stream information is given on some charts, and Tide Tables also appear in *Brown's Nautical Almanac* and *Reed's Nautical Almanac*. The latter contains charts showing tidal streams round the British Isles.)

The Nautical Almanac, issued annually, contains the astronomical data required for finding the ship's position and the error of the compass by observations of heavenly bodies. Much of it is reproduced in *Brown's* and *Reed's Nautical Almanacs*, both of which contain much other useful information.

Nautical and *Navigation Tables* contain permanent (as distinct from the ephemeral tables published yearly in *The Nautical Almanac*) mathematical tables for working navigational problems. These can be classed as tables for general use: *Bowditch's* (U.S.); *Burton's*; *Inman's*; and *Norie's*. Other tables have been compiled for facilitating the solution of special problems: for fixing position; *Sight Reduction Tables for Marine Navigation* (HO 229) in six volumes, each volume covering 15° of latitude; *Sight Reduction Tables for Air Navigation* (A.P. 3270) in three volumes, Vol. I for selected stars, Vol. II for other heavenly bodies, covering latitudes 0° to 39°, and Vol. III covering latitudes 40° to 89°. These tables are considered by the Royal Yachting Association as the most suitable for use in yachts, and Vols. I and III are used in solving the problems set in Chapters 17–19. *Azimuth Tables* for finding the deviation of the compass.

Extracts from the *Tide Tables*, Vol. I; *The Nautical Almanac*, and *Radio Signals*, Vol. V, are included, by permission, at the end of this book.

PAPER I—PRELIMINARY NAVIGATION

1. Define and illustrate with a diagram: Latitude, Great Circle, a Meridian, a Knot.

2. What is the length of a sea-mile or nautical mile? How long is a cable?

3. What is the:
   (*a*) variation and
   (*b*) deviation of the compass?

   What causes them to change their value?

4. Correct the following bearings taken by compass so that they can be laid off on the chart:

(*a*) Bearing 152° C.	Deviation 1° E.	Variation 2° W.
(*b*) Bearing 165° ½ C.	Deviation 2° W.	Variation 9° W.
(*c*) Bearing 200° C.	Deviation 3° W.	Variation 3° E.
(*d*) Bearing 309° C.	Deviation 3° E.	Variation 2° E.
(*e*) Bearing 109° C.	Deviation 2½° E.	Variation 14° W.

5. What courses must I steer by compass to make good the following true courses?

(*a*) Course 217° T.	Variation 5° W.	Deviation 2° W.
(*b*) Course 345° T.	Variation 15° E.	Deviation 5° W.
(*c*) Course 095° T.	Variation 12·5° W.	Deviation 2·5° W.
(*d*) Course 003° T.	Variation 18° E.	Deviation 4° W.
(*e*) Course 011° T.	Variation 15° E.	Deviation 5° E.

6. Using the 'Deviation Card' (*see* page 8), find the deviation and hence the compass course to steer in each of the following examples:

(*a*) True Course to make good, 051° T., Var. 11° W.

(*b*) True Course to make good, 269° T., Var. 11° W.

(*c*) True Course to make good, 114° T., Var. 4° E.

(*d*) True Course to make good, 001° T., Var. 6° E.

7. EXERCISE: Complete the following table:

True Course	Variation	Mag. Co.	Deviation	Comp. Co.	Comp. Error	Bearing of Object	
						Comp.	True
041°	11° E.		6° W.			153°	
			2° E.	312°	20° W.		149°
	15° W.			237°	10° W.	075°	
271°	19° E.		4° E.			306°	
116°		110°		113°			180°
207°		217°	5° W.			022°	
	12° E.			030°		266°	280°
		200°	6° E.			043°	030°
357°			9° E.		21° E.	168°	

Considerable emphasis has been laid on the simple matter of correcting courses and bearings because of its extreme importance. A correction applied in the wrong way doubles the original error. Many vessels have been lost through steering a wrong course.

8. On a mercator chart:

(*a*) Which scale is used to measure distance?

(*b*) How do great circles appear?

(*c*) 'A line joining two places (unless running North and South) is not the shortest distance between them.' Does this matter? What is such a line called?

(*d*) Why is the vertical latitude scale on a mercator chart not the same as the horizontal scale (except at the equator?)

9. (*a*) Ship's Head is N.W. by compass. Compass error 5° W. A lighthouse is abeam to port. What bearing should you lay on the chart?

(*b*) A beacon is bearing 085° relative when you are heading 216° T. What is the true bearing of the beacon?

CHAPTER 2

CHART WORK

'What's the good of Mercator's North Poles and Equators, Tropic Zones and Meridian Lines?'
. . . 'They are merely conventional Lines!'

Edward Lear.

Required for this chapter: Chart instruments, Burton's.

Admiralty charts are at present being gradually transformed in appearance, though as there are some 4000 to be replaced, it will be some years before the process is completed. The chief change noticeable is the use of colour: land being buff, shallow water light blue, and where shoals dry between high- and low-water, the colours merge to a distinct olive-green. A most important difference from the present charts is that vertical measurements—heights and depths—are shown in metres, or metres and decimals. 4·5 metres, for example, being printed 4_5 (as fathoms and feet are shown at present). The legend *Depths in Metres* is boldly displayed. Depths are measured from the Lowest Astronomical Tide, so that a sounding will always show the least possible depth at the place.

The remarks that follow refer principally to the older charts.

Reading the Chart. The various symbols and abbreviations used on charts are shown on Chart No. 5011, now published as an attractive booklet, reproduced in end pocket. Some important points are mentioned below.

Soundings. The figures printed over areas depicting the sea show the depth of water, or soundings. Depths of water are usually shown in fathoms (1 fathom = 6 ft.). On large-scale charts they are sometimes given in feet. The units used are clearly printed under the title of the chart. Depths below 11 fathoms are given in fathoms and feet, the number of fathoms being written first and then the feet on a slightly lower level.

Depths are given below *Chart Datum*, which is the level of the lowest predictable tide under average meteorological conditions. The actual depth exceeds that charted by a safety margin called 'the height of the tide', which height can be found from the tide tables. (*See* Chapter 5.)

Fathom Lines are printed on the chart to enclose all soundings of or below certain depths, banks, shoals, etc.

Heights. Underlined figures on banks which dry at low water are used to show the height of the bank above chart datum. These heights are always in feet, irrespective of the units used for soundings.

Heights of lights, hills, etc., are given in *feet* above the level of Mean High Water Springs, which is the average height of high water at all spring tides. Heights of small islands are shown beside the island to which they refer and are enclosed in brackets so that they cannot be mistaken for soundings.

The nature of the seabed is indicated by abbreviations, capital letters being used for nouns and small letters for adjectives, e.g. M means soft mud; fS, fine sand.

Charted positions of light vessels, buoys, etc., are shown by the small circle on their waterline. Lighthouses are at the centre of the star.

Lights are identified by their characteristic phases: *A Fixed light* (F.) is a continuous steady light. *A Flashing light* (Fl.) shows at

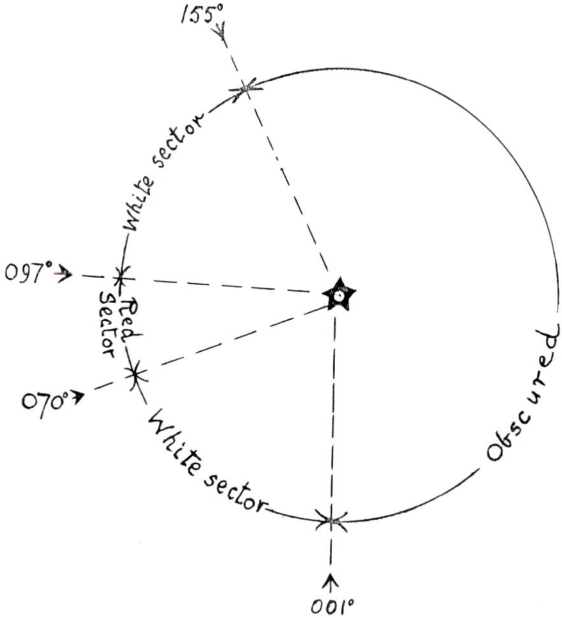

FIG. 2.1

regular intervals. The duration of light is less than the period of darkness separating the flashes.

An Occulting light (Occ.) is a steady light which is eclipsed at regular intervals, the duration of darkness being less than the period of light.

An *Isophase light* (Iso.) shows at regular intervals and the duration of light is equal to the duration of darkness.

A *Quick Flashing light* (Qk. Fl.) flashes continuously more than 60 times a minute.

An *Interrupted Quick Flashing light* (Int. Qk. Fl.) is a quick flashing light which is totally eclipsed at regular intervals.

An *Alternating light* (Alt.) is one which changes colour, e.g. Alt. *W.R.* ev. 5 sec. means a light changing from white to red and red to white alternatively, every five seconds.

Some lights show different colours in different sectors. The sectors are shown only on large-scale charts. Full details are given in the Admiralty List of Lights.

Lights may flash Morse letters, e.g. Mo (K) means that Morse letter K is flashed.

Most of the characteristics given above can be arranged in a group or groups, e.g. Gp. Fl. (3) 10 sec. means a group of 3 flashes given every 10 seconds. Gp. Fl. (3, 4) means two groups, one of 3 and the next of 4 flashes.

The visibility of lights given upon the chart assumes the observer's eye to be 15 ft. above sea level. (*See* Fig. 2.19, page 39.)

All bearings referring to lights (descriptions of sectors, etc.) are given from *seaward.* (*See* Fig. 2.1)

Character of Fog Signals.

Warning. Fog signals are heard at greatly varying distances because the behaviour of a sound wave in air depends upon the state of the air. Fog signals cannot, therefore, be implicitly relied upon.

Diaphone emits a powerful low note ending in a sharp fall in pitch, called the 'grunt'.

Tyfon. Powerful note of medium pitch, similar to a ship's whistle.

Siren. Medium power, high or low note, or combination. In the latter case one of the notes may be inaudible.

Reed and *Nautophone.* Low power, high note.

Whistle. Low power, low note.

Bell. Worked either mechanically or by the action of the sea. In the latter case the sound is irregular and, if the sea is calm, the bell may not be heard.

Tidal Streams.

On some charts the direction in which the tidal streams flow are indicated by arrows. The set (the direction of flow) of the ebb stream being shown by arrows having no feathers (⟵) and that of the flood stream by arrows which have feathers on one side only (⟶).

Sometimes the directions and rates of tidal streams are tabulated at hourly intervals from High Water at Dover (or the nearest

Standard Port). Separate tables are given for individual localities, each of which is marked by a number or letter in a diamond, thus ⟨E⟩.

Tidal Streams

Hours	Position ⟨A⟩ Lat. 58°30′N. Long 1°14′E. (G.15)			Position ⟨B⟩ Lat. 58°30′N. Long 4°14′E. (G.13)			Position ⟨C⟩ Lat. 58°09′N. Long 1°24′W. (G.17)			Position ⟨D⟩ Lat. 57°10′N. Long 0°0′ (G.05)			Position ⟨E⟩ Lat. 57°10′N. Long 3°34′E. (G.02)		
	Direction	Rate Sp.	Rate Np.	Direction	Rate Sp.	Rate Np.	Direction	Rate Sp.	Rate Np.	Direction	Rate Sp.	Rate Np.	Direction	Rate Sp.	Rate Np.
Before H.W. Dover 6	001°	0·4 kn.	0·2 kn.	340°	0·3 kn.	0·2 kn.	351°	0·7 kn.	0·4 kn.	021°	0·5 kn.	0·3 kn.	007°	0·4 kn.	0·2 kn.
5	000°	0·4	0·2	330°	0·3	0·2	000°	0·7	0·4	012°	0·7	0·4	008°	0·5	0·3
4	002°	0·4	0·2	350°	0·2	0·1	357°	0·6	0·3	019°	0·7	0·4	028°	0·5	0·3
3	033°	0·3	0·1	025°	0·1	0·1	080°	0·1	0·1	026°	0·5	0·3	028°	0·4	0·3
2	095°	0·2	0·1	066°	0·1	0·1	189°	0·4	0·2	022°	0·3	0·2	070°	0·3	0·2
1	149°	0·3	0·1	147°	0·2	0·1	173°	0·6	0·3	Slack			120°	0·3	0·2
H.W.	179°	0·4	0·2	157°	0·3	0·2	175°	0·7	0·3	188°	0·4	0·3	174°	0·4	0·2
After H.W. Dover 1	184°	0·5	0·2	158°	0·2	0·1	180°	0·7	0·4	196°	0·7	0·4	189°	0·5	0·3
2	194°	0·4	0·2	177°	0·2	0·1	174°	0·6	0·3	188°	0·7	0·4	204°	0·5	0·3
3	203°	0·3	0·2	192°	0·2	0·1	177°	0·3	0·2	193°	0·8	0·4	220°	0·4	0·3
4	245°	0·2	0·1	255°	0·1	0·1	270°	0·1	0·1	208°	0·4	0·2	260°	0·3	0·2
5	000°	0·3	0·2	334°	0·2	0·1	327°	0·5	0·3	Slack			306°	0·3	0·2
6	001°	0·4	0·2	340°	0·3	0·1	355°	0·7	0·4	030°	0·4	0·2	350°	0·4	0·2

Currents which are more or less constant in direction are shown by arrows with feathers on both sides (⋙——➤) or a wavy arrow with no feathers (∼∼➤).

Using the Chart. As a general rule, the largest scale chart available should be used, because small-scale charts may omit details which overcrowd them.

In measuring distance, remember to use the latitude scale at the sides of the chart.

Use of Parallel Rulers. This needs some practice. In order to measure the direction of a line, place one edge of the rulers along it and then, holding that edge firmly, open the rulers. Next press the opened edge firmly, and close the other to it. Repeat as necessary. The angle between the line and the meridian can be measured either at a compass rose or by moving the index of the parallel rulers to a meridian and noting the reading against the same meridian on the other edge. Since there will be two readings to choose from (e.g. 140° or 320°), be careful to select the direction you want and not the reciprocal. When the angle has been measured move the rulers back to the line to check that the rulers have not slipped during the transfer.

If using set squares, place one edge along the line, then place the other set square along another side of the first set square and slide the first one towards a compass rose or meridian. With a straight-edge and protractor, slide the protractor along the straight-edge till it reaches a meridian.

Shaping the Course. Symbols used in plotting:

True Course or True Heading (——◄——) refers to the ship's fore-and-aft line and is the direction in which the ship is pointing, irrespective of the direction of her movement.

True Course Made Good or Track (——←←——) is the true direction of the ship's track over the seabed.

The effect of current, or a current or tidal vector, (——←←←——).

It should be noted that the ship's speed always refers to her speed through the water. Effective speed or ground-speed is her speed made good over the seabed and is *measured along the track*.

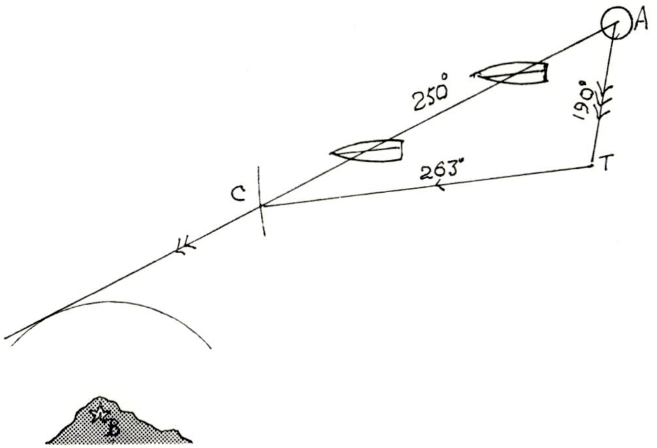

FIG. 2.2—Shaping the Course

Referring to Fig. 2.2, it is desired to set course from position *A* to pass 2 *M* off point *B*. Proceed as follows:

With centre *B* and radius 2 *M*, draw a circle. From *A* draw a line tangential to the circle; this is the True Course to Make Good or the required Track. Measure its direction against the nearest meridian or the outer graduated circle of the Compass Rose. Mark the track with two arrow-heads.

Next, consult the chart or Atlas of Tidal Streams to determine how the tide will set during the first part, or the whole, of your passage from *A* to *B*. If the rate of the tide is navigationally significant the course to steer must be adjusted so that the ship will maintain her required track.

Suppose the required track to be 250° T., the ship's speed 6 knots and the tide setting 190° at 2½kn. Then: from *A* plot the tidal vector *AT*, 190° 2½ M, to any convenient scale. *T* is then the position the ship would have reached had she been stopped but allowed to drift with the tide. However, her speed is 6 *k* and she could, therefore, reach any point 6 *M* away from *T* at the end of one hour. Since it is desired to move along the track, a distance of 6M is taken on the dividers or compasses and *C* marked on the track 6M from *T*. The *direction* of the line *TC* is the true course to steer. The length *AC* is the effective speed, or speed made good.

Note that the vessel will move along the line *AC*, but her head will be pointing in the direction *TC*.

To get an accurate result the constructed speed-triangle, or triangle of velocities, must be of a good size. Often its sides may be the distances gone in one hour (as described above), but this is not essential, and the same answer will be obtained if, for example, a speed-triangle of 2 hr., 3 hr. or half-an-hour is drawn.

The tidal stream varies from hour to hour and, if desired, several hourly tide vectors may be plotted, in sequence, from *A*, and the distance sailed by the ship in the same number of hours is marked from *T* on the track and the triangle completed to find the course to steer. (*See* Fig. 2.3.)

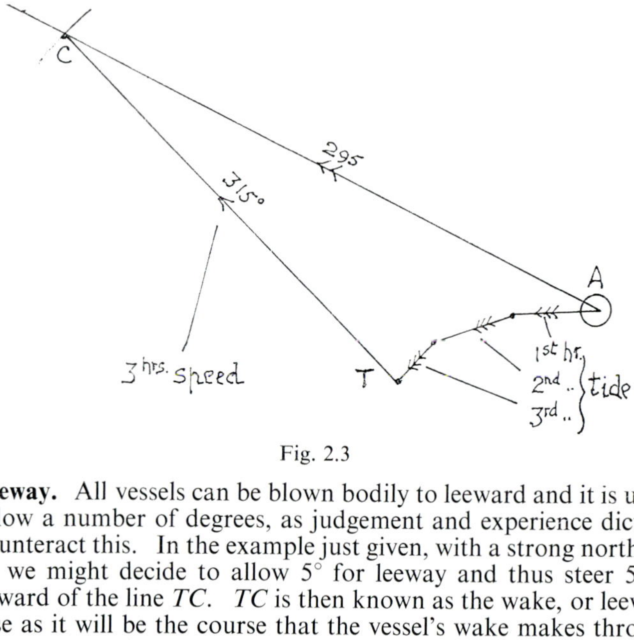

Fig. 2.3

Leeway. All vessels can be blown bodily to leeward and it is usual to allow a number of degrees, as judgement and experience dictate, to counteract this. In the example just given, with a strong northerly wind we might decide to allow 5° for leeway and thus steer 5° to windward of the line *TC*. *TC* is then known as the wake, or leeway, course as it will be the course that the vessel's wake makes through the water even though her head will be pointing 5° to windward.

Remember (*a*) That if you wish to *make good* a certain course you must steer to windward by the amount of the leeway. (This course is not drawn on the chart.)

(*b*) That if you have been steering a certain course and wish to know your course through the water, the leeway must be allowed to leeward of the true course steered.

Dead Reckoning and Estimated Positions. The D.R. position is that obtained from the course steered and the speed through the water, i.e. the position the ship would be in if she experienced no leeway, tidal stream, or current. It is written on the chart as +1100 (the cross indicating the position and 1100 the time she was supposed to be there).

The Estimated Position (E.P.) is that obtained after allowances have been made for the estimated effects of leeway, current and tide. It is the best estimate we can make of where we should be. It is written on the chart as △ 1100.

Always write the *time* and *log reading* against the position.

To find the Estimated Position at a given time.

EXAMPLE: Referring to Fig. 2.4. The ship is in position *A* at 1320, the log then reading 13 M. Course is 259° C., compass error 6° E. At 1600 the log reads 39·7 M. Leeway caused by a southerly wind is estimated as 5° and, according to the tidal information on the chart, the tide sets 040° at 2 kn throughout the period. It is required to find the E.P. for 1600.

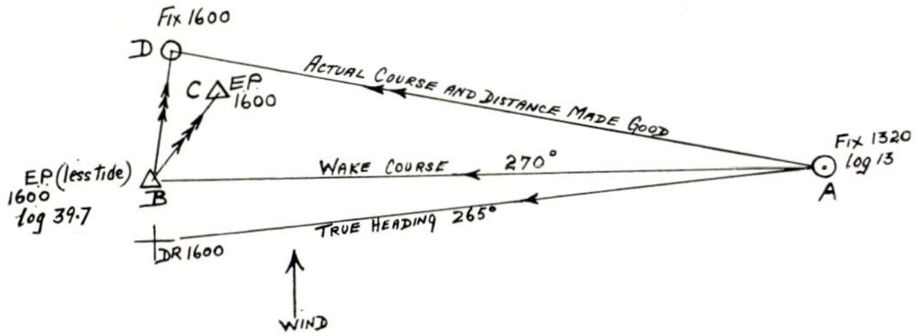

FIG. 2.4

The true course (heading) and D.R. position for 1600 are shown here only for illustration. They would not usually be plotted on the chart. If a fix, *D*, was obtained at 1600 *BD* would be the actual set and drift experienced.

(i) Apply leeway to the true course to obtain the wake course. (*Never apply leeway to a compass course.*)

Course 259° C., Error 6° E. True Course = 265°
leeway 5°

Wake Course 270° T.

(ii) Distance run by log = (39·7—13) = 26·7 M.

From *A* lay off *AB* in direction 270°, distance 26·7 M. Position *B* would be the E.P. if there was no tide or current.

(iii) The tide is estimated to have set 040° at 2 kn for 2 hr. and 40 min., which makes the drift 5·3 M. From *B* lay off *BC*, 040°, 5·3 M. Point *C* is the E.P. for 1600.

Although we have taken care in finding the right course to steer we must not assume that our vessel will arrive precisely at the Estimated Position, for we cannot know with certainty the strength and direction of tides or currents, the error in the log or in steering, and we may be uncertain of the compass error. We must, therefore, fix the ship's position on the chart at regular intervals if it is possible to do so. Various methods of obtaining a fix will now be discussed.

Position Lines. A position line (P.L.) is a line, drawn on a chart, on which the ship is known to have been at a particular time. It is distinguished by an arrow-head at one or both ends and the time at which the observation was made. (*See* Fig. 2.5.) There are many ways of obtaining position lines, those commonly used are:

(i) *By observing two fixed objects in transit.* The P.L. is drawn on the chart simply by producing the line joining the objects. The accuracy of this method is high.

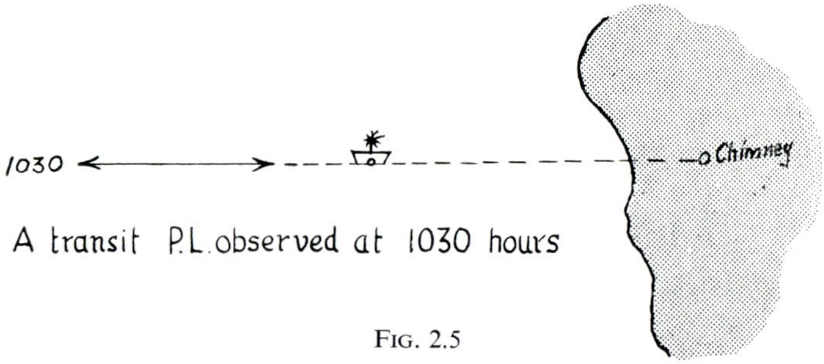

Fig. 2.5

(ii) *A compass bearing of a fixed object.* The P.L. is drawn by plotting, from the object, the reciprocal of the *true* bearing. (When converting the compass bearing to true, remember to take the deviation for the ship's head.) The accuracy of this method depends on how closely the compass error is known and the distance of the object from the ship. If the compass error is uncertain the bearing of a near object will will give a more reliable P.L. than a distant one. An error of 1° in the bearing of an object 6 M away gives an error of about one cable in the P.L.

(iii) *A relative bearing of a fixed object*, when added to the true course gives the true bearing. The procedure then is the same as in (ii). (*See* also *Pelorus* page 78.)

(iv) *A vertical sextant angle of an object of known height.* Suppose that a light has a charted height of 156 ft. This means, precisely, that the height of the light (not the top of the lighthouse) is 156 ft. above high water at spring tides. If we measure the altitude of the lantern above the level of the water with a sextant (described in Chapter 7), we can find our distance from the lighthouse by referring to a table, 'Distance by Vertical Sextant Angle' (*Burton's*, page 129). The table is entered with the height of the object and the distance off is read opposite the angle measured. The distance given if the table is computed for observer's eye at water-level but the effect of height of eye is navigationally insignificant except in rare cases.

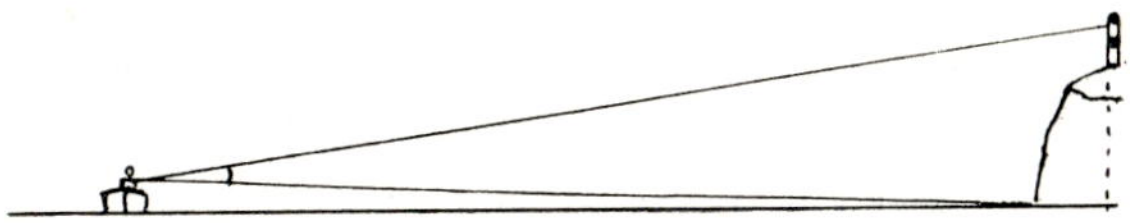

FIG. 2.6

Although the height of the light is given above Mean High Water Springs, the actual height of the tide can usually be ignored because any error from this cause puts the ship nearer to the light than her true distance.

To plot the P.L., place the centre of the compasses on the object and with radius equal to the distance off, draw an arc. The accuracy of this method is high.

EXAMPLES FOR PRACTICE:

	Object	*Corrected Vert. Sextant Angle*	*Distance off (M)*
(*a*)	C. St. Vincent Lt. (221 ft.)	1° 23′	1·5
(*b*)	Yarborough Monument (312 ft.)	1° 58′	1·5
(*c*)	Duncansby Head Lt. Ho. (180 ft.)	0° 28′	3·6

(v) *By Soundings.* When a fathom line is particularly well defined it can be found by taking a line of soundings and noting the time at which the observed sounding (corrected for the height of the tide) agrees with the fathom line on the chart. This fathom line, itself, then constitutes an approximate position line.

At other times, when the sea bottom is sufficiently sloping, a P.L. parallel to the course made good can be obtained. Mark on the edge of a strip of paper or on the edge of the ruler, a line of soundings spaced the proper distance apart according to the scale of the chart and the time interval between soundings, place the first sounding near the Estimated Position and then move the paper about parallel to the course expected to be made good and decide where the soundings agree best with those marked on the chart.

(vi) *By Radio Bearings.* Radio Beacons are established for the same purpose as lighthouses, sending out identifiable radio signals far beyond visibility range, so that vessels equipped with Direction Finders can obtain bearings of them. There are Radio Beacons on most of the coast lines of the world, including about 80 on the coasts of the United Kingdom. A list of them is contained in the *Admiralty List of Radio Signals*, Vol. II, and for the coasts of the Mediterranean to the entrance of the Baltic, *Reed's Nautical Almanac*, showing the grouping, identification signals, range and frequency. A useful chart of Radio Beacons (275 (2a) is also published.

Characteristics: All beacons are allocated a 2-letter Identification Signal, which is followed by a long dash, during which the bearing is taken. The complete signal, repeated every 6 min., is as follows:

Identification Signal, made 3 to 6 times;

Long dash for 25 sec.;

Identification Signal made once or twice;

Silent period for 5 min.

Radio Beacons are grouped, each member of the group using the same frequency and transmitting in turn, so that bearings can be taken in sequence from each. The group transmission cycle being 6 min.

EXAMPLE: In mid-English Channel from what Radio Beacons can I obtain bearings?

Referring to *Reed's Nautical Almanac*, it is seen that some of the following group might be used:—St. Catherine's Point Lt. H. (Identification Signal CA), Start Point (SP), Casquets Lt. H. (QS), Le Havre Lt. V. (LH), Roches Douvres Lt. H. (RD), Pointe de Barfleur (FG). Range of all these is 50 *M*, their frequency 291·9 kc/s. Only beacons well within the listed range should be used.

Operation. The correct frequency having been set and the beacon identified, while the long dash is being transmitted the aerial is turned until the signal fades out, then turned a few degrees more until it is again just audible. This is done several times until the mean position of the null is found and the bearing of the radio beacon is then written down. It is most important for the yacht's head to be steady while the bearing is being taken and the exact heading noted when the null is being measured if the bearings taken are relative.

The radio bearings are subject to several errors: (*a*) Quadrantal; (*b*) Coastal effect; (*c*) Sky-wave or Night effect; (*d*) Half-convergency; (*e*) 180° Ambiguity; (*f*) (If compass bearings) Compass error. The greater the distance of the beacon, the greater the effect on the accuracy of the bearing.

(*a*) *Quadrantal Error.* Steel rigging acts as an aerial and may deflect the signal towards the fore and aft line of the vessel. To allow for this the D.F. must be calibrated for quadrantal error (so called because it is greatest on the bow and quarter and zero on ahead, astern and beam bearings). A curve or table is made and used in much the same way as a Deviation curve or table is for the compass.

Calibrating the D.F. for Quadrantal Error. The yacht must be in sight of a Radio Beacon, and the weather calm. Any other radio set should be switched off and the aerial disconnected and a good position chosen for the D.F. aerial, as far away from steel rigging as possible. The yacht is then slowly swung through 360°, a series of radio bearings of the Radio Beacon being taken and, simultaneously, bearings by hand-bearing compass while the ship's head is held steady. These compass bearings corrected for deviation and converted to relative bearings as necessary, are then compared with the radio bearings, the difference being the quadrantal error. These are plotted on a graph and a fair curve drawn. Note that the correction refers to the *Relative bearing* and not (as for the compass) to the ship's head. For example, a relative bearing of 215° having been observed, the quadrantal error, found from the curve below, is $1\frac{1}{2}+$. The correct bearing is $216\frac{1}{2}°$.

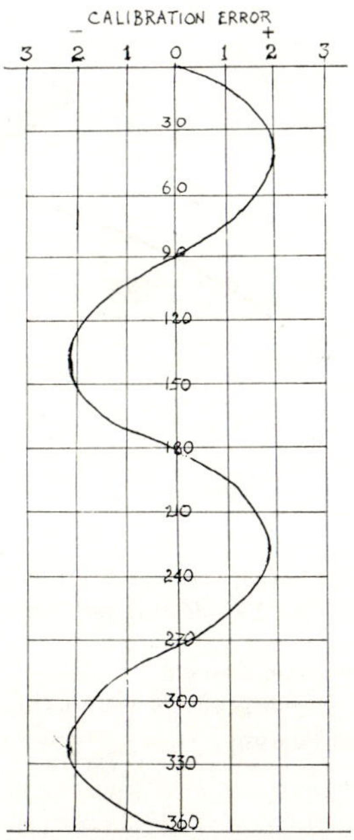

FIG. 2.7—Quadrantal Errors

(*b*) *Coastal Refraction* (Land Effect). When radio waves from a shore station cross the coastline they are bent towards it. The resulting error is not measurable and, though negligible when the angle made with the coast is large, it can be considerable at small angles. It will always place the ship nearer to the coast than she really is.

(*c*) *Night Effect.* Near the times of sunrise and sunset great changes are taking place in the ionosphere, and radio bearings should not be relied upon unless within 25 miles of the radio station.

(*d*) Radio bearings are arcs of great circles which appear as curves on a Mercator chart and, since the plotting of great circles is not practicable, they must first be converted to rhumb line bearings by applying a correction called *Half-convergency.* This correction is tabulated in *Burton's* (page 128). The table is entered on the left with the mean (average) latitude between radio station and ship, and their difference of longitude at the top. *Half-convergency is always applied towards the equator.*

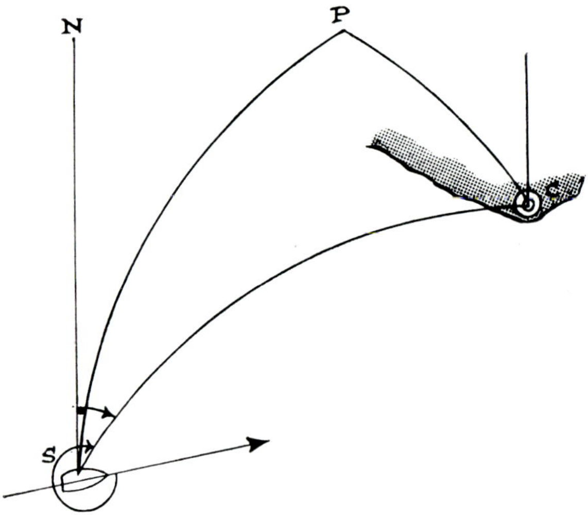

FIG. 2.8—Radio Bearing

∠*PSC* represents the great circle bearing
∠*NSP* represents the half convergency of the meridians
∠*NSC* represents the chart bearing

EXAMPLE: St. Catherine's Point (Lat. 50° 34·5′ N., Long. 1° 17·8′ W.) bears 315° by D.F., ship's head 090° C., compass error 9° W. D.R. position 49° 50′ N., 3° 8′ W. Find the rhumb line bearing to plot on the chart.

Radio Beacon Lat.	50° 34·5′ N.	Long.	1° 17·8′ W.	
D.R.	49° 50′ N.		3° 8′ W.	
Mean Lat. (approx.)	50° N.	Diff. Long.	2° W.	
Ship's head	090° C.			(approx.)
Compass error	9° W.			
Ship's head	081° T.			
D.F. Bearing	315° relative			
Sum	396°			
Subtract	360° (whenever the sum exceeds 360°)			
Great circle bearing	036°			
*Half-convergency+	1° (to nearest ½° in practice)			
Mercator bearing	037°			

*Towards the equator is plus in this case.

(vii) *By Consol Bearings.* Consol radio stations send signals which can be received by ordinary Medium Frequency receivers, such as D.F. receivers, at long range and which give postion lines. They can be received up to 1000 miles from the station, more at night, but must not be used within 25 miles of it and they are not sufficiently precise to be relied upon for making a landfall or for coastal navigation.

Accuracy is greatest on the normal (at right angles to), the line of aerials where, by day, the error seldom exceeds $\pm \frac{1}{2}°$. At 60° from the normal the error may have doubled and will increase rapidly as the line of aerials is approached.

The consol aerials radiate a pattern of alternate sectors of dots and dashes, each sector being divided by an 'equisignal' where the dots merge with the dashes into a continuous tone. The pattern rotates and passes over the navigator who can tell his angular position in the sector by counting the number of dots or dashes heard before the equisignal. Details of all Consol stations appear in the *Admiralty List of Radio Signals*, Vol. 5.

Operation. Having tuned to the frequency and identified the long, continuous signal interrupted with the station's call-sign, the navigator counts the number of dots (or dashes) which precede the equisignal and then the number of dashes (or dots) which follow it. In each cycle there is a total of 60 characters, but some will be merged in the equisignal. Both dots and dashes should be counted and the total subtracted from 60. Half of the characters missed should then be added to the characters heard first; this number is that which should have been heard, it being assumed that an equal number of dots and dashes was lost.

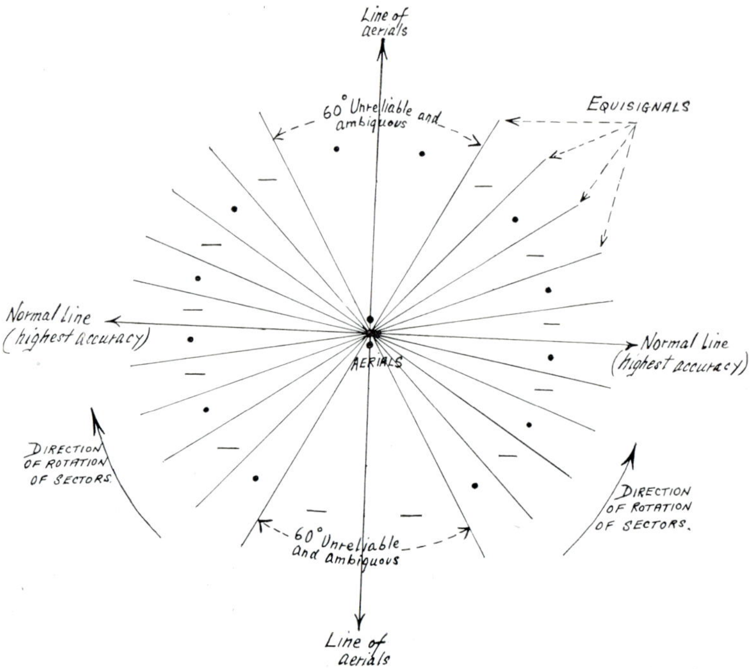

FIG. 2.9—Radiation Pattern of a Consol Station

Sequence of operation:—

(1) A long continuous note interrupted by the station call sigh.
(2) A short silence.
(3) The keying cycle consisting of dots and/or dashes.
(4) Another short silence before the cycle of operations is repeated.

During the keying cycle each sector moves in the direction indicated so as to occupy the position of the adjacent sector. They return to their former positions during the silence which follows.

EXAMPLES:

(1) Count of 38 dashes—equisignal—18 dots

Number of characters	38+18 = 56
Lost characters	60−56 = 4
Corrected count	38+ 2 = 40 dashes

Several counts should be taken and the average used.

(2) Observed count. Equisignal—54 dots—silence. Lost characters 6, i.e. 3 dashes and 3 dots. Dots were counted last, therefore the corrected count is 3 dashes.

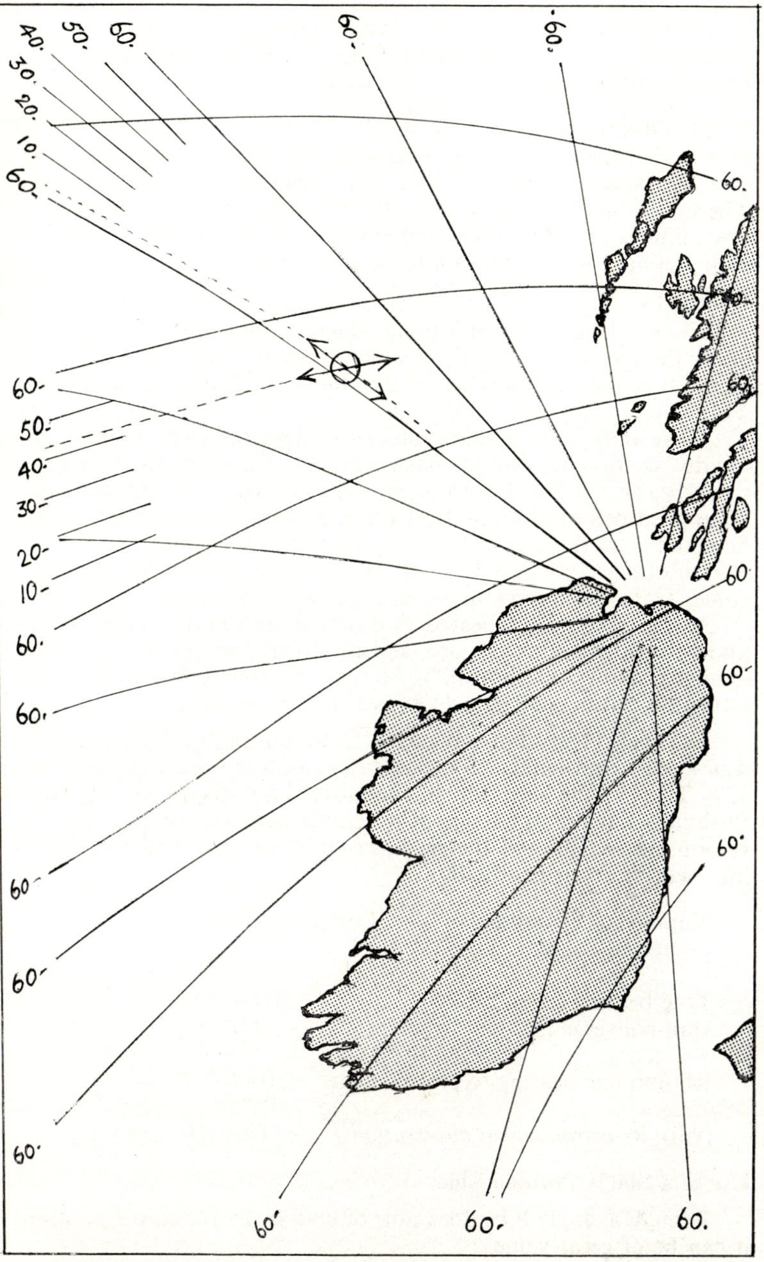

FIG. 2.10

Representation of a part of a Consol Chart showing how a fix is obtained. Bushmills bears approximately 127° T., distant 140 *M*. A count of Bushmills gives 4 dots and a count of Stavanger gives 43 dashes.

(3) Observed count 54 dashes—equisignal—silence. Lost characters 6, i.e. 3 dashes and 3 dots. Dashes were heard first, therefore the corrected count is 57 dashes.

(4) Observed count Equisignal—52 dots—Equisignal. If the two equisignals heard were of exactly the same duration the corrected count would be 60 dots, but since there is no means of knowing whether or not this is the case, the usual practice is to wait for a few minutes and then take further observations, by which time the ship will have moved to a more positive count.

EXAMPLES:

To find the position line along which the observer is located, use either the special Consol chart (which is usually the most convenient method) or the Consol tables in the *Admiralty List of Radio Signals*, Vol. 5.

These tables are divided into two pages for each station, one for dot sectors and one for dash sectors. They give the true great circle bearing of the ship from the station. Half-convergency must be applied before plotting it on the chart and this can most readily be taken from the special tables at the end of the Consol tables, but *Burton's* (page 128) may be used. Remember that the bearing given comes from the *station* and is not taken by the ship.

As the pattern is repeated in different sectors it is necessary to know in which sector you are, and the rough bearing of the station must be known (within 10°). It can be found from the ship's estimated position or by a D.F. bearing of the station.

EXAMPLE: The observed count from Bushmills Consol Station was 1 dot—equisignal—53 dashes. The station bore approximately 127° T. from the ship which, therefore, bore about 307° T. from Bushmills. If the difference of longitude between the ship and the station was 4° 30′ and the mean latitude was 56°, find the rhumb line bearing from Bushmills.

Number of characters = 54. Lost characters = 6.
Corrected count = 1+3=4 dots

True bearing from Table	311·1°
Half-convergency	− 1·9°
Rhumb line bearing to plot on chart	309·2°

(viii) *By astronomical observations. See* Part III.

Use of a Single Position Line.

Though a single P.L. does not, of course, fix the ship's position, it can be of great value.

(i) *To confirm track.* A P.L. which is parallel or nearly parallel to the required track will indicate whether or not the ship is deviating from that track.

(ii) *To check effective speed.* A P.L. which cuts the track at or near 90° gives a very good indication of effective speed, providing the course has not been altered since the last fix was obtained.

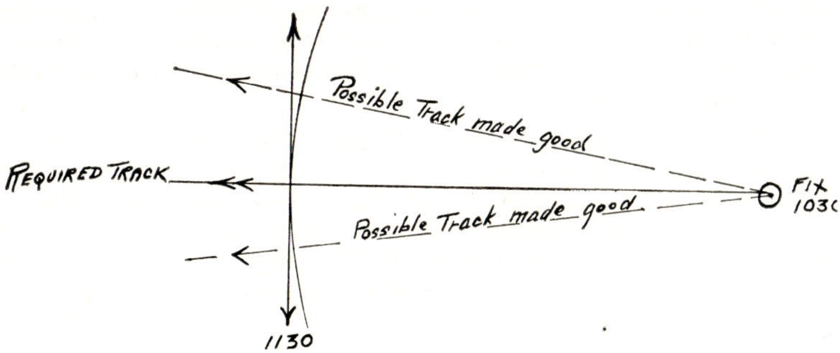

FIG. 2.11

(iii) *Leading Lines and Clearing Lines.* Two beacons or other objects may be placed to form leading marks so that, when in transit (symbol ø), they may be used to lead a ship through a channel or clear of a danger. They may also be placed to act as clearing marks, the transit line usually leading just clear of the danger. In this case the ship should keep the marks 'open' and not get on the wrong side of the line.

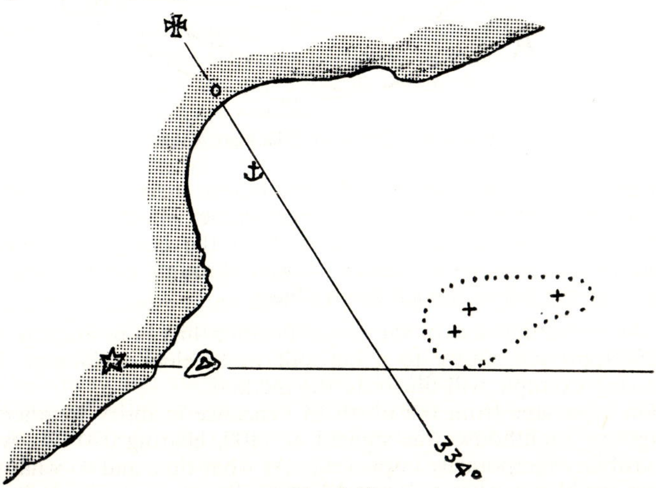

FIG. 2.12

Leading Marks—Church in transit with beacon. Leads to anchorage.
Clearing Marks—Lighthouse with island open to the north leads clear of shoal.

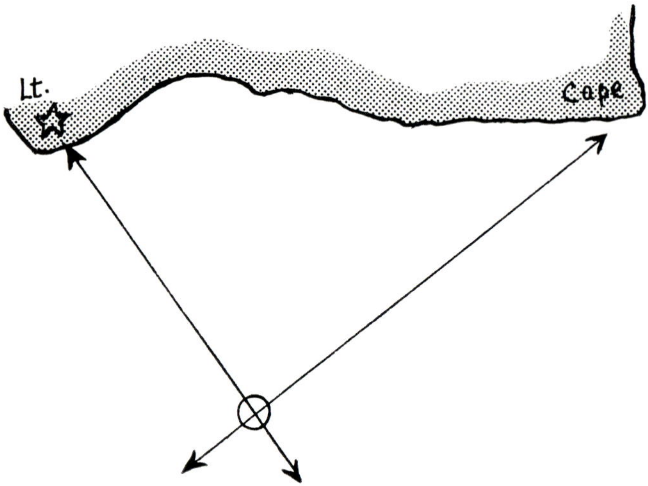

FIG. 2.15—Position by Cross Bearings

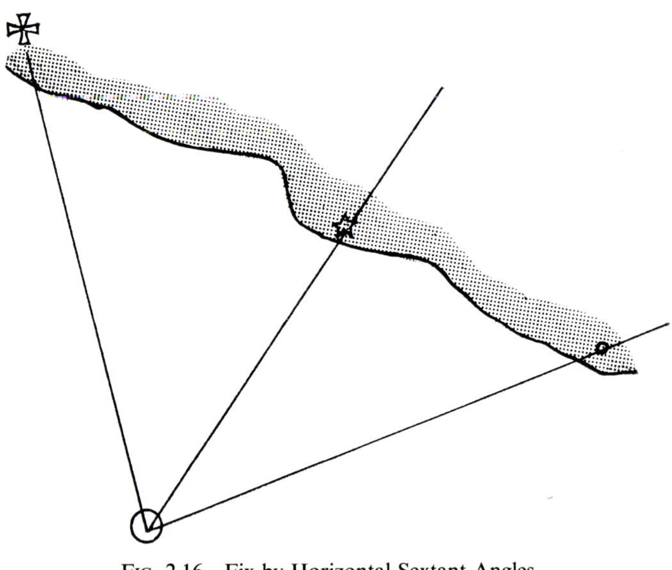

FIG. 2.16—Fix by Horizontal Sextant Angles

(iii) *Fix by Horizontal Sextant Angles.* This method has a high degree of accuracy, is completely independent of compass error, but requires three identifiable objects. If the horizontal angles between the three objects are measured with a sextant, a fix can be obtained immediately. In the example illustrated the information would be written: 'Church 47°, Lighthouse 34°, beacon'. The corresponding position on the chart can be found by using a Station-pointer, an instrument resembling a circular protractor with three long arms; the central one fixed at 0° and one movable arm at

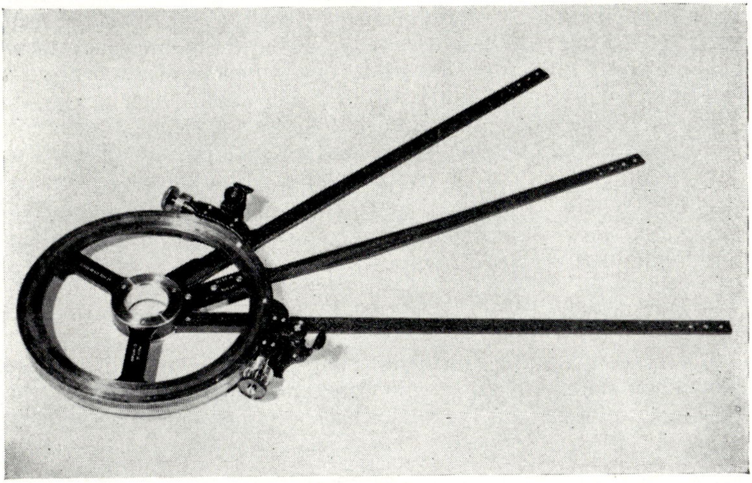

PLATE I—Station Pointer

each side. (*See* Plate 1, above.) In this case the left-hand one would be clamped at 47° and the right-hand one at 34°. The station-pointer is now manoeuvred on the chart until the edge of each arm lies over the appropriate object, the ship's position is then marked through a hole in the centre of the instrument.

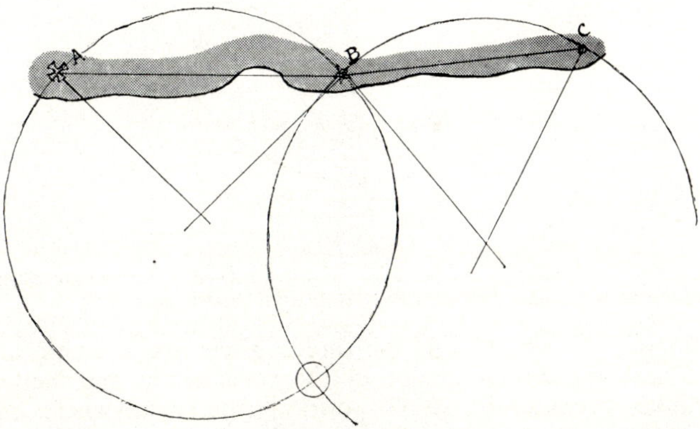

FIG. 2.17

When no station-pointer is available the same result can be obtained by drawing the lines on either tracing paper or a Douglas protractor.

If preferred, a geometrical construction can be employed to obtain a fix.

Method: Join the three points *A*, *B* and *C*. (Fig. 2.17). At *A* and *B* draw an angle equal to the complement of the angle observed between them (in the figure, 90°—47° = 43°). From the point where these Lines intersect draw a circle with its circumference passing through *A* and *B*. This is a circular position line upon which any point will subtend the angle observed (47°). Draw the same construction from *B* and *C* and obtain the second circular position line. The ship's position is where the two circular position lines intersect.

Note: Where an observed angle is greater than 90°, proceed as usual, but draw the complementary angles on the *far side* of the line joining the observed points.

The same method can be employed to get a fix when the compass error is unknown.

Three compass bearings are taken and, since they cannot be plotted on the chart, the angles between them are found and the position is plotted as if the angles had been measured by a sextant.

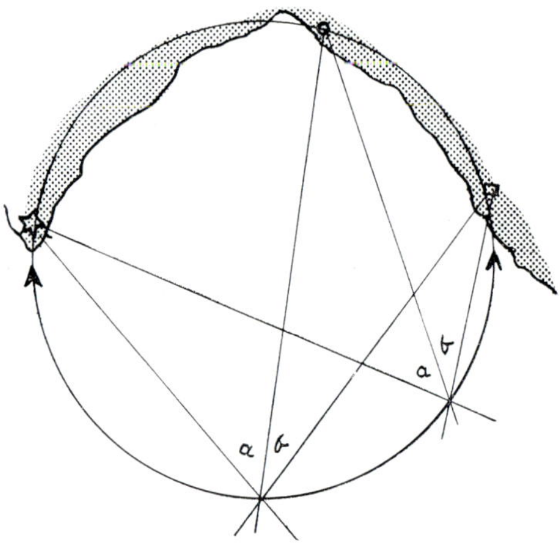

FIG. 2.18

Case where horizontal angles fail. If the observer is anywhere along the circumference the angles between the objects will be the same.

Choosing the objects. If the three objects and the ship are all on or near the circumference of the same circle, the method is unsuitable, because the ship's position can be anywhere on the circumference on the seaward side of the objects.

The objects chosen should thus be:

(*a*) All on, or nearly on, a straight line or

(*b*) With the centre object nearer the ship than the other two, or

(*c*) With the ship inside the triangle formed by the three objects.

(iv) *Fix by Bearing of a Rising or Dipping Light.* When visibility is good a powerful light will often show its loom (reflection) in the sky when the actual light is a little below the horizon. When the loom is approached and the light appears on the horizon it is said to 'rise'; conversely, when bound away from the light it is said to 'dip' at the instant it disappears below the horizon.

The distance at which a light will rise or dip depends on its height and that of the observer's eye. The visibility of any light given on the chart has been calculated for an observer whose eye is 15 ft. above sea-level. At this height the horizon is 4·5 M distant. To find the rising or dipping distance for any other height of eye it is necessary only to subtract 4·5 M from the visibility given on the chart and then add the distance to your own horizon. *See* the 'Distance of Sea Horizon' Table in *Burton's*, page 137.

It is thus sometimes possible to obtain a fix by taking a bearing of a light at the instant or rising or dipping. Great accuracy should not be expected of this method, as the distance at which the light appears or disappears over the horizon is affected by the scend (rise and fall with each wave) of the vessel and also by the existing atmosphere condition.

FIG. 2.19—Distance by Rising and Dipping Bearings

EXAMPLES FOR EXERCISES: Find the distance at which each of the following lights will rise or dip.

(1) Duncansby Head Lt.	Visible 20 M.	Height of eye 35 ft.
(2) Noss Head Lt.	Visible 18 M.	Height of eye 25 ft.
(3) Smalls Lt.	Visible 17 M.	Height of eye 20 ft.

Answers: (1) 20—4·5+6·9 = 22·4 M.
(2) 18—4·5+5·8 = 19·3 M.
(3) 17—4·5+5·2 = 17·7 M.

(v) *Running Fix* is used when only one object is available for bearings and it is not possible to get a vertical sextant angle. A bearing is taken and the P.L. plotted on the chart; later, when the bearing has changed sufficiently (not less than 20°) to ensure a satisfactory angle of cut, another bearing is taken and plotted. The first P.L. is then transfered to the time of the second P.L. The point where they intersect constitutes the running fix. Its certainty depends not only upon the accuracy of the P.L. and their angle of cut, but also on the accuracy of the course and distance run during the interval between them.

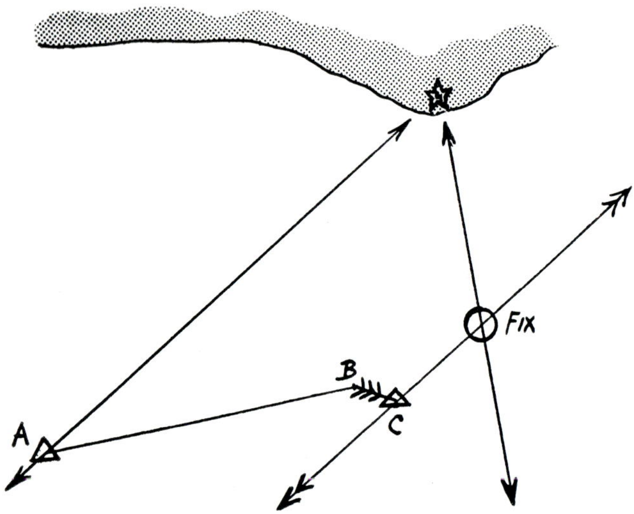

FIG. 2.20—Position by Running Fix

To transfer the position line. (Fig. 2·20.) From any point (*A*) on the first P.L., draw the course and distance (*AB*) travelled through the water during the interval. From point *B* draw the set and drift of the tide, if any (*BC*). Through *C* draw the transferred position line parallel to the original position line, distinguishing it by two arrowheads at each end. Since the vessel was somewhere on the first P.L. at the time of the first observation she will be somewhere on the transferred P.L. at the time of the second observation. As she is also somewhere along the P.L. obtained by the second observation, she must be at the position where they cross.

When there is no wind or tide acting across the course, a running fix can be made without plotting the position lines on the chart.

Doubling the Angle on the Bow. The angle which the first bearing makes with the vessel's heading is noted and the time taken. When this angle has been doubled the time is noted again. The distance travelled in the interval is the distance off the object at the time of the second observation.

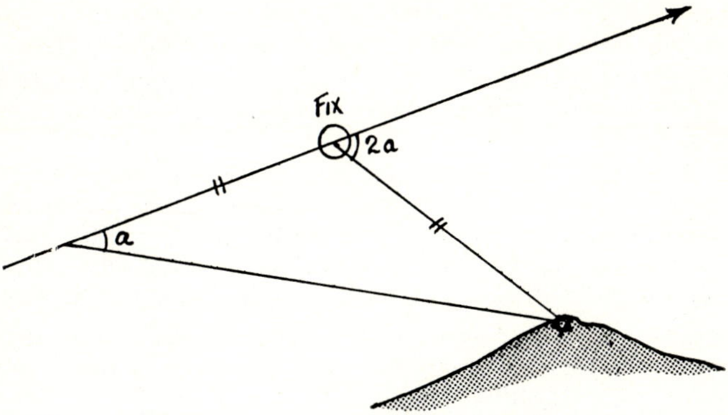

FIG. 2.21—Doubling the angle on the bow

The 4-point Bearing. The time is noted when the object is exactly 4 points on the bow, that is when the bearing makes an angle of 45° with the course steered, and again when it is abeam (90° to the course steered). The distance run in the interval is the distance off the object when abeam. This is the simplest of all running fixes but it gives the distance from the object only when the ship is nearest to it.

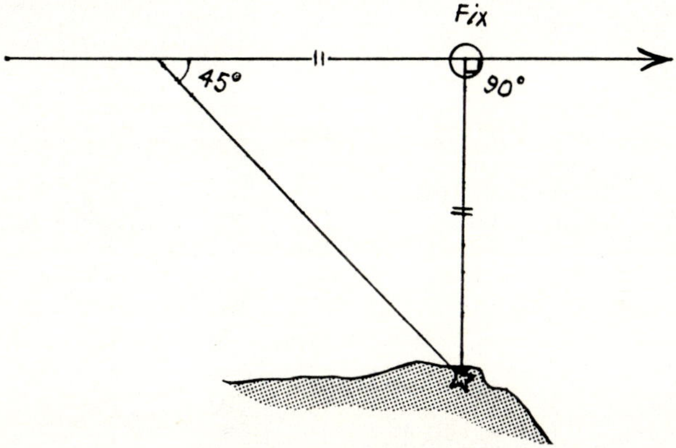

FIG. 2.22—Four-point bearing

Once again, the accuracy of any running fix depends upon the accuracy with which the run between bearings is known.

To Find the Set and Drift of Tide or Current experienced. When the ship's position has been fixed the actual set and drift of the tidal stream experienced may be found. The E.P. (found by applying leeway, if any, to the course steered) is plotted and any difference between this position and the fix must be due to the tide or current. (*See* Fig. 2·4.)

Worked Example: The worked example on the inset chart should be carefully followed. It can be worked on the Practice Chart.

Note: (1) In practice a chart of a larger scale would be used and this would give more accurate results than are possible on a small-scale chart.

(2) The work falls into two parts. (*a*) The preparation of courses to be steered and distances to be run to pass certain distances off (which would vary according to visibility, etc.). This may be done before starting. (*b*) The periodical fixing of the ship's position and adjustment of the course which takes place during the passage.

EXAMPLE 1: On a voyage from Torbay to Falmouth find the true and compass courses to steer and distances to run from an anchorage with Orestone (50° 27·3′ N., 3° 28′ W., height 108 ft.), bearing 035° T. and subtending a vertical angle of 0° 40′ to pass Berry Head not less than 1 M, Start Point 3 *M* and 3 *M* south of Eddystone Lighthouse. On the course from Start Point to Eddystone the tide was expected to set 224° at 2 kn. The ship's speed is 12 kn. Variation (for 1974) 7° W. Use Deviation Card on page 8.

From anchorage with Orestone 035° T. $1\frac{1}{2}$ *M* to 1 *M* off Berry Head	Co.	151° T. distance $2\frac{1}{2}$ *M*
	Var.	7° W.
		158° M.
	Dev.	$3\frac{1}{2}$° W.
	Co.	$161\frac{1}{2}$° C.
From 1 *M* off Berry Head to 3 *M* off Start Point	Co.	200° T. distance 14 *M*
	Var.	7° W.
		207° M.
	Dev.	4° E.
	Co.	203° C.
From 3 *M* off Start Point to 3 *M* south of Eddystone	Required track	264° T.
	Co. to steer	270° T.
	Var.	7° W.
		277° M.
	Dev.	$1\frac{1}{2}$° W.
	Co.	$278\frac{1}{2}$° C.

2. What angle should be set on the sextant to round Berry Hd. Lt. Ho. (191 ft.) at a distance of 1 M? (*Answer*: 1° 47′.)

3. At 1650, co. 203° C., Start Pt. Lt. Ho. bore 250° C. and Berry Hd. Lt. Ho. 010° C. Required the Lat. and Long. of the fix. (*Answer*: 50° 15½′ N., 3° 30½′ W.)

4. At 1711, co. 203° C., Prawle Pt. ø Bolt Hd., and Start Pt. bore 300° C. Required the position. (*Answer*: 50° 11′ N., 3° 33′ W.)

5. At 1748 Bolt Tail 61° Prawle Pt. 29° Start Pt. Required the Lat. and Long. of the fix. (*Answer*: 50° 9½′ N., 3° 45′ W.)
Note that this is obtained by the horizontal angle method.

6. If the height of the observer's eye is 25 ft., at what distance would you expect to see the Eddystone Light? Range of the light given on the chart is 17 M (*Answer*: 18·3 M.)

7. At 1845 Eddystone Lt. bore 294° C. and at 1915 it bore 320° C. Ship's estimated course and speed made good is 264° T., 13¼ kn. What distance will you pass abeam of Eddystone Lt.? (*Answer*: About 3 M, but 26° is a small angle for a Running Fix.)

Exercises in Chartwork.

The following six questions may be answered by drawings to scale (1 cm to 1 M is convenient). No chart is needed. Use variation 5° W.

1. Ship steering 083° C., Dev. 2° E. speed 10 kn, current 180° T. at 2 kn. What is:

(*a*) The course and speed made good?

(*b*) The true course steered?

2. Ship steering 083° C., Dev. 2° E., speed 10 kn, current 180° T. at 2 kn, wind N., leeway 10°. What is:

(*a*) The course and speed made good?

(*b*) The compass course steered?

3. Course to make good 350° T., Dev. 6° W., current 270° T., rate 3 kn, wind W., leeway 3°. What are the true and compass courses to steer if the ship steams at 10 kn?

4. Track 070°, ship's speed 6 kn. Compass error 5° W. Tide set in first hour 350°, 3 kn; in second hour 055°, 2½ kn; in third hour 070°, 2 kn. What is the compass course to steer to make good the track course and what is the speed made good in three hours?

5. *A*, *B* and *C* are three land-marks to the northward of the ship. *A* bears 270°, 5 M from *B* and *C* bears 090°, 4 M from *B*. The horizontal angles between them are: *A* 37°, *B* 52° C. What is the ship's distance and bearing from *B*?

6. *A*, *B* and *C* are three landmarks to the westward of the ship. *A* bears S.W., 5 M from *B*. *C* bears N., 2 M from *B*. Horizontal angles between them are: *A* 25°, *B* 47½° C. What is the bearing and distance of *B* from the ship. If *B* bore 270° C., what is the compass error and deviation?

Practice Chart 5103.

Positions and Courses

7. Plot position *A* in Lat. 50° 10½′ N., Long. 1° 20′ W. The ship steams 063° T. for 59 M. What is the position arrived at?

8. A vessel in Lat. 50° 8′ N., Long. 1° 10′ W. steams 080° T. for 50 M, then a/c (alters course) to 191° and runs another 40 M. Find the final position.

9. Beachy Head Lighthouse bears 057° T., distant 7 M. Find the Latitude and Longitude of the ship's position.

10. From the position in Question 8, find the true course to steer and the distance to a position 10 M S.E. of St. Catherine's Point.

Fixes by Cross Bearings and Courses

11. At 1300 Radio D.F. bearings were taken of Start Pt. 264° T., Casquets 168° T., Shambles 036° T. Set course to pass 6 M abeam of Beachy Head Lt.

Find the ship's position at 1300 and true course to steer and distance to go.

12. At 2151 Nab Tower bore 310° T. and Owers Lt. Vessel (Gp. Fl., to eastward of Nab), bore 045° T.

Find the ship's speed and new course to steer to the position off Beachy Head in 8.

Fixes by Vertical Sextant Angles

13. S. Foreland Lt. is 374 ft. above high water (the tower is 69 ft. high). The vertical angle subtended is 2° 4′. What is the distance off?

14. N. Foreland is 188 ft. above high water. What angle would you set on the sextant to round it at 3 M?

Where would you find the heights of these lights?

15. Beachy Head Lt. (100 ft.), bearing 016° C. Compass error 11° W. Vertical angle 0° 14′. Find the distance off and position.

Practice Chart 5067.

Running Fixes

(A convenient Time, Speed and Distance Table is in *Burton's* Inside Front Cover.)

Note: In the next six questions bearings and courses are all given magnetic. Variation 8° W.

16. Course 058° M. Trevose Head Lt. H. bore 070° M. log 23. Later, Trevose Head Lt. H. bore 089° M. log $28\frac{1}{2}$° M. What is then the latitude and longitude of the ship's position? How far off will Trevose Head be when abeam?

17. Course 243° M. Speed 6 kn. At 0735 Carnbrea Monument, bore 174° M. At 0835 Carnbrea Monument, bore 141° M. What is the Latitude and Longitude at 0835? How far off is the Whistle Buoy?

18. Course 269° M. Speed 8 kn. Tide setting 190° T., $1\frac{1}{2}$ kn. At 0315 Sevenstones Lt., bore 230° M. and at 0345, 208° M. What was the distance off at 0345?

19. Course 110° M. Wolf Rock Lt. H., bore 4 points on the port bow, log 37. Later, Wolf Rock Lt. H. was abeam to port, log 41. Tidal stream considered to be negligible. What was the distance off, when abeam?

A lighthouse is abeam when it is at right angles to the ship's head (not necessarily her track), but this is generally about the closest point of approach.

20. Course 127° M. 0110 Round Island Lt. 144° M., log 32. Tidal stream 060° at 2 kn. 0215 Round Island Lt. 165° M., log $38\frac{1}{2}$. What is the distance off at 0215 and when Round Island is abeam?

21. Course 066° M. 2115 Lizard Lt. 035° M, log 62. Tide as at ◇C, 5h after H.W. Springs at Devonport. 2215 Lizard Lt. 019° M, log 70. What is the distance off at 2215 and when abeam? At what time should the Lizard Lt. be abeam?

Finding the Estimated Position

22. From 50° 25′ N., 6° 31′ W. the ship steered 093° C. (Compass error 7° W.) at $10\frac{1}{2}$ kn. The tide was expected to set 045° at $1\frac{1}{2}$ kn. Find the E.P. 2 hrs later and the course and speed expected to be made good.

23. At 1100 in 50° 36′ N., 5° 15′ W., course was 230° C. (Dev. from the Deviation Card, Var. 8° W.). Leeway 10°, wind N.W. Ship's speed $5\frac{1}{2}$ kn. Tide setting S.W. at $\frac{1}{2}$ kn. Find the E.P. at 1500.

24. 1200 Lizard Lt. H. 352°, distant 9 M. Course 077° T., speed 5 kn, wind S., leeway 3°. Tide until 1400, set 225° at 2 kn and afterwards slack. Find the E.P. at 1500 and the course, distance and speed expected to be made good.

25. 0830 Bishop Rock Lt. H. 320° T., distant 5 M, steering 075° C. (Deviation from the card, variation 8° W.). Leeway 5°.

wind S. Speed 7 kn. Expected tide N., 1 kn. Find the E.P. at 1200. What should be the bearing and distance of the Longships Lt. H. at this time?

Exercises on Practice Chart 5103.

Chart Paper A

1. Plot the following positions and give them in the form of a bearing from and distance off the nearest lighthouse.

A	Lat. 50° 39′ N., Long.	0° 10′ E.
B	49° 36½′ N.	0° 15½′ W.
C	49° 45′ N.	2° 33′ W.
D	49° 53′ N.	5° 55′ W.

2. Plot the following positions and give them in the terms of Latitude and Longitude.

E	Lundy Is. North Lt. bears	047° T., distant	10¾ M.
F	Lizard Lt.	029½°	11 M.
G	Les Hanois Lt.	103°	13¾ M.
H	St. Catherine's Lt. Ho.	314°	8 M.

3. What is the true course and distance from a position where Ushant main Lt. bears 206° T. and Vierge Lt. 098° T. to position *D*?

4. What is the true course and distance on each leg from position *D* to position *E* passing 5 M off the Longships?

5. What is the true course and distance from *A* to *C*?

6. If, on the above course, C. de la Hague Lt. should be sighted at a distance of 22 M, on what true and on what relative bearing should it rise?

Practice Chart 5067.

Chart Paper B

Using Var. for 1974 and Dev. from the Deviation Card,

1. Find the Lat. and Long. of the following fix: Trevose Hd. Lt. Ho., bore 049½° C., Park Hd., bore 089° C., Kelsey Hd, bore 191½° C., Ship's Head 039° M.

2. Find the Lat. and Long. of the following fix: Course S.W. magnetic. Manacle Pt., bore 002° C., Lizard Pt. Lt. Ho., bore 266° C. Coverack Hotel, bore 338° C.

3. At 0920, Sennen Church and Longships Lt. Ho. were in transit and bore 095° C., and Cape Cornwall bore 054° C.

(i) What is the deviation?

(ii) What is the Lat. and Long. of this position?

4. Course 330° T. Speed 10 kn. Slack Water. Wolf Rock Lt. Ho. bore 303° C. 30 min. later Wolf Rock Lt. Ho. bore 269° C. How far off Wolf Rock were you at the time of the second observation?

5. 2115, Bishop Rock Lt. (Scilly Isles) bore 297° C. Course 262° T., speed 6 kn 4 hrs before High Water at Devonport. At 2215, Bishop Rock Lt. bore 319° C.

(i) What is your distance off at 2215?

(ii) Assuming the Course and Tide to be unchanged, how close will you get to the Bishop Rock Lt.?

6. What is the chart symbol for a rock awash at the level of chart datum?

Practice Chart 5103.

Chart Paper C.

Variation 12° W., deviation from the Deviation Card.

1. At 1000 ship steaming 248° C., at 16 kn Casquets Lt. Ho. bore 207° C. At 1200 Les Hanois Lt. Ho. bore 124° C. Required the 1200 position.

2. 0700 L. de Vierge Lt. Ho. bore 160° T., distant 10 M, Course 280° C., log 40. 1000 a/c 310° C., Log 85. 1200 log 102. Tide set 240° T. at 2 kn from 0700 to 1000 and 270° at $1\frac{1}{2}$ kn from 1000 to 1200. Wind N., leeway 5° on first course and 3° on second course. Find the E.P. at 1200.

3. At 1305 Wolf Rock Lt. Ho. bore 005° C. Sounded: 44 fathoms. Required the approximate distance from Wolf Rock.

4. Plateau des Triagoz and Les Sept Iles Lts. were in transit bearing 093° C. and at the same time I. de Bas Lt. bore 193° C. Give the ship's position as a bearing and distance of I. de Bas Lt.

5. At 1600, Godrevy Lt. Ho. bore 120° T., 15 M. Course 049° C., speed 18 kn. At 1900, Hartland Pt. Lt. bore 127° C. and Lundy Is. South Lt. bore 056° C. Find the position at 1900 and the set and drift since 1600.

6. How is the 6-fathom line marked on the chart?
How is the 10-fathom line marked on the chart?
How is the 100-fathom line marked on the chart?

Practice Chart 5067.

Chart Paper D

1. 1006. Horizontal sextant angles were observed as follows:
St. Antony Head 42°, Nare Head 70°, Dodman Point (East end).
Required the Lat. and Long. of this position.

2. 1209, Lizard Radio Beacon bore 0° Relative and Round Island Radio Beacon bore 250½° Relative. Ship's head 060° C. Quadrantal error as Quadrantal Error Curve on page 27. What is the Lat. and Long. at 1209?

3. Rounding the eastern edge of the Scilly Isles, it is desired to keep 1 M east of the rocks between Eastern Island and St. Martins. What angle should you set on the sextant, using the Daymark on St. Martins and Eastern Island as marks?

4. In misty weather, speed 5 kn, at 0536, Runnel Stone buoy was sighted bearing North magnetic. What course can be safely steered now, and to what course and when could you safely alter to approach Penzance Bay?

5. 0450, Bishop Rock bore 153° C. Sounded: 44 fathoms. Required the Lat. and Long. and the distance off Bishop Rock. Course 222° C.

6. What is the difference between a flashing and an occulting light?

Practice Chart 5103.

Chart Paper E

Variation 9° W. throughout, Deviation from Deviation Card.

1. 0130, Ushant Lt. rose bearing 111½° C. Height of eye 35 ft. Course 020° C. Required the distance off Ushant and compass course direct to the Needles.

2. 0322, Ushant Lt. dipped bearing 176° C., course 052° T. Required the Lat. and Long. of the ship's position.

3. 0730, Radio bearings of Casquets 022½°, Ushant 168°, Round Island 234° all relative. Ship's head 052° T. Required the position. Are you making good the course?

4. 1300, Portland Lt. Ho. 293½° C., Anvil Pt. Lt. Ho. 350½° C. The ship has made good her course and logged 15 kn, find the Observed Position. Find the set and drift of the tide since 0730.

5. From the fix at 1300 a/c to a position 1 M west of the Needles. If the tide is now setting 080° at 2 kn, find the compass course to steer and the E.T.A.

6. (*a*) What sort of sound does a diaphone fog signal make? (*b*) What colour are lightvessels painted?

Practice Chart 5067.

Chart Paper F

1. At 0730, Black Head bore 044° C. and Lizard Pt. Lt. Ho., 230 ft., 010° C. Compass error unknown. Lizard Pt. Lt. Ho.

subtended a vertical angle of 0° 41′. What is the distance off Lizard Pt. Lt. Ho., the Lat. and Long. of this position, and the compass error and deviation?

2. From the 0730 position find the Courses to steer and the total distance to go to a position with the Seven Stones Lt. V. abeam to port, distant 2 M. Wind S.W., expected leeway 5°, speed 6 kn. For the first 2 hours the tide is expected to set north in to Mounts Bay at 1 kn. Assuming the Runnel Stone buoy will be visible at 2 M, when would you expect to see it?

3. In Mounts Bay, deviation unknown, the following bearings were taken: Lizard Pt. Lt. Ho. 116° C., St. Michael's Mount 012° C., Gwenapp Hd. 279° C. What is the ship's position and the deviation of the compass?

4. Co. 063° C., speed 5 kn, steering parallel to the coast in poor visibility, Pendreen Lt. Ho. was observed bearing 183° C. at 0912. At what time and to what Course should you alter to anchor in St. Ives Bay?

5. 0915, Round Island Lt. Ho. (Scilly Isles), bore 190° C., log reading 97. Course 073° C. Speed 5 kn. Tide estimated to be setting 230° at 1·6 kn. 1027 Seven Stones Lt. V., bore 117°, log 103.

(i) What is the distance off Seven Stones Lt. V. at 1027?

(ii) What is the Lat. and Long. of this position?

(iii) What will be the distance off the Seven Stones Lt. V. when abeam and at what time will this be, assuming your course and the tidal stream to be unchanged?

6. Wolf Rock Light is described as Alt. Fl. W.R., and Godrevy Light as Fl. W.R. How do these two lights differ in character?

Practice Chart 5103.

Chart Paper G

Use variation given in the questions and deviation from the Card, quadrantal corrections from the Curve.

1. Ship steering 080½° T. at 15 kn. At 2210 relative radio bearing of Land's End beacon was 273°. The set of the tide was estimated at 235° T., 1½ kn, for the next 3 hours and the visibility was about 5 M. If you do not see Lizard Head, when should you alter course for Falmouth to pass 3 M off Manacles buoy?

2. 1700, Ship's head 178° T. Lundy Is. Radio Beacon, bearing 251°, Land's End Radio Beacon 351° relative. What is the ship's position and would you consider it reliable?

3. Assuming position at 1700 to be 50° 55′ N., 05° 58′ W., the ship continued to steer 178° at 15 kn. If there is no tide, at what time would you alter course to pass 5 M off Lizard Lt. Ho., passing

midway between Wolf Rock and the Runnelstone whistle buoy? Give the compass course and E.T.A.* off Lizard. Variation 9° W.

4. At 2250, Lizard Hd. Lt. bore 000° T., 5 M. Find the compass courses and distance to Banc de Seine Lt. Vessel passing 10 M off Pt. de Barfleur on each course. Speed through the water 15 kn. Tide estimated to set 245° at $1\frac{1}{2}$ kn for first 3 hours afterwards negligible. Variation on first course 8° W. and on second course 7° W. Required E.T.A. at alter course position and at destination.

5. On another voyage at 1440 in D.R. 49° 54′ N., 3° 22′ W. when bound for Cherbourg in foggy weather, course 080° T., speed 9 kn, tide estimated to set 190° at $1\frac{3}{4}$ kn. At 1640 began to take hourly soundings: 1640, 38 fathoms; 1740, 34 fathoms; 1840, 44 fathoms; 1940, 65 fathoms. When would you expect to be north of Alderney Lt. Ho. and how far off?

6. What do the following chart symbols mean:
(*a*) P.A.; (*b*) E.D.; (*c*) (U), against a light?

*Expected time of arrival.

CHAPTER 3

THE VOYAGE

'A voyage—a victory!' *Arab saying.*

Preparing a Cruise. Before cruising on an unfamiliar coast a considerable amount of homework should be done. Besides considering the time at your disposal, and your aims: to go as far as you can or see more of fewer places; the experience or otherwise of your crew—all outside the scope of this book, there is much navigational information which you will require.

Get to know the coast you intend to cruise on. The chief reason for this is, of course, the safety of your ship; but with this knowledge you are more likely to make right decisions on your day-to-day programmes. Buy one or two small-scale charts of the region to be visited, but do not get any others at this preliminary stage, because the later you purchase them the fewer corrections to them you will have to make (*see* page 13). You will also need the appropriate *Admiralty Pilot* (*see* page 13) and the best *Sea Guide* (*see* page 14). There are also the *Handbook of the Cruising Association* and the Clyde Cruising Club and Irish Cruising Club. The Royal Cruising Club compiles a *Foreign Port Information Folio*, corrected each year, and copies are held by The Little Ship Club and the Cruising Association, both of which can supply members with photostat copies of either individual sheets or the whole folio.

Spread out the chart and, with the *Pilot* and *Sea Guide*, work along the coast, compiling a list of anchorages and ports as you go.

An example follows:

Pilot page	Sea Guide page	Place	Miles from previous port	Approach	Shelter	Mooring	Remarks
201	35	None-such	22	D & N but not in strong Northerly winds	Good	a/s YC pontoon	Hospital (XIII century Cathedral)

Abbreviations: D day, N night, a/s alongside, f/a fore-and-aft, s/t stern-to

Ring page numbers if they should be read again before approach to the place.

You will now know what other charts you will need. Usually Admiralty charts are the best, but there are places, where intricate inshore waters, inaccessible to large ships but a fine cruising ground for yachts, are better covered by local government charts. It is better to have too many charts than too few, for even though a visit to a certain place may not be planned, stress of weather may drive you there.

Work through the *Pilots* and large-scale charts, noting in the margin of the former anything you think would need reading again, and on the charts pencil arrows or ring inconspicuous (inconspicuous on the *chart*) dangers or useful marks which might otherwise be overlooked. At the end of the *Pilot* books are a series of views. Any that illustrate the places on your list should be noted.

Lights.

Reed's Nautical Almanac gives a complete list of lights from Denmark to Brest, beyond consult the *Admiralty List of Lights.*

Tides.

At this stage note only the range at Springs and Neaps of ports which are deep enough for entry (*Reed's* or *Admiralty Tide Tab les*)

Radio Weather Forecasts.

Local reports are particularly useful and since the vocabulary used is small, even some foreign language broadcasts may be listened to with advantage.

Time Signals.

Time Signals will be needed if sights are to be taken during passage (*See Reed's* and *The Radio Times* for B.B.C.) *Admiralty List of Radio Signals*, Vol. IV (all-world).

Stores. Note where fuel, fresh water and stores are obtainable and repairs may be carried out.

Port of Entry. In some countries entry must be first made in certain Ports of Entry. Here a document permitting other places to be visited and showing any prohibited areas will be issued.

Some Yacht Clubs appoint Honorary Port Officers who can advise on special problems, such as leaving a yacht for a few weeks, or laying-up.

Documents. A *Certificate of Registry* or, if the yacht is not registered, a *Certificate of Ownership*, which can be obtained from the Royal Yachting Association, Cruising Association or Motor Boat Association.

The Royal Yachting Association have an agreed International Certificate for Pleasure Navigation with many countries and this can be issued to unregistered yachts to be used as 'ship's papers' upon application to the R.Y.A.

An Overseas Helmsman's Licence and Certificate of Registration can also be issued to those who wish to use small boats on the inland waters of Europe, where the local government require certificates of 'proficiency', especially for small powered-craft.

If the yacht is on charter, a copy of the *Charter Party*. If going foreign, a *Licence* under the Small Craft Regulations 1953 should be obtained from the Custom House.

A yacht should be insured and it is as well to check that your cruising ground is covered by the insurance policy.

Crew. Each member should have a valid *passport*, if going foreign. Few countries now require visas.

Making a Passage.

In confined waters, navigation is by *pilotage*; directions and distances being estimated by eye, though the compass and lead (or depth meter) may be used when necessary and the chart briefly consulted. When a vessel has moved off shore, pilotage gives place to *coastal navigation*; courses being steered and bearings being taken by compass and the ship's position being fixed on the chart as described in the previous Chapter. When out of sight of land recourse is had to *astro-navigation*. (*See* Part III.)

Preparation. Preparation is an important part of any voyage, even if only a short passage. It is not generally possible to memorise all the details necessary, so they should be entered in the navigator's notebook before starting, which is usually much easier to consult than the chart when under way. The specimen page shows how this may be done. (*See* Fig. 3.1.) Most navigators prefer to start from the bottom of the page, when, if in a channel, the buoys, etc., to be passed to starboard are drawn on the right and those to be passed to port on the left. The time of high water should be noted and, with the aid of the tidal atlas, the direction of the tidal stream shown for each hour once the time of departure has been decided: indeed the direction of the tidal stream may well decide at what time you will sail. In the example given the yacht is sailing from the Solent to Poole harbour. The navigator is familiar with the Solent, and so starts his notes from Hurst Castle. Characteristics of lights have not been shown in this case since this passage is to be made in daylight, but fog signals are noted. So, of course, is the tide. Although it is not intended to enter Christchurch harbour, details may be noted if the yacht's draft permits her to enter in case it became necessary to do so. Details of the entrance to Poole harbour and the channels within it would also be noted down in the same way.

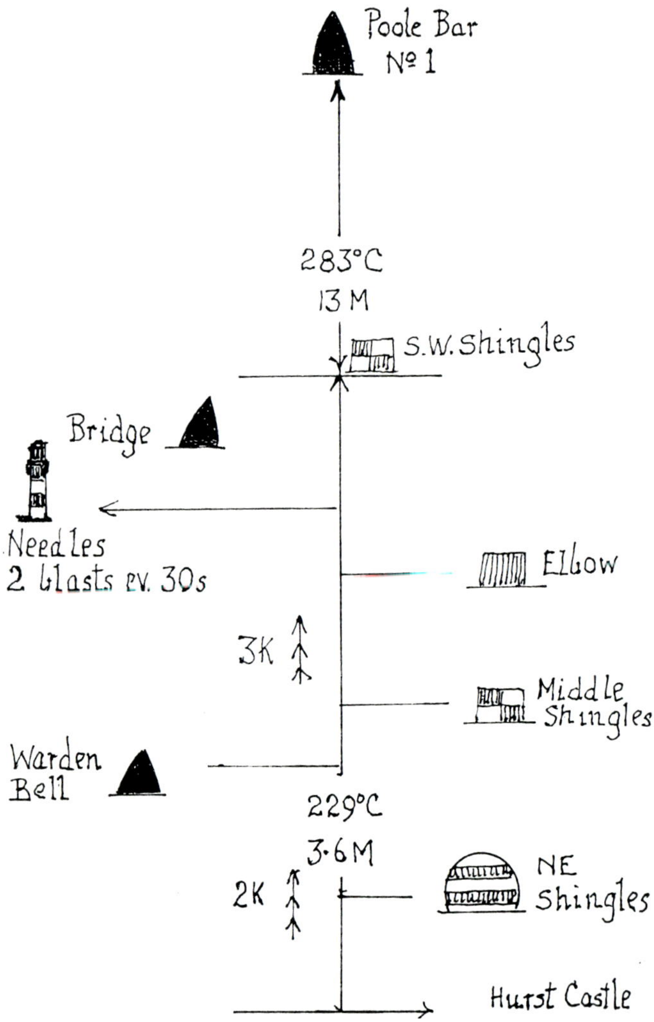

FIG. 3.1
A page from a navigation notebook.

Allowing for the Tide. The basic method of allowing for the set of the tidal stream has already been explained in Chapter 2. Here we are concerned with the effect of the tide over a considerable period. Take the Tidal Atlas. The set of the tidal stream is shown for every hour before and after high water at Dover. Look up the time of high water for the day of sailing and pencil this time on the chartlet marked H.W. Dover, adding an hour on the subsequent pages, so that at any time during the passage we can easily see what the tide is doing.

On a coastwise passage, unless we are steering across a bay, in and out of which the tide sets, the tide is likely to be either against or with us, and only our ground-speed will be affected, but on a cross-channel passage with the tide on one side or the other, allowance must be made in the course to be steered. We could draw a series of tide-triangles for each hour, but it is often sufficient to list the hourly sets, take the average, and allow this. Providing that our speed remains constant during passage, this is accurate enough but, should our speed be much altered by, for instance, the change in direction or strength of the wind the problem will have to be worked out again.

Take, for example, a passage from the Needles, Isle of Wight, to Cherbourg, approximately 180°, 60 M distant. High water at Dover on this particular day is at 1600 and our E.T.A. off the Needles is 1800. We estimate our average speed at 6 kn and our T.O.P. (Time on Passage) will, therefore, be about 10 hrs. Mark the chartlet '2 hrs. after H.W. Dover', 1800 and the following 10 chartlets 1900, 2000, etc. On the edge of a piece of paper mark your hourly progress and move this scale from one chartlet to the next, listing the set and drift for each hour, estimating the set by eye and noting that, in this case, Spring rate is applicable.

Time	From H.W. Dover	Set	Drift
1800	2 hrs. after	250°	2 M
1900	3 hrs. after	260°	2 M
2000	4 hrs. after	260°	2 M
2100	5 hrs. after	260°	2 M
2200	6 hrs. after	Change	
2300	5 hrs. before	095°	1½ M
0000	4 hrs. before	095°	1½ M
0100	3 hrs. before	090°	2½ M
0200	2 hrs. before	095°	1½ M
0300	1 hr. before	095°	1½ M

If these vectors are drawn, the resultant tide for the 10 hrs. will be found to be 155°, 1 M, and it can be seen from the table drawn up that the flood and ebb streams have nearly cancelled each other. The tide in the English Channel repeats its movement about every

12 hrs, and the position reached after that time will be on track, if that course has been steered, though the vessel will have followed a serpentine path over the ground to arrive there. In practice, therefore, we need concern ourselves only with the effect of the tide beyond the 12-hr. position.

In the following exercises in Cross-Channel Tidal Problems, proceed as follows: On the navigation chart join the points of departure and arrival and measure the track and distance to go and, allowing the expected speed, find the approximate time on passage (T.O.P.).

Using the tidal chartlets, and starting from the appropriate hour before or after H.W. Dover, list the set and drift of the tide for each hour of your passage. Draw these tidal vectors joining one to the next, on paper to a scale of, e.g. 1 cm to 1 M, and find the resultant set and drift of the tide for the passage. Transfer this tidal vector to the navigation chart and join the end of it to the arrival position. This is the wake course required.

As the tidal stream generally reverses in direction, one can often assess the total effect of the tide without drawing all the vectors. In Exercise 1, for example, the ebb sets about 265°, 9½ M and the flood about 110°, 3 M. A total tidal vector of 250°, 7 M would be a close estimate.

If your speed varies and you take a longer or shorter time on the passage, the effect of the tide will, of course, not be quite the same.

Exercises.

1. On a voyage from Portsmouth to Le Havre, find (i) the track; (ii) the compass course to steer; (iii) the distance to go; and (iv) approximate time of passage from the Nab Tower to the Banc de Seine Lt. V. Expected speed 10 kn. Departure from the Nab Tower 1 hr. after H.W. Dover. Spring tides. Wind S.W. Expected leeway 3°.

2. On a passage from Jersey to Dartmouth find (i) the track; (ii) the compass course to steer; and (iii) the distance to go from a position where Les Hanois bore 045° distant 5 M to the entrance of the River Dart. Ship's speed 6 kn. Spring tides. Departure Les Hanois 5 hrs after H.W. Dover. Wind S.W., expected leeway 5°.

3. Bound to Poole from Cherbourg, find (i) the track; (ii) the compass course to steer from Cherbourg to 2 M N.E. of Anvil Point; and (iii) the distance to go. Speed 8 kn. Departure 5 hrs. after H.W. Dover. Spring tides.

On passage, if nearly becalmed or stopped, the set and drift should be plotted hourly in order to keep a reasonably accurate E.P. for, of course, the slower the ship's speed the greater the effect of the tide upon her progress.

Cruising at High Speed.

The foregoing remarks about planning apply (except with regard to tides) with particular force when proceeding at a very high speed. While under way no chart work involving rulers and dividers can be carried out, or Direction Finder used, because of the violent motion. Quick and certain recognition of navigation marks–buoys, W/T masts, beacons, etc., is essential. Views of the coast should be studied and descriptions of landmarks memorised as far as possible.

Before the passage it is as well to cut charts to a constant size, securing them to a board in the order in which they will be wanted under a perspex cover, and discarding each chart when the vessel runs off it. Courses and expected times of sighting and arrival should be marked *boldly*. Buoys and inconspicuous, but important marks, should be ringed: so should hazards. In short, anything navigationally important should be emphasised.

The navigation notebook is made up as described. Tidal streams will have little appreciable effect on the course steered, and can usually be ignored; but rough water may require a reduction of speed, and it will probably pay to keep in sheltered waters if possible, even though this means going further. Avoid known overfalls and tide races.

On passage, if in doubt, or in fog, *stop* to check your position and to listen for fog signals. A watch should be kept on the helmsman to see if he is steering a good approximation to his course and not, for example, crabbing to windward. The time of arrival at each mark must be noted in the navigation notebook and, if different from the E.T.A., future E.T.A.s must be corrected. A Time, Speed & Distance Table can be used for this.

Sailing to Windward.

When the wind is adverse we must tack towards our destination. It will probably be easy to see which tack will allow us to lie closest to the direct course, but if there is any doubt, put the tidal stream, if there is one, on the lee bow, which will help to push you to windward.

When to Tack. The wind is unlikely to remain steady indefinitely, and if we spend too long on one tack we may find we should have done better to tack earlier.

The Windward Line. From the destination draw 'the windward line', a line dead to windward, passing through the destination. (*See* Fig. 3.2.) Draw also an angle of 5° on each side of it. Tack when the yacht reaches these border lines—that is, when the bearing of destination is 5° beyond the windward line. If we are within this 10° sector when the wind shifts, whether it veers or backs, it will be to our advantage. If the wind is *expected* to shift in one direction we should keep in the sector on the side of the windward line which lies towards where the wind is expected to change. Reassess the situation when the wind or tide has altered.

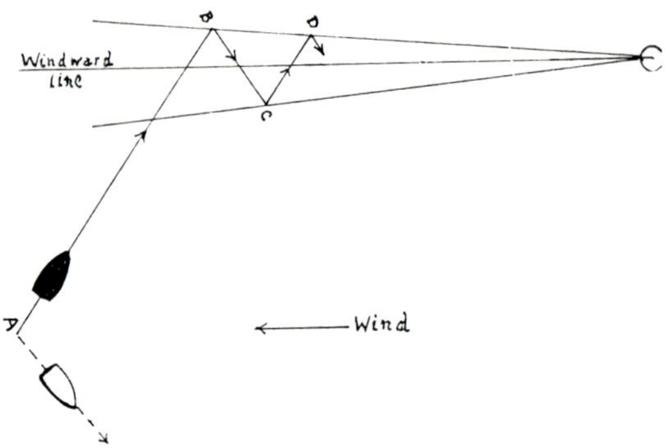

FIG. 3.2

The wind is East, so the windward line is drawn 270° from the destination. The starboard tack is clearly the tack to start on. We should tack when we reach *B* and again at *C* and *D* and so on. But if the wind is likely to veer we should keep in the sector which lies to the south of the windward line, tacking as soon as it is reached.

Buoys and Buoyage.

Buoys are used to mark the edges of navigable channels and shoals, submerged, isolated dangers, or the entrances to fairways or channels. It is useful to note how far inshore it is safe for a yacht to sail, for, usually, the buoys mark the edge of the deep water channels and there may be plenty of room inshore for shallow-draft vessels.

There are two systems of marking dangers: the Lateral and Cardinal, the former being used around the British Isles at present.

The Lateral System of Buoyage. Buoys marking the starboard side of a channel, when entering harbour (or going in the direction of the flood-tide*) are called starboard-hand buoys: buoys marking the port side of a channel when entering harbour (or going with the flood stream) are called port-hand buoys. In general, starboard-hand buoys are painted black and port-hand buoys are painted red.

Starboard-hand Buoys, on starboard when entering harbour, are conical; black or black and white chequered, and if lit, flash groups of 1, 3 or 5 white flashes.

Port-hand Buoys, on port when entering harbour, are can-shaped, red or red and white chequered, and if lit, flash groups or 2 or 4 white flashes or, occasionally, red flashes.

* For the general direction of the flood stream *see* Fig. 5.1.

Middle-ground Buoys. Buoys marking middle-grounds or the junction of channels which may be passed on *either hand*, are spherical in shape and painted in red and white or black and white horizontal bands. *Red* and white if the main channel lies to the right, or when the channels are of equal importance; *black* and white if the main channel lies to the left. If lit, they show a light distinct from neighbouring lights.

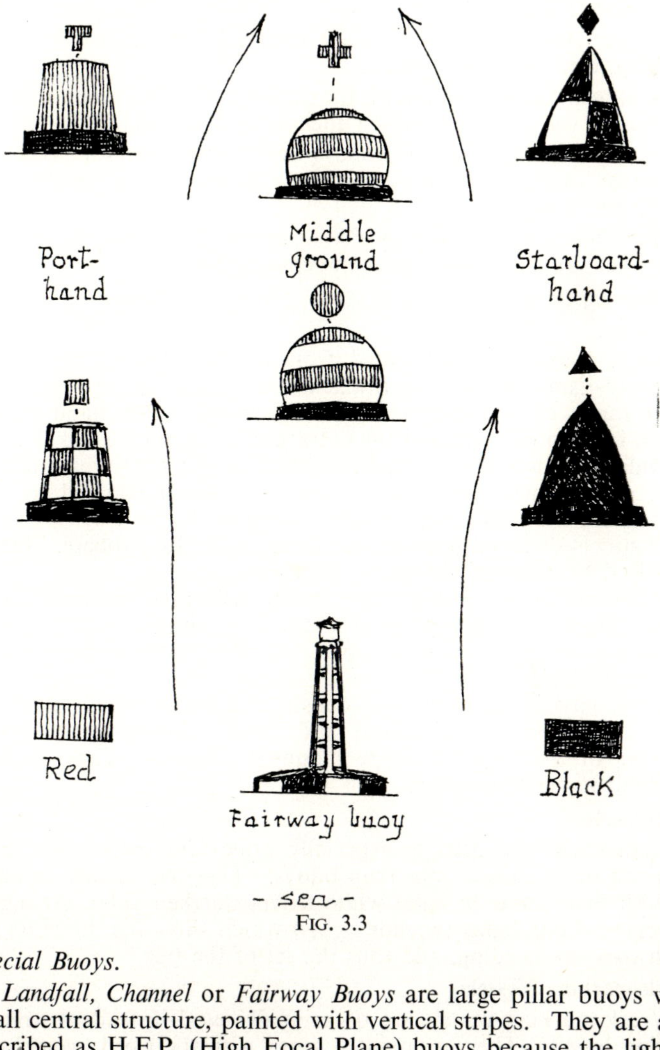

FIG. 3.3

Special Buoys.

Landfall, Channel or *Fairway Buoys* are large pillar buoys with a tall central structure, painted with vertical stripes. They are also described as H.F.P. (High Focal Plane) buoys because the light is on a higher level than other buoys.

F

Isolated Danger Buoys, Spherical. Wide black and red bands separated by a narrow white band.

Wreck-marking Buoys are all coloured green and no other buoys are painted this colour. Their shape indicates the side on which they should be passed. At night: leave to port: a double green flash every 10 secs. Leave to starboard: a triple green flash every 10 or 15 secs. Pass on either side: an Interrupted Quick Green Flash every 5 secs.

Danger Area Buoys mark areas reserved for practice firing or bombing, etc. Red and yellow.

Telegraph Cable Buoys mark the position of submarine cables of any kind in the vicinity of which anchoring is prohibited.

Outfall and Spoil-ground Buoys mark the position where a sewage or other pipe discharges into the sea or the discharging ground for dredgers.

Spar Buoys, *Dan Buoys* or *Pricks* at sea may mark the end of fishing-nets or the limits of minesweeping areas.

Mooring Buoys are of many sizes and various shapes.

Topmarks.

Buoys may be differentiated from one another by topmarks. Starboard-hand buoys carry a black cone, point up, or (except at the entrance of a channel) a black diamond. Port-hand buoys carry a red can or (except at the entrace of a channel) a red 'T'.

Middle-ground buoys: Main channel to the left: Outer end—Black cone, Inner end—Black diamond. Main channel to the right: Outer end—Red can. Inner end—Red 'T'.

Channels of equal importance: Outer end—Red sphere. Inner end—Red St. George's cross.

Important buoys usually have radar reflectors mounted upon them. These double the range at which they can be detected. They should not be confused with topmarks, which at a distance, they may resemble.

Large navigation buoys are constructed on a cylindrical base which floats almost submerged and have a lattice-work superstructure of the required shape mounted upon it. They usually bear their name or number.

Lightvessels.

Lightvessels are used at especially important marks, for they are much more conspicuous than buoys. They are usually painted red with their name in large white letters on their sides. At night they carry riding-lights (anchor lights) which show the direction in which they are heading, and thus the set of the tide.

Wreck-marking Vessels.

Wreck-marking vessels are occasionally used instead of wreck-marking buoys. They are painted green with the word 'wreck' on their side. They carry the following signals:

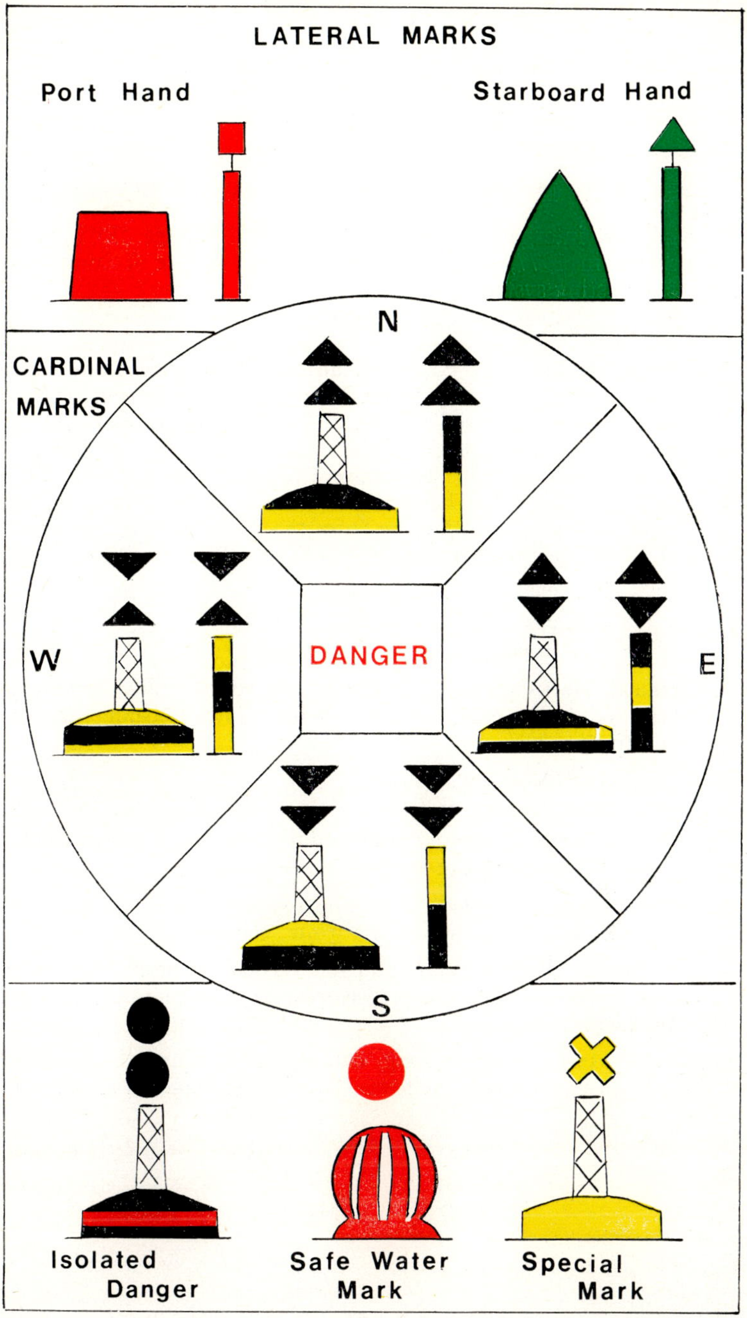
LATERAL MARKS
Port Hand
Starboard Hand
CARDINAL MARKS
N
W
DANGER
E
S
Isolated Danger
Safe Water Mark
Special Mark

THE NEW INTERNATIONAL SYSTEM OF BUOYAGE

The new international buoyage system (*see* page 58) is to be introduced in 1977, though it will be long before it completely replaces existing systems.

Five classes of marks are used:—

1. **Lateral Marks.** The colour of all port-hand marks and lights are red and starboard-hand marks and lights are green. (Green will no longer be used for wreck-marking purposes.) At the same time a new concept of the '*General Direction of Buoyage*' will replace the present 'Direction of Main Stream of Flood Tide' (Fig. 5.1, page 92). The general direction of flood tide past the British Isles is assumed to be from the south-west to the north-east. Changes in direction (Fig. 5.1) therefore result on:

   The east coast of England;
   The east and north coast of Scotland;
   The north-east coast of Ireland.

2. **Cardinal Marks** (page 61). Black and yellow buoys with top-marks as in Fig. 3.4. Lights: White quick flashing or very quick flashing:

   North: uninterrupted;
   East: groups of 3 (at 3 o'clock on a clock face);
   South: groups of 6 followed by 1 long flash (at 6 o'clock);
   West: groups of 9 flashes (at 9 o'clock).

3. **Isolated Danger Marks.** Black and red horizontal bands. Top-marks: 2 black spheres. Light: white, 2 flashes.

4. **Safe Water Marks.** White and red vertical stripes. Topmarks: 1 red sphere. Light: white, isophase, occulting, or a *single* long flash.

5. **Special Marks** (marking, e.g. cables, spoil-ground, anchorages, etc. Colour and light: yellow. Topmarks: diagonal cross.

Leave to Port: 2 green lights or shapes, vertically arranged. Fog signal: 2 strokes on a deep-toned bell.

Leave to Starboard: 3 green lights or shapes arranged vertically. Fog signal; 3 strokes on a deep-toned bell.

Pass Either Side: 4 green lights or shapes, 2 arranged vertically from each end of a horizontal yard. Fog signal: 4 strokes on a deep-toned bell every 30 secs.

Dredgers.

Dredgers may use long warping wires which obstruct part of the fairway. The international signal (to be introduced probably in 1976) will be:

(i) 3 all-round lights in a vertical line, the top and bottom ones red and the middle one white, or 3 shapes in a vertical line, the top and bottom ones balls and the middle one a diamond;

(ii) 2 all-round red lights or two balls in a vertical line on the side on which the obstruction exists;

(iii) 2 all-round green lights or 2 diamonds in a vertical line on the side on which vessels may pass.

The Cardinal System of Buoyage.

This system is used abroad where the lateral system is inapplicable. Buoys are placed to the north or east or south or west of the danger they mark.

The shapes, colours and topmarks are not the same in every country, though generally, North and West are black and white, and South and East are red and white.

The French, Belgian and Dutch system is shown in Fig. 3.4.

The various buoyage systems in Europe are to be harmonised, beginning in 1977. The main features are:

Lateral System: red buoys and red lights are used exclusively for port-hand lateral marks and green buoys and green lights exclusively for starboard-hand lateral marks.

Cardinal System: North buoys; black above yellow, Very Quick Flashing or Quick Flashing. South buoys; yellow above black, VQF(6) or QF(6). East buoys; black with yellow band, VQF(3) or QF(3). West buoys; yellow with black band, VQF(9) or QF(9). Topmarks as in Fig. 3·4.

Before Sailing.

Clear at the Customs;
Call the Weather Centre;
Check your Charts, etc.;
Cards to the Coastguard.

If you are bound abroad, the British Customs like you to clear outwards before sailing. This can usually be done by telephone a day or two before leaving.

Have a check-list of navigation (and other) gear which you may take ashore between voyages: charts, parallel rulers, dividers, pencil, rubber, pilot-books, almanac, tide tables, binoculars, etc., and check that you have everything on board.

Obtain the latest weather forecast. Set the alarm clock to warn you when to listen. A local forecast can be obtained by telephone from Meteorological Offices in London and many other towns in the British Isles.

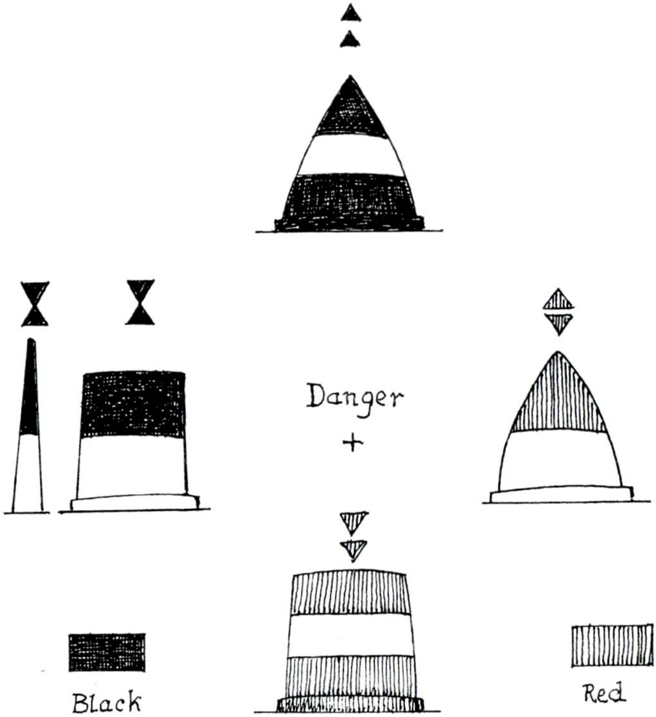

FIG. 3.4

H.M. Coastguard is responsible for the co-ordination of search-and-rescue measures for all vessels in distress around Great Britain and Northern Ireland and are in close co-operation with search and rescue organisations of neighbouring Western European countries. They call upon, as appropriate, local boats, Royal National Life-boat Institution lifeboats and inshore rescue boats, Coastguard, Royal Navy or R.A.F. helicopters or search aircraft and cliff-rescue parties. To do this effectively they need details of your yacht and of her movements.

For normal sailing activity—boating, racing, cruising or fishing, you should complete and send in a **Card CG 66A** (white) (*See* Fig. 3.5.) This shows details of the craft and her safety equipment, and is instantly available should you be in distress. It is lodged with the Coastguard by a boat wishing for surveillance in a *Local Area.* The Coastguard are responsible for declaring the boat overdue and for taking action.

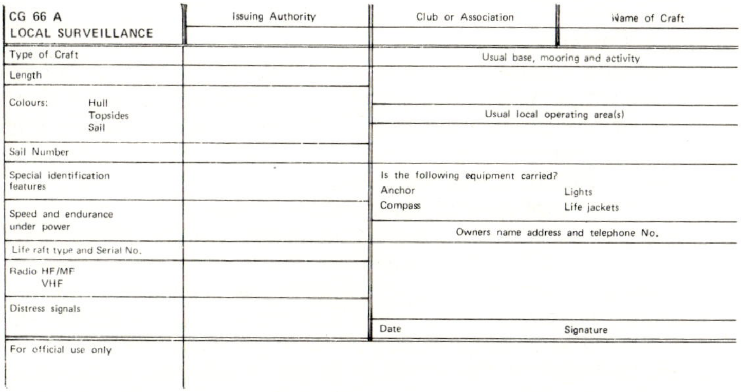
CG 66 A
LOCAL SURVEILLANCE
Issuing Authority
Club or Association
Name of Craft
Type of Craft
Usual base, mooring and activity
Length
Colours: Hull
Topsides
Sail
Usual local operating area(s)
Sail Number
Special identification features
Is the following equipment carried?
Anchor
Lights
Compass
Life jackets
Speed and endurance under power
Owners name address and telephone No.
Life raft type and Serial No.
Radio HF/MF
VHF
Distress signals
Date
Signature
For official use only

FIG. 3.5

For coastal passages or a long voyage you should complete and send in a **Card CG 66B** (pink) (*See* Fig. 3.6), together with a copy (white) to an 'agent'—such as a relative or friend—ashore, who will then alert the Coastguard if there is reason to believe that you are overdue. By despatching further CG 66B Cards during your voyage you will keep the Coastguard and your 'agent' up to date with your progress. It is the responsibility of the Agent to decide when the boat is overdue and to inform the Coastguard, who will then take action.

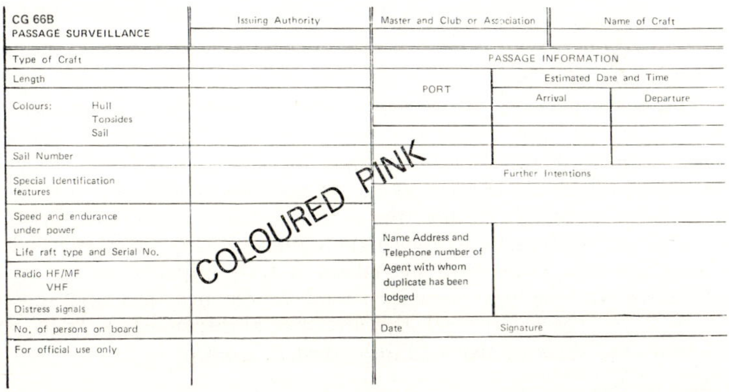
CG 66B
PASSAGE SURVEILLANCE
Issuing Authority
Master and Club or Association
Name of Craft
Type of Craft
PASSAGE INFORMATION
Length
PORT
Estimated Date and Time
Arrival
Departure
Colours: Hull
Topsides
Sail
Sail Number
Further Intentions
Special Identification features
Speed and endurance under power
Life raft type and Serial No.
Name Address and Telephone number of Agent with whom duplicate has been lodged
Radio HF/MF
VHF
Distress signals
No. of persons on board
Date
Signature
For official use only
COLOURED PINK

FIG. 3.6

Should your agent—or anyone else—urgently require the services of the Coastguard they should dial 999 and ask for the Coastguard.

CG Cards can be obtained from the Coastguard, Yacht Clubs, Harbour Masters and Marinas. *Addresses and telephone numbers of H.M. Coastguard Rescue Headquarters Stations are given in the Appendix*, page 264.

During a Passage.

Navigation is easy when you know where you are: a full and careful record of your progress should, therefore, be kept in the log-book or navigation notebook. The entries may be quite brief. Useful abbreviations are: a/c = alter course; ←| = abeam to port; |→ = abeam to starboard.

Make quite sure of the identity of each buoy as you pass it. If the name cannot be seen, compare its shape, colour and topmark, if any, with the details shown on the chart. Note which way the tide is setting as you pass buoys or lightvessels—they will appear to be steaming against the tide. (If a lightvessel or an anchored ship is to be passed at a short distance, pass it on the down-stream side to avoid fouling it if the strength of the current has been underestimated.) At definite intervals the helmsman should enter the course he has been steering, assessing it as accurately as possible. The log reading should be noted at half-hour intervals. Logs not infrequently get fouled by weed, and regular readings at short intervals will show if this has happened, and the log can be cleared at once without much loss in the mileage recorded.

The time and log reading should be entered whenever any navigational mark is abeam; when course is altered; when the yacht goes about, and the new course steered. The barometer reading and direction and force of the wind is entered at the change of the watch. A good E.P. can thus be found at any time. When opportunity presents itself, the ship's position is fixed and marked on the chart—not forgetting to add the time and log-reading at the fix.

Changing the Watch.

During a passage when the watch is changed it is important (particularly at night) that the relieving watch is properly briefed before taking over. When coasting, the relief should first consult the chart and satisfy himself as to the position, course and speed of the vessel; the set of the tide; the compass error; and characteristics of lights or landmarks likely to be sighted during his watch. He should not take over until his vision is fully adjusted to the light conditions and he has been informed of the bearings and movements of any vessels in sight and any special information about, for example, the state of the visibility, past or present.

Safety at Sea.

Safe navigation is largely the result of good navigational *habits* which need to be formed by regular and punctilious repetition. For example, the course to be steered when passed verbally to your relief should always be repeated by him to show that it has been *correctly heard*; *times* always noted against positions plotted on the chart; log-book entries regularly made. It has been said that one advantage of keeping a systematic log is that it makes one observant.

Navigating in Fog.

'I have looked at a wall of fog only to realise a few moments later that it was covered with rivets!' *J.S.*

The incidence of fog at night may be detected simply by not seeing expected shore lights, or an apparent absence of passing vessels. If it can be seen that the yacht will soon enter fog (*a*) get a fix at once, if possible; (*b*) note the bearings and courses of any vessels in sight; (*c*) hoist your radar reflector if it is not already aloft; (*d*) post a lookout where his hearing will not be affected by engine-noise.

If in doubt as to your position, alter course parallel to, or away from, the land. Note the time that you pass a buoy and when it disappears from sight: this will enable you to estimate the visibility and the distance at which you may expect to sight the next buoy.

When outward bound there is no advantage in hugging the coast if you are to make an ocean passage, so stand out at once into open water and get clear of traffic-lanes, but if coasting or approaching narrow waters the situation is different. The longer we navigate by dead reckoning the greater the probable error in the D.R. position. We should, therefore, navigate from point to point, passing each navigational mark close enough to sight and identify it (if this can be done safely) or, next best, to hear it clearly.

A depth-meter is a great help. One can often run into a certain depth, aiming to keep the next navigational mark not ahead, but on one bow or the other. Then when you have reached the desired depth, but have not yet seen or heard the lightvessel, or whatever it may be, there is no doubt which way to turn, and course is maintained to keep the required depth until the lightvessel is heard or sighted.

In buoyed channels, small craft drawing little water should keep literally to the starboard side of the channel, steering directly from buoy to buoy. In this way you will be clear of ships which will be towards the centre of the channel, and unlikely to meet other yachts on an opposite course. It is improbable that the visibility will be so bad that a small vessel will be unable to avoid the buoy when it is sighted.

It may be safer, and in the long run perhaps save time, if the vessel is anchored in dense fog. Do not do so in a frequented channel, but in shallow water to one side. Allow sufficient depth for the fall of the tide and remember that the tidal stream may run strongly, so that plenty of cable should be used.

Landfall.

If the coast you are approaching is well lighted, but unfamiliar, it is usually easier to make a landfall at night, because the range at which lights may be seen and identified is often greater than the distance at which you will be able to recognise navigational marks in daytime. The best time is just before dawn, so that the position may be fixed by lights, but the entry into harbour be made in daylight. Whether by day or night it is better to make for some landmark or light which you know can be seen and identified at a good distance, rather than directly to your destination, if this is not likely to be recognisable from far off.

If it is necessary to make a landfall by day on a coast which has few prominent landmarks it may be best to steer to a position 5 or 10 M to one side, and probably to windward of, the required position. When the coast is sighted, course is altered towards your destination and, while running nearly parallel with the coast, landmarks may be identified and a fix obtained.

At night the loom of town lights can be helpful and perhaps a D.F. bearing of a local radio station may be obtained. When first sighting a distant light, check its characteristics carefully, for the scend of the sea may obscure some flashes and make it difficult to identify.

Arrival.

Before arrival study the chart and pilot to familiarise yourself as much as possible with the entrance, etc., and make notes in the navigation notebook.

In a foreign port do not forget to hoist the Courtesy Flag (the flag of the country visited), at the starboard yardarm or speader, and 'Q' flag at the port spreader: the club burgee at the masthead and the ensign aft.

Pay a personal visit to the Custom House or Port Office as soon as possible, taking with you the ship's Certificate of Registry or Ownership and all passports. In France the Customs will then issue the Green Card (Passeport du Navire étranger). This will be needed for obtaining duty-free stores and must be stamped by the Customs before leaving France. Regulations vary, of course, from country to country.

Arrival at Home.

Hoist Q flag or a red light over a white light until visited by a Custom House officer.

The Ocean Passage.

When finding the course and distance for an ocean passage it is not always sufficient to draw a line from departure to destination on a mercator chart. The Great Circle track should be drawn. The Great Circle is the shortest distance between two places and thus always shorter than the rhumb line distance found from the mercator chart. The difference is greatest when the distance between the ports is large; when the difference of Long. between them is great; and when the Lats. are high. For example, on a passage across the Pacific Ocean from Bass Strait (between Australia and Tasmania), and Callao in Peru, the mercator or rhumb line distance is 7321 and the Great Circle distance is 6772, a saving of 549 M.

Whether one can follow the Great Circle track is another matter, for the direction of the prevailing winds and the weather likely to be met with must be taken into account. Monthly Weather or Pilot Charts may be consulted.

It is possible to calculate the Great Circle distance and courses, but simpler to plot the track upon a chart drawn on polar gnomonic projection.

The Polar Gnomonic Chart is mentioned earlier on page 12. In this projection the observer is supposed to be situated at the earth's centre, looking directly at one of the poles. The meridians then all appear as straight lines radiating from the pole and the parallels of Lat. as circles of increasing radius, centred on the pole. (*See* Fig. 3.7.) *All Great Circles appear as straight lines*, so all that one has to do to find a Great Circle track is to plot the positions of departure and destination and join them with a straight line. The disadvantage of the gnomonic chart is that one cannot measure courses and distances, and for this, the Great Circle track must be transferred to a mercator chart.

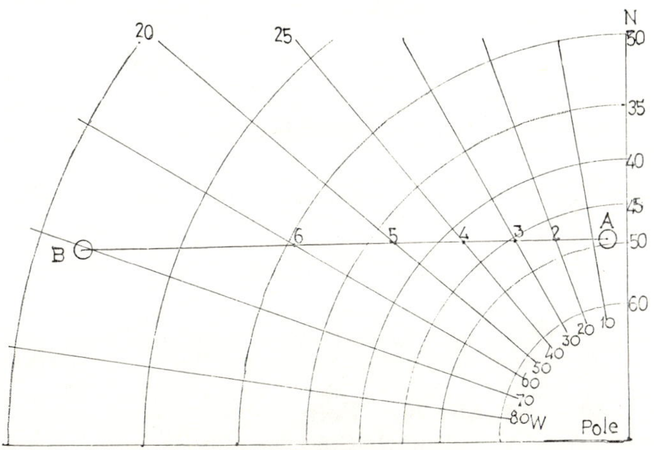

FIG. 3.7

To do this the Lats. at which the Great Circle track cross each 5° or 10° of Long. are noted. These positions are then plotted on the mercator chart and joined by straight lines (rhumb lines). (*See* Fig. 3.8.) The ship sails along these from position to position; a process known as *Great Circle Sailing*.

In some cases, since Great Circles on a mercator chart appear to curve towards the poles,* the latitude reached by the Great Circle track may enter high latitudes where ice may be encountered. In this

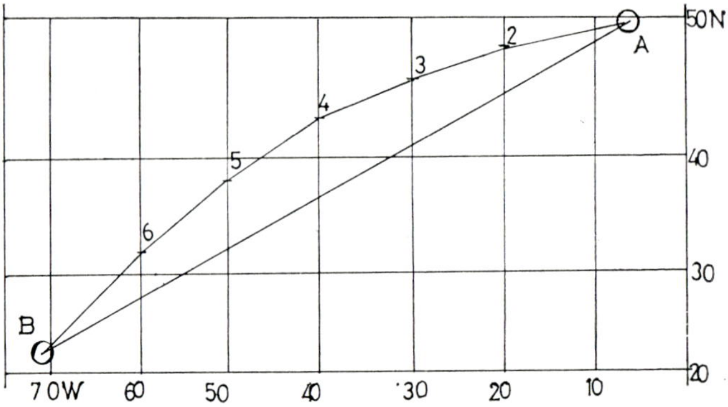

FIG. 3.8

case the ship sails along a Great Circle track as described above, to the highest safe parallel of latitude. She then sails east or west until a Great Circle track can again be followed. This is known as *Composite Sailing*. (*See* Fig. 3.9.)

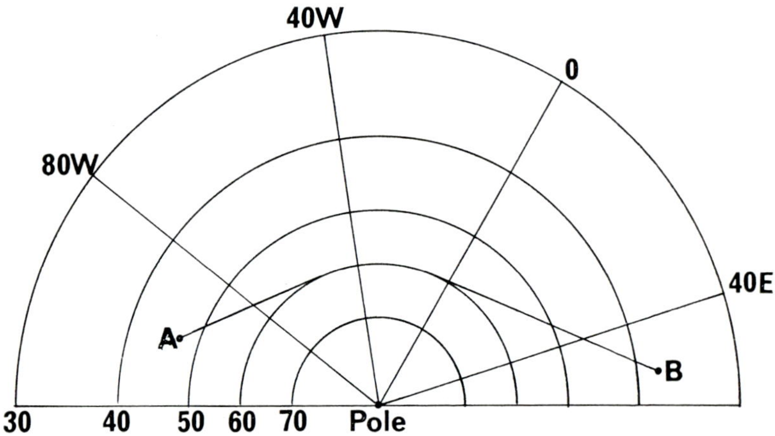

FIG. 3.9—Composite Sailing

It is not necessary to pass exactly through any of these positions or strictly to follow the Great Circle track, for a few miles either side will make no appreciable difference to the distance. So if a gnomonic chart is not available one can be drawn for the purpose.

* The point of nearest approach to the pole is known as the vertex of the Great Circle.

Constructing a Polar Gnomonic Chart. (*See* Figs. 3.7 and 3.9.) Draw a long line on the back of a chart, and from its centre-point, representing the north or south pole, draw angles every 5° or 10° apart, representing meridians. Then draw semi-circles representing parallels of Lat. The radius of these circles is found by the formula: radius of the parallel = scale × cotangent Lat.; the scale being the length chosen to represent the radius of the earth; here 6 in. Their value is shown below.

Lat.	Radius	Radius 15 cm
60°	3·46 in.	8·63 cm
55°	4·20 in.	10·50 cm
50°	5·03 in.	12·58 cm
45°	6·00 in.	15·00 cm
40°	7·15 in.	17·88 cm
35°	8·57 in.	21·42 cm
30°	10·39 in.	25·98 cm
25°	12·87 in.	32·17 cm
20°	16·49 in.	41·21 cm

A draughtsman would draw circles of such large radius with a beam compass, but they can be drawn as follows: Cut a narrow strip of stiff, strong paper about 18 in. long from the edge of an old chart. Near one end pierce a tiny hole. From this zero point measure the length of the radii, at each point pierce a hole just large enough to take the tip of a pencil. Pin the zero point to the pole on the chart, insert a pencil successively at each hole and draw in the semi-circles, or quarter-circles, if the d. long. between the places concerned does not exceed 90°.

Fig. 3.7 shows a chart drawn in this manner to find the Great Circle track between positions off the Bishop Rock (Lat. 49° 42′ N., Long. 6° 27′ W.) and the Bahama Islands (Lat. 22° N., Long. 71° W.). The track was found to cross the meridians in these Lats.:

Long. 10° W. (omitted as being close to the departure position)
Long. 20° W. crossed in Lat. 48·4° N.
Long. 30° W. crossed in Lat. 46·1° N.
Long. 40° W. crossed in Lat. 42·8° N.
Long. 50° W. crossed in Lat. 38° N.
Long. 60° W. crossed in Lat. 32° N.

These positions were then transferred to a mercator chart (Fig. 3·8.) Although the Figs. are too small to allow exact measurement to be obtained from them, they show clearly where the Great Circle track lies in comparison with the mercator track. To calculate the Great Circle distance and initial course, *see* Chapter 17, and the distances and courses between the Great Circle positions, *see* Chapter 8.

It will be noticed that there is a considerable difference between the Great Circle courses and the rhumb-line course (the latter always cuts the meridians at the same angle and thus remains the same; the former changes continually). In the example we have used, this difference is 24° (265½° Great Circle—241½° mercator). In head-winds this difference may affect the tack we should choose to sail on. If, in the example used, we had a S.W. wind—practically dead ahead of the mercator course—we might decide to sail either on the port tack or on the starboard tack. But the starboard tack will put us some 75° from the Great Circle—our most advantageous course—while on the port tack we shall be only 25° from it. Since the Great Circle track always lies on the polar side of the mercator rhumb line, the tack which lies to the polar side of the rhumb line is usually to be preferred.

It can be realised, too, that since the Great Circle track is the shorter, there must be a track of equal length to the rhumb line on the other side of the Great Circle track, though it *looks* much longer on the mercator chart. There is, then, except at the beginning and end of the voyage, a zone of considerable breadth in which the distance to travel is much the same.

CHAPTER 4

THE COMPASS

'The Mariner's Compass . . . Pre-eminently the instrument upon which the safety of the vessel depends.'

Captain S. T. S. LECKY.

The Earth as a Magnet. The earth's magnetic field is like one that would be produced by a huge magnet near the earth's centre. Because the axis of this magnet does not coincide with that of the earth the two areas known as the earth's north and south magnetic

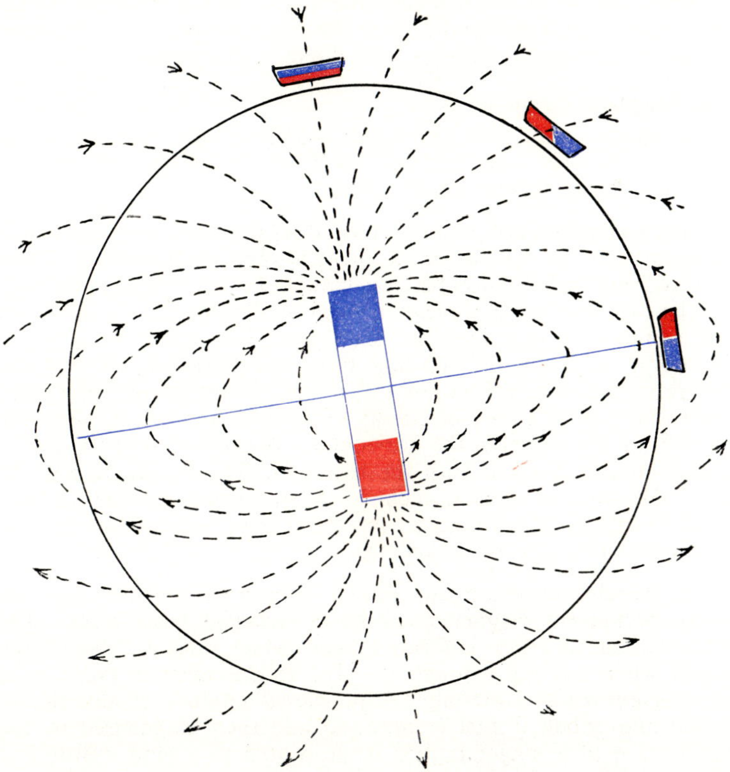

FIG. 4.1

poles do not lie at the north and south geographical poles, but are in the vicinity of Hudson Bay and South Victoria Land respectively. The magnetic poles move very slowly, gradually encircling the geographical poles.

The earth's magnetic poles are not diametrically opposite one another, so the magnetic equator, or aclinic line, does not form a great circle but is a sinuous curve which twice crosses the equator and has a latitude range from about 11° N. to 14° S.

It can be seen from Fig. 4.1 that the earth's lines of force are horizontal only on the aclinic line and that they are vertical at the magnetic poles. Consequently, the north-seeking end of a freely-suspended magnetic needle will point vertically downwards at the north pole, vertically upwards at the south pole, and remain horizontal only on the aclinic line. A *compass* needle is prevented from leaving the horizontal.

Horizontal Force is the horizontal component of the earth's total magnetic force. It is the *Directive Force* which will hold the compass needle in the magnetic meridian. It is maximum on the magnetic equator line and zero at the magnetic poles. A magnetic compass is, therefore, useless near the magnetic poles.

The Magnetic Compass.

Magnetic compasses are today usually of the type known as liquid compasses, the compass card, to which the needles are attached, being submerged in a liquid.

The Compass Card, often of mica, does not float, but is so nearly buoyant that the weight on the pivot is very small indeed. There is, thus, very little friction between card and pivot. Besides buoying up the card the liquid prevents it from oscillating and the action of the compass is described as 'dead beat', i.e. when disturbed it returns to and remains pointing along the meridian. A sapphire cap rests upon the iridum point of the pivot and the pivoting point itself is above the centre of gravity of the card and this prevents the card from tipping from the horizontal. A yacht's compass card is usually boldly marked with the letters **N, NE, E, SE, S, SW, W** and **NW,** and in degrees, each 5th degree being marked and the final 'O' being often omitted from the tens of degrees for the sake of clarity (thus 150° is marked 15).

The Needles, usually a pair, placed on each side of the pivot, are short, permanent magnets enclosed in nickel or brass cases. The latest design, however, is a ring magnet which does not disturb the liquid when moving through it. The ring magnet is made by a process known as sintering. A powdered mixture of aluminium, nickel and cobalt is first compressed and then magnetised by the application of a magnetic field while heated to a temperature just below melting point.

The bowl, (Fig. 4.2) of brass or copper, has a glass top (and bottom if it is illuminated from below); a corrugated expansion disc or chamber to allow for the slight expansion and contraction of the liquid; and a leaden ballast ring. It is completely filled with an anti-freeze

PLATE II.—Ship's Liquid Compass in Binnacle

mixture. There is a good margin between the edge of the card and the bowl to prevent eddies affecting the card when the bowl turns. A black wire lubberline, accurately aligned in the direction of the ship's head, indicates the compass direction in which the ship is pointing.

Should an air bubble form, it can be removed by tipping the bowl on one side, so that the filling plug is uppermost. The plug is unscrewed and a little distilled water added, the bowl being gently shaken to help the air to escape, and the plug then replaced.

PLATE III.—Yacht Compass (Sestrel Major) in Binnacle

In some compasses (*see* Plate III) the bowl is spherical and the gimbals (*see* under 'binnacle') are placed within it. The curved

PLATE IV.—Grid Compass (Sestrel Medina)

perspex cover magnifies the divisions on the card, making it easy to read at a distance.

Grid Compasses. In order to help the helmsman, particularly when he is steering by tiller and some distance from the compass, a grid compass may be used. This has a grid on the face of the bowl, which can be rotated. (*See* Plate IV.) To steer a given course, the grid is first rotated until the course to be steered is against the lubber line and then locked there by a small lever. The helmsman then alters course until the **N-S** line, or lines, on the grid are over and parallel to the **N-S** line or lines conspicuously drawn on the card. It is easier to keep the lines parallel than to keep the lubber line against the number of degrees to be steered.

The Binnacle. The bowl rests in the binnacle in two gimbal rings, one within the other. The axis of the inner being athwartship and the axis of the outer being fore-and-aft the compass bowl remains horizontal although the ship may roll or pitch. The outer ring is secured to the binnacle by damped springs to absorb vibration.

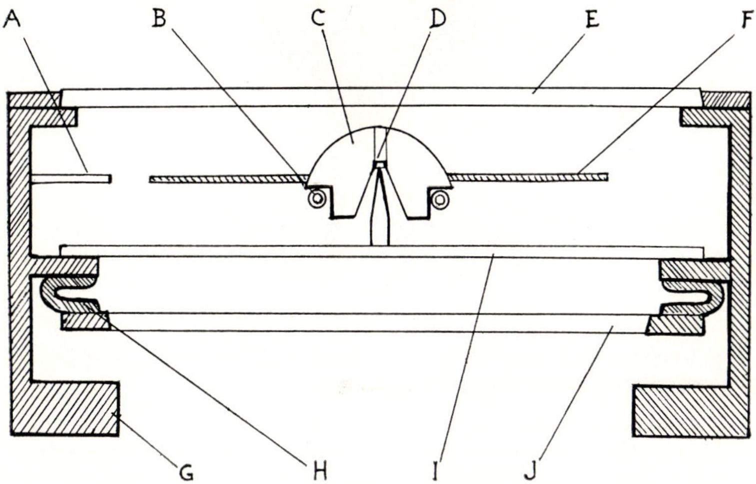

FIG. 4.2.—Compass Bowl

A—Lubber line
B—Compass needles
C—Float
D—Jewel cap, resting on pivot
E—Verge glass
F—Card
G—Balancing ring
H—Corrugated ring for expanded bottom
I—Bridge supporting pivot
J—Bottom glass

Besides elevating the compass to a convenient height, the binnacle houses the magnets and soft-iron correctors (if any) and the lighting arrangement. Twin electric wiring must be used, and held in place by non-magnetic clips to avoid the possibility of the current affecting the compass.

A removable helmet protects the bowl and the Azimuth Circle or Azimuth Mirror which rests upon it.

Hand-Bearing-Compasses. (Plate V.) The steering compass is not usually conveniently placed for taking bearings and, where no standard compass is mounted for this purpose, a hand-bearing compass is very useful. The compass is 'aimed' at the object observed and the compass bearing read in the reflected, and magnified, image of the small part of the card seen in the prism. In Plate IV, the compass bearing is 253½°, every second degree being marked on this card.

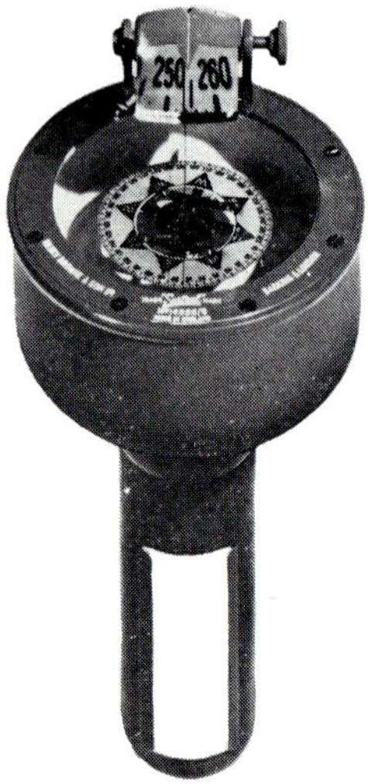

PLATE V.—Hand-Bearing Compass

Hand-bearing compasses are usually illuminated by a torch battery which fits inside the handle.

The Bino Compass is a development of the hand-bearing compass. It enables the navigator to use his binoculars to find a landmark and, when found, to read off its bearing. It is a small compass which can be attached to any standard 50 mm binoculars, and when this is done the observer sees the compass markings superimposed upon his field of vision when he looks through the binoculars, both by night or by day. The Bino compass can also be used separately from the binoculars as a simple hand-bearing compass.

When taking bearings with any type of hand-bearing compass, the observer must be careful to stand as far away as possible from any steel fittings, such as wire backstays and guard rails.

The Azimuth Circle. To facilitate the observation of bearings the *standard* (i.e. principal) *compass* is fitted with either an Azimuth Circle or an Azimuth Mirror. The Azimuth Circle has a ring carrying a prism in which the compass card graduations are reflected. To take the bearing of an object on the level of the horizon the observer turns the Azimuth Circle till the prism is in the right direction and the object appears in the small v-sight above it. The compass bearing is seen where the vertical line on the prism crosses the reflection of the card.

In a yacht with only a small steering compass, the bearing of an object can be taken by pointing the yacht directly at the object and then reading the direction of the ship's head.

If the object to be observed is the sun above the horizon the black reflecting mirror is adjusted so that the sun's reflection appears over the v-sight. Shades can be placed in front of the mirror so that the observer is not dazzled.

The Azimuth Mirror is used for the same purpose. Here, the prism is mounted on an inclined tube in which a magnifying lens is fitted. When observing objects on the horizontal plane the prism is turned so that the arrow engraved at one side of it points

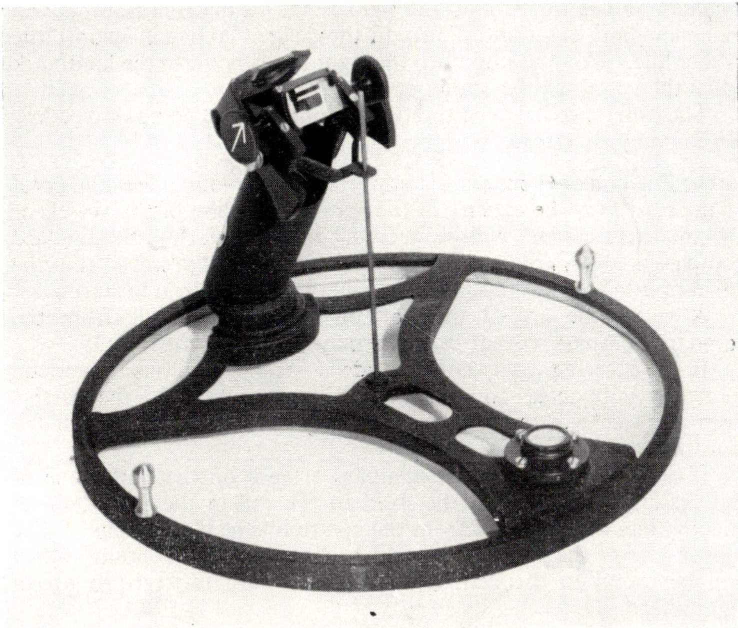

PLATE VI.—Azimuth Mirror

downward. The reflection of the compass card then appears in the prism while the observer sights the object above it.

If the sun above the horizon is to be observed the arrow is turned to point upwards and the prism adjusted so that the reflected sun appears in it. The compass card reading is seen below it through the tube.

In Plate VI the vertical *shadow pin* can be seen. This is fitted in some compasses and is used to obtain an accurate reciprocal bearing of the sun (required, as will be seen later, for finding the deviation of the compass).

The Pelorus is a compass-repeater; a dial without needles, which can be rotated and clamped in any direction. Sights are fitted so that bearings can be taken with it. It can be used in two ways.

(i) *To take a bearing of an object which is not visible from the compass.* The pelorus is set with the ship's course clamped to its lubber line.

N.B.—The pelorus must be lined up exactly in the fore-and-aft line of the vessel. The helmsman reports when the ship is steady on her course and the bearing is then observed with the pelorus.

(ii) *To set a true course by compass when the true bearing of a landmark is known.* Clamp the sight vanes on the true bearing and then slew both sight and dial until the desired course is against the lubber line. Con the ship until the sight vanes point to the landmark when the vessel will be on her course.

The difference between the true course and the compass heading is the compass error.

The Position of Compasses is important. The lubber line and pivot of the card must be *exactly* in the fore-and-aft line of the vessel, or else incorrect courses will be steered. In larger yachts the standard compass is always mounted on the centre-line of a vessel so that the effect of iron to port is counteracted by the effect of iron to starboard.

A very wide arc of the horizon should be visible from the standard compass so that bearings may be directly observed.

If the steering compass is inside a steel wheelhouse magnetic screening will occur and only a small proportion of the earth's lines of force will reach it because they tend to pass through the iron instead of the air.

If the steering wheel and compass are not on the centre-line of the vessel, a mark should be fixed to the rail in the bows off-set exactly the same distance from the centre-line as the compass. The lubber line of the compass should be lined up to this point, and it will be useful to the helmsman when steering to keep an object right ahead.

Two steering compasses, which can be quite small, fitted one on each side of the cockpit, eliminate parallax when the helmsman is on one side or the other. They also provide a useful check on compass deviation.

Magnetism in a Ship. A vessel built of steel will, of course, have an effect upon the compass, and engines and other iron or steel fittings in glass-fibre or wooden yachts will also affect the compass if they are close enough. The strength of a magnetic field diminishes as the square of distance removed. (Move a magnetic object twice its distance away and the effect will be $\frac{1}{4}$ of what it was, not $\frac{1}{2}$.) So that distance is important. Magnetic force will pass through a wooden or glass-fibre bulkhead, so that an iron pan hung in the cabin would affect a compass mounted upon the other side in the cockpit.

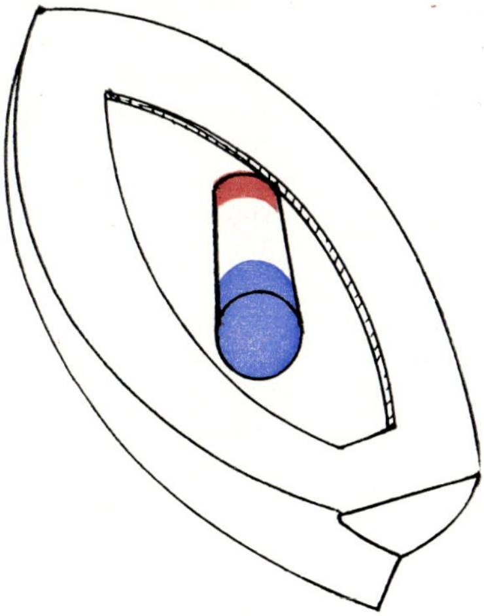

FIG. 4.3.

The combined effect of all the iron in the ship produces an effect similar to having a single large magnet placed in the hull

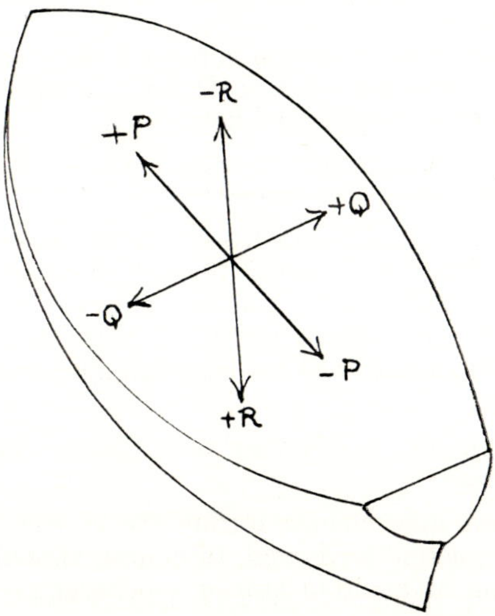

FIG. 4.4.

(*see* Fig. 4.3). Theoretically, its influence could be counteracted by placing a small magnet near the compass pointing in the opposite direction. In practice the ship's magnetic force is counteracted by resolving the total force into a fore-and-aft, athwartship, and vertical components (called *P*, *Q* and *R*), (*see* Fig. 4.4), and small magnets are placed fore-and-aft, athwartships, in the horizontal plane, with another placed vertically beneath the compass, at appropriate distances.

The compass needle cannot be tilted from the horizontal, so that the vertical component does not cause any deviation when the ship is upright, but it can do so when she is heeled over, either while sailing on a tack or when rolling. This error is named **Heeling Error.** The heeling error changes its direction when the ship

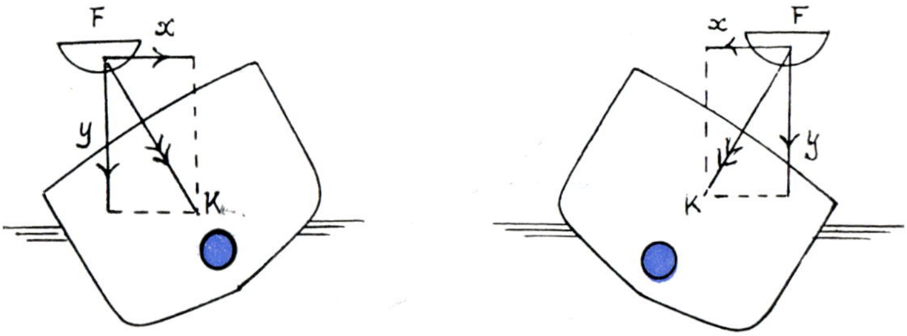

FIG. 4.5.

changes her list from one side to another (*see* Fig. 4.5). On northerly and southerly courses, if the ship is rolling, this will cause an unsteady compass.

The art of compass correction is beyond the scope of this book, and is usually carried out by a professional compass adjuster, but a few points must be emphasised here:

(i) A compass which may have an unknown error is not an aid to navigation; it is a danger;

(ii) A compass may be adjusted so that the amount of deviation is negligible, but usually there is a small residue which cannot be eliminated. This can, however, be discovered and recorded on a Deviation Card (*see* page 81) or Curve (page 83) and allowed for when shaping courses and taking bearings;

(iii) The deviation is different for different directions of the ship's head;

(iv) Deviation may change during the course of time; if the Lat. is much altered; and, of course, should any structural alteration be made in hull or superstructure involving iron or steel work.

The remainder of this chapter describes how a Deviation Card or Curve can be constructed, and then how the deviation can be further checked.

Making a Deviation Card.

To find the deviation on all points of the compass the vessel must be out of the fairway, underway, upright, and within sight of a clear landmark not closer than three miles. The sea must be calm.

Swinging Ship. Where possible, swing your vessel close to a charted buoy. The true bearing of the landmark from the buoy can then be taken from the chart and the variation for the year applied to it to obtain the magnetic bearing. Observe the compass bearing of the landmark on at least eight different headings. Keep the buoy between yourself and the landmark. It will be easier to remain in the same position if the observations are made alternatively on opposite headings: e.g. N., S., then N.E., S.W., then E., W., then S.E., N.W.

The record might begin to look like this:

Bearing of Landmark from buoy	117° T.
Variation	6° W.
	123° M

Head by Compass	*Compass Bearing*	*Magnetic Bearing*	*Deviation*
000°	120°	123°	3° E.
180°	125½°	123°	2½° W.
045°	117°	123°	6° E.
225°	129°	123°	6° W.

(It will be noticed that the deviations on opposite courses are usually equal and opposite.)

We can swing the ship without the aid of a buoy.

The procedure is then to circle tightly at very slow speed, at slack water, observing and recording the compass bearing of the landmark at regular intervals, say every 30° or 20° of the compass. This must not be hurried, for it is essential for the yacht to be steadied on the new course before the bearing is taken.

The result might be as follows:

Head by Compass	*Compass Bearing*	
180°	335°	
210°	336°	
240°	337°	
270°	336°	
300°	335°	and so on.

When a complete circle has been made, all the bearings are added together and the sum divided by 12 or 18, as the case may be, to get the average compass bearing—and this should be the correct magnetic bearing of the landmark. In this example it might turn out to be 335°. The difference between the magnetic bearing and the compass bearing on every course is the deviation on that course, so the table could be completed:

Head by Compass	*Compass Bearing*	*Magnetic Bearing*	*Deviation*
180°	335°	335°	0°
210°	336°	335°	1° W.
240°	337°	335°	2° W.
270°	336°	335°	1° W.
300°	335°	335°	0° etc.

A Deviation Card or Curve is then drawn up as on page 7 and page 8.

EXAMPLE: Construct a Deviation Card and Curve from the following compass bearings taken of a distant lighthouse:

Ship's Head by Compass	*Compass Bearing*
000°	120°
030°	117½°
060°	116°
090°	117½°
120°	120½°
150°	123°
180°	125½°
210°	128°
240°	129½°
270°	128°
300°	125½°
330°	123°

The Deviation Curve for these observations can be seen in Fig. 4.6.

It should not be supposed that two compasses in the same vessel will necessarily have the same deviations. Since it is often impossible to take bearings from a steering compass, its deviation should be found either by using a pelorus (*see* page 78), or by comparison with another compass whose deviation is already known. The helmsman steadies the ship on her course and the heading is then noted on the other compass. When the deviation, if any, has been applied to this heading and the magnetic course found, the difference between it and the compass course steered is the deviation.

EXAMPLE:

Course steered 280°	Heading by other Compass 276° C.
	Deviation 5 E.
281° M	281° M
Deviation 1° E.	

A complete Deviation Card for the Steering Compass may be drawn up by this method.

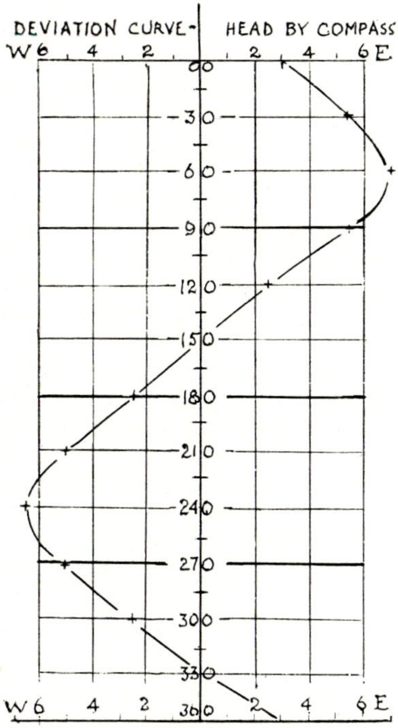

FIG. 4.6.

As before stated, it is desirable to check the deviation of the compass from time to time.

Deviations by Transit bearings.

One good and quick method of obtaining the deviation is to observe the compass bearing of two fixed objects in transit, then comparing it with the true bearing measured from the chart. The disadvantage of this method is that it can be used only when transit bearings happen to be available.

In Fig. 2.12 the leading marks bear 334° T. when in transit, if their compass bearing is 338°, the compass error is 4° W.

Study Fig. 1.4 again; the rule still holds good:

'Compass best—error west
Compass least—error east.'

If the variation is 7° W., the deviation is 3° E.

Deviation by Amplitude of the Sun.

Twice a day, when the Sun rises and sets, it is possible to find the deviation with very little calculation, the Sun's bearing by compass being compared with the true bearing at that time. We need to know only the ship's approximate position and the Sun's declination (for which we require the G.M.T.).

Declination is the angular distance that a heavenly body is north or south of the celestial equator.

The amplitude is the amount that the Sun is north or south of the east point at the moment of rising, and north or south of the west point when setting. Twice a year, at the vernal and autumnal equinoxes, the declination is 0°, and the Sun rises bearing exactly east and sets exactly west. In summertime, when the Sun is north of equator, the declination and amplitude are always north, in the the winter both are south. The amplitude is thus always named the same as the declination. When the amplitude has been found it is converted to true bearing by adding or subtracting it from 090° at sunrise, or 270° at sunset.

The amplitude is calculated for the moment when the Sun's centre is on the horizon. Owing to refraction (the bending of the light rays by the atmosphere) this actually occurs when the Sun's centre appears to be about *its own diameter above the horizon.* (*See* Fig. 4.7). It is at that moment that the bearing should be taken by compass. Though in low latitudes the path of the Sun appears nearly vertical, in high latitudes it slopes, so it is important that the observation should be made at the right time.

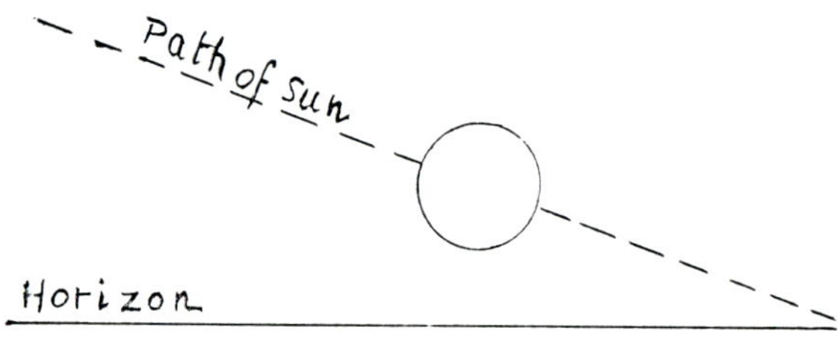

FIG. 4.7.—The time at which the sun should be observed for amplitude

A table of Amplitudes is included in most Nautical Tables and one from *Burton's Nautical Tables* is reproduced on page 262. It is entered with (1) the ship's latitude and (2) the Sun's declination. The Sun's declination is tabulated in the *Nautical Almanac* for every hour of the year. All times in the *Nautical Almanac* being G.M.T., we need to know the date and time at Greenwich in order to find the declination.

WORKED EXAMPLE: On Jan. 3rd, at 1550 G.M.T., in Lat. 54° N., the Sun set bearing 241° C. Find the Sun's amplitude and thence its true bearing. The variation was 17° W., find also the compass error and deviation.

Declination for Jan. 3rd. 1600 = S. 22° 47·6′, that is nearly S. $22\frac{3}{4}°$.

Amplitude	W41° S.
	270°
Bearing	229° T.
Bearing	241° C.
Compass Error	12° W.
Variation	17° W.
Deviation	5° E.

Exercise

Deviation and Compass Error by Amplitude

In each question find the compass error and deviation.

1. On Jan. 2nd, 1973, in Lat. 45° 25′ N., Long. 17° 40′ W., the Sun set bearing 245° C. The correct G.M.T. was 18hr. 2 min. Variation 9° W.

2. On Jan. 5th, in Lat. 50° N., Long. 14° W., the Sun rose bearing $144\frac{1}{2}°$ C. at 0900 B.S.T. Variation 18° W.

3. On Jan. 1st, in Lat. 50° N., Long. 0°, the Sun set bearing $241\frac{1}{2}°$ C. Variation 8° W. G.M.T. 1555.

4. On Sept. 23rd, in Lat. 19° 55′ N., Long. 79° 45′ W. at 1815 Z.T. (Zone +5) the Sun set bearing 272° C. Variation 2° E.

5. On Sept. 19th, in Lat. 20° N., Long. 157° 15′ W. at 1827 Z.T. (Zone +10) the Sun set bearing 265° C. Variation 10° E.

6. On Sept. 21st, in Lat. 0°, Long. 82° W., the sun rose at 0528 Z.T. (Zone +5) bearing 086° C. Variation 9° E.

For Zone Time (*see* page 00).

Deviation by Azimuth of the Sun.

The compass bearing of the Sun can also be observed when the Sun is above the horizon (though it is difficult to do so accurately at high altitudes) and is then compared with the true bearing to find the compass error. The true bearing is found from Tables. Before the Tables can be used it is necessary to extract certain data from the *Nautical Almanac* for the G.M.T. of the observation.

The Geographical Position of the Sun. A straight line joining the earth's centre to the Sun's centre cuts the earth's surface at a point called the *Geographical Position* (G.P.) of the Sun.

As the Sun travels through the heavens on its daily path from east to west, so the G.P. moves westwards along a parallel of Lat. which is the same as the Sun's *Declination*.

The Long. of the G.P., if measured from 0° to 360° *westwards* from the Greenwich meridian, is called the Sun's *Greenwich Hour Angle* (G.H.A.).

The Sun's Declination and Greenwich Hour Angle are listed on the daily pages of the *Nautical Almanac*, and thus the Sun's Geographical Position can be found for any instant of time. The declination gives the Sun's Lat. and the G.H.A. gives its Long. (If the G.H.A. exceeds 180° the G.P. will be in east Long., e.g. 270° west of Greenwich is the same as 90° east Long.)

The Sun's Azimuth and True Bearing. The bearing of the Sun is the same as the bearing of the G.P. from the observer.

A Great Circle passing through the observer and the Sun's G.P. cuts the observer's meridian at an angle called the *azimuth*. The azimuth can be named either from north or south, depending on which Tables are used; but whichever way it is expressed, each gives the same true bearing when converted to 360° notation. For example, N. 130° W. or S. 50° W. are both equal to 230°.

To determine the azimuth of the Sun the following data is required: Observer's Lat.; Sun's declination and Local Hour Angle.

The Local Hour Angle (L.H.A.) of the Sun is equal to the difference of Long. between the observer and the Sun's G.P. and is always measured westward (0° to 360°) from the observer's meridian.

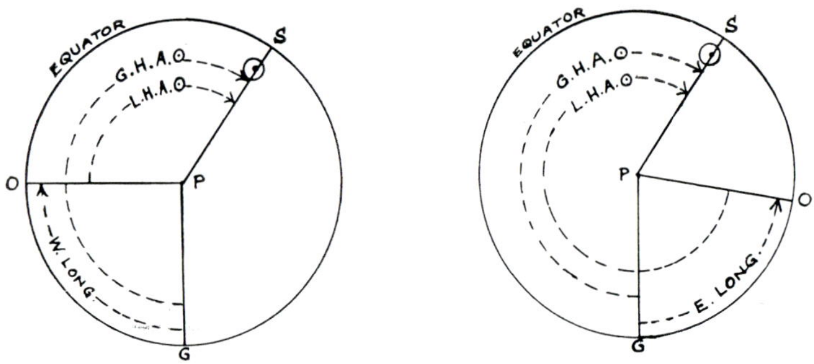

FIG. 4.8.

Both figures represent the earth on the plane of the equator. We are looking down upon the northern hemisphere from above the north pole.

P represents the North Pole
PG „ the Greenwich Meridian
PO „ the Observer's Meridian
PS „ the meridian of Sun's Geographical Position

To find the Sun's Local Hour Angle and Declination.

(1) Enter the *Nautical Almanac* at the appropriate date on the right-hand page of opening.

(2) Write down the G.H.A. and Declination of the Sun tabulated against the nearest whole hour of G.M.T. *before* the G.M.T. for which the data is required. This entry is close enough for the Declination, but the G.H.A. must be found for the exact minute and second.

(3) Turn to the Increment and Correction Table near the end of the *Almanac* and extract the Sun's increment for the minutes and seconds of G.M.T. over the whole hour. The increment is always added to the G.H.A.

(4) To the corrected G.H.A. of the Sun apply the ship's Long. to obtain the Local Hour Angle of the Sun. Add for east Long.; subtract for west Long.:

'Longitude west; Greenwich is best,
Longitude east; Greenwich is least.'

If necessary, add 360° to the G.H.A. when the Long. has to be subtracted.

If necessary, subtract 360° from the L.H.A. when the Long. has been added.

Worked Example 1: It is required to find the L.H.A. and Declination of the Sun at G.M.T. 09hr. 08 min. 46 sec. on Jan. 3rd, 1973. Ship's position is in Lat. 46° 13′ N., Long. 18° 35′ W.

Jan. 3rd G.M.T.	G.H.A. 9 h =	313° 52·3′	Dec. S. 22° 49·3′
Inc.	8m 46s	2° 11·5′	
Correct G.H.A.		316° 3·8′	
Long. W.		− 18° 35′	
Required L.H.A.		297° 28·8′	

Worked Example 2: D.R. position 28° 40′ S., 158° 50′ W. Find the Sun's L.H.A. and Declination at Zone Time 17 hr. 12 min. 30 sec. on Sept. 20th, 1973.

Z.T.	20d 17h 12m 30s	
Zone	+ 10h	
G.M.T.	21d 3h 12m 30s	(Note the Greenwich date)
G.H.A. 3h	226° 42·0′	Dec. N. 0° 48·0′ d 1·0′
Inc. 12m 30s	3° 7·5′	− 2′
G.H.A.	229° 49·5′	N. 0° 47·8′
Long. W.	−158° 50′	
L.H.A.	70° 59·5′	

Examples for Exercises: In each question find the L.H.A. and Declination of the Sun.

1. Date and time: Sept. 19th, 1973, G.M.T. 07 hr. 06 min. 45 sec. Ship's position: Lat. 36° 19′ N., Long. 30° 27′ W.

2. Date and time: Jan. 2nd, at 0910 Z.T. (Zone +9). Ship's position: 36° 50′ S., 137° 20′ W.

3. Date and time: Jan. 6th, at 0812 Z.T. (Zone −6). Ship's position: 32° 16′ S., 156° 36′ E.

4. Date and time: Sept. 21st, at 2205½ Z.T. (Zone +6). Ship's position: 50° 10′ S., 90° 15′ W.

Tables for finding the Azimuth of a heavenly body.

All the standard *Nautical Tables* contain A, B and C Tables for this purpose; *Sight-Reduction Tables* present the azimuth often by a single entry; *Reed's Nautical Almanac* has a very concise 'Time–Azimuth Table' with full instructions for its use; the *Weir Azimuth Diagram* (Admiralty Chart No. 5000) provides a diagramatic method for finding the azimuth.

To find the Azimuth by use of A, B and C Tables.

WORKED EXAMPLE: In Lat. 50° 30′ N., Long. 3° 15′ W., at 10hr. 5 min. 30 sec. G.M.T. on Jan. 1st, the Sun bore 140° by compass. The local variation was 8° E. What was the deviation? Proceed as follows:

(1) Find the L.H.A. and declination of the Sun as previously shown. The Declination to the nearest decimal of a degree is accurate enough.

(2) Enter Table *A* with *Lat.* on the left and *L.H.A.* along the top or bottom of the page. Extract the value for *A*, interpolating as necessary. For practical purposes it is accurate enough and will assist interpolation if Lat. is also used to the nearest half of a degree.

Label *A* as + or −, according to the rule given at the top of the page.

(3) Enter Table *B*, on the facing page, with *Declination* and *L.H.A.* Again interpolate to the nearest decimal of a degree. Write down the value of *B* and the sign according to the rule given at the top of the page.

(4) Deduce *C* by taking the *algebraic* sum of *A* and *B*. Label *C* with the appropriate sign.

(5) Enter Table *C* with arguments *Lat.* and *C*. Extract the azimuth from the top of the appropriate column.

(6) The azimuth should be named according to the rules at the top of the page. Note that quadrants are numbered clockwise, beginning with *NE*.

G.M.T. 21d 10h 05m 30s

G.H.A. 10h	329° 6·0′	Dec. S. 23°
Inc. for 5m 30s	1° 22·5′	
G.H.A.	330° 28·5′	*A*+1·87
Long. W.	− 3° 15′	*B*+ ·78
L.H.A.	327° 13·5′	*C*+2·65
True Bearing	149¼°	
Compass Bearing	140°	
Compass Error	9¼° E.	
Variation	8° E.	
Deviation	14½° E.	

There are other rules for naming the azimuth which some navigators find preferable to those given in Burton's.

1. *A* is named the same as the latitude when the L.H.A. lies between 90° and 270°. Otherwise it is named opposite to the latitude.

2. *B* is always named the same as the declination.

3. *C* is named N. or S., whichever sign predominates when the two are combined. Thus *A* and *B* are added when of the same name and the difference is taken when thev are of opposite names.

In the example given above: *A* = 1·87 S.
B = ·78 S.
C = 2·65 S.

4. The azimuth is always west when the L.H.A. lies between 0° and 180°, otherwise it is east.

To find the Azimuth by using Sight-Reduction Tables.

Using the same WORKED EXAMPLE and *Sight Reduction Tables for Air Navigation* (*A.P. 3270*), Vol. 3, open Vol. 3 at page 'Lat. 50°, Declination (13°—29°) CONTRARY NAME to Latitude'. Under column headed 23° (the Declination) and line 327 (the L.H.A.) find the value of Z (the Azimuth): 149°. Interpolating for ¼°, L.H.A. makes it 149¼°.

Since, in this example, the Lat. is 50½°, a second opening under Lat. 51°, Declination 23° CONTRARY NAME, and L.H.A. 327° must be made (4 pages on). This also gives 149° and interpolation 149¼°.

At the top of the page is the direction: N. Lat., L.H.A. greater than 180° . . . Zn = Z. Zn is the angle measured from north, i.e. the bearing. If the L.H.A. had been less than 180°, following the given rule, we should subtract the Azimuth (Z) from 360° to get the Bearing (Zn). Directions for South Lat. are given at the foot of the page.

Compass Error by Azimuth

In each question find the compass error and deviation

1. On Jan. 5th, 1973, at 1200 G.M.T., the Sun bore 178° C. D.R. Position: Lat. 51° 28′ N., Long. 40° 12′ W. Variation 35° W.

2. On Sept. 23rd at about 1336 at ship, Chronometer 1 hr. 2 min. 6 sec. Error 6 min. 7 sec. slow of G.M.T. Position: Lat. 57° 47′ N., Long. 8° 41′ E. Sun bore 235′ C. Variation 27½° W.

3. On Jan. 3rd at 10 hr. 6 min. 42 sec. G.M.T., in Lat. 48° 30′ N., Long. 5° 32′ W. Sun bore 161° C. Variation 11° W.

4. On Sept. 24th at 1003 B.S.T., in Lat. 35° 17′ N., Long. 2° 47′ E., the Sun bore 139° C. Variation 10° W.

5. In the morning of Sept. 20th, in Lat. 38° 52′ N., Long. 30° W. Variation 27° W. Sun bore 142½° C. at 0812 Z.T. (Zone +2.)

6. On Jan. 2nd at 0612 Z.T. (Zone −7) in Lat. 2° 5′ S., Long. 99° 5′ E., the Sun bore 115° C. Variation 4° W.

Every 34 mB reduction in pressure ≡ ·3M.
Mean standard pressure in UK ≡ 1016 mB
Scotland 1011 mB

CHAPTER 5

TIDES

'The heart-throb of the mighty ocean.'

Less poetically, the tide is the rising and falling of the sea, which usually takes place twice a day.

The oceans are free to move and are attracted by the moon (and to a less degree) by the sun.

The tide-producing forces cause the oceans to 'heap-up' on the sides of the earth nearest to and farthest from the moon, leaving low-tide areas between them. As the earth turns on its axis, these tidal waves follow the moon but, while the earth is rotating from west to east, the moon is also moving along its orbit in the same direction, and consequently crosses each meridian about 50 min. later each day. Hence, the time interval between two successive high tides averages about 12 hr. 25 min.

The Sun also attracts the sea but, being so much farther away, its influence is far less than that of the moon. At the New and Full Moon, when the moon and Sun are nearly in line with the earth, their combined effect gives higher high waters and lower low waters. These are called Spring Tides and occur at intervals of about 15 days. When the moon and Sun are pulling at right angles to one another, the Sun reduces the effect of the moon causing lower high waters and higher low waters, which are known as Neap Tides. Spring and neap tides, therefore, follow one another on alternate weeks.

The highest Spring Tide is usually not upon the day of Full or New Moon but a day or two later ($1\frac{1}{2}$ days later round the British coasts) because the continents obstruct the free passage of the tidal waves. A fuller explanation of the tide generating force is given in the *Admiralty Manual of Navigation*, Vol. 1.

The Tidal Stream. The tidal wave approaches the British Isles from the west, floods up the English and St. George's Channels, round the north of Scotland, where it attains a rate of 10 kn at Springs in the Pentland Firth, and southward down the North Sea, meeting the east-going stream off the coast of Kent. The ebb stream runs from thence in the contrary directions. The direction and speed of the tidal stream, obviously of importance to navigators, is shown in the *Admiralty Tidal Atlas of the British Isles*. On each page the direction of the tidal stream for a certain hour before or after high water at Dover is shown. To use the Atlas, the time of high water at Dover is found from the Tide Table and the Atlas is then opened at the appropriate page. These charts are reproduced on a smaller scale in *Reed's Nautical Almanac*.

Tidal information is also printed on many charts in tabular form, showing the set and rate (i.e. the direction and speed) of the tidal stream for every hour before and after High Water at Dover. Each column refers to a different locality on the chart shown by capital letters or numbers placed in a diamond, thus: ⟨A⟩

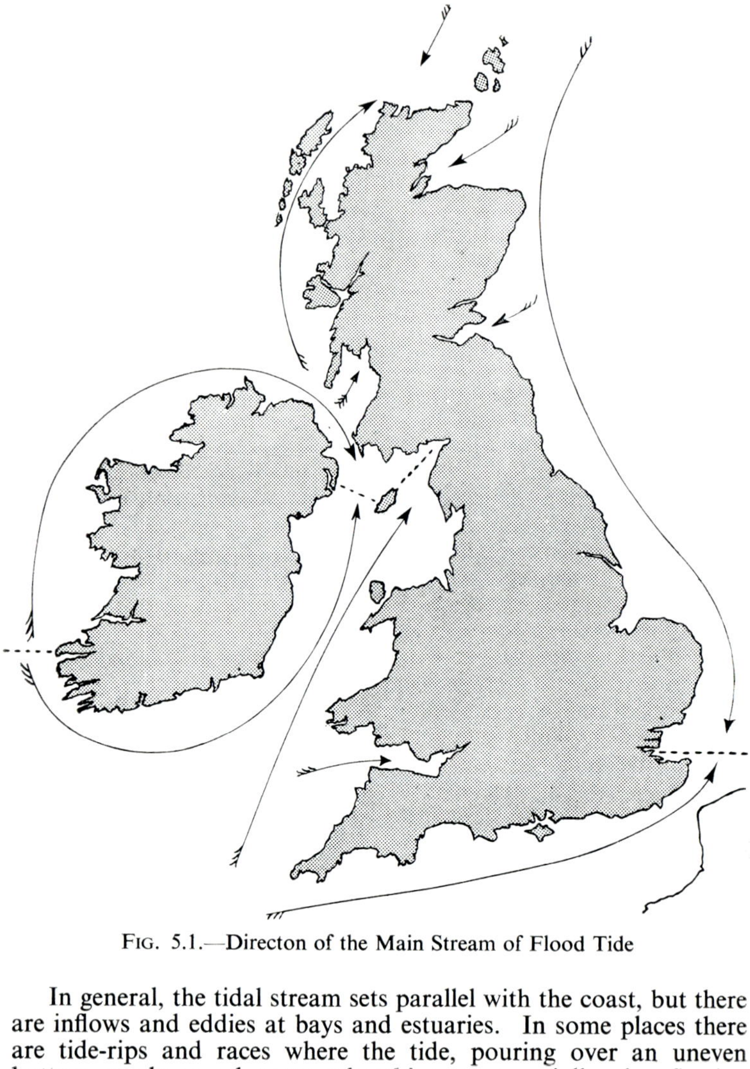

FIG. 5.1.—Directon of the Main Stream of Flood Tide

In general, the tidal stream sets parallel with the coast, but there are inflows and eddies at bays and estuaries. In some places there are tide-rips and races where the tide, pouring over an uneven bottom, produces a dangerous breaking sea, especially when flowing against the wind.

Tides. The vertical movement of the water is shown in the *Admiralty Tide Tables*. *Reed's* and *Brown's Nautical Almanacs* also contain a large section on Tides and Tidal Streams.

The *Admiralty Tide Tables*, Vol. I, deals with European Waters. In Part I the times and heights of High and Low Water at *Standard Ports* are listed for each day of the year.

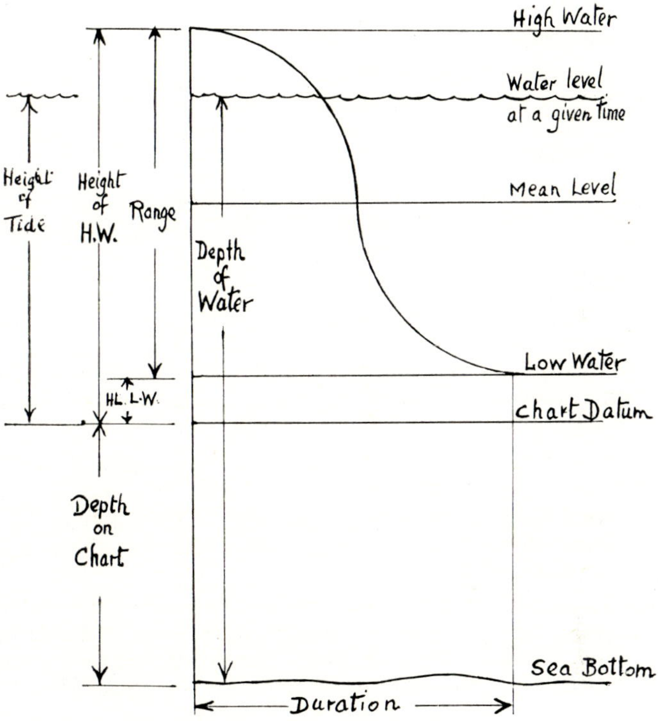

FIG. 5.2

Note that the tide does not rise and fall at a uniform rate. Both tidal stream and rate of rise and fall are usually fastest at half tide

An example of a Standard Port prediction is shown on page 25. On Jan. 7th, for instance, Low Water occurs at 34 min. past midnight (B.S.T.), with a height above chart datum of 1·9 m or 6·4 ft.; first High Water at 0635, height 7·1 m or 23·4 ft.; second Low Water at 1252, height 1·7 m or 5·7 ft.; second High Water at 1051, height 6·9 m or 22·6 ft. It will be noticed, by comparing the height of High Water with other days, that this is a day of Spring Tides.

In Part II are listed the differences of time and height for predicting the tides of *Secondary Ports*. The term 'Secondary Port' does not imply that the place concerned must be of secondary importance, but is a device for keeping the Tide Tables to manageable proportions. Each Secondary Port is associated with a nearby Standard Port.

All heights and height differences are given in metres and feet.

All times are given in B.S.T. (Zone −1). If G.M.T. is required subtract 1 hr.

Tidal Predictions at Secondary Ports.

The times and heights of H.W. or L.W. at a secondary port are found by applying time and height *differences* to predictions given for a selected Standard Port. A specimen page of Secondary Ports on Brest (height in feet) is given on page 259.

Time Differences are tabulated under certain times at the Standard Port. When the actual time falls between those given the difference may be interpolated.

Height Differences are given only for certain heights at the Standard Port and may be interpolated or extrapolated, as appropriate, for the actual height.

It is usually sufficient to do the interpolation by eye, but if greater precision is required it can be rigorously interpolated as shown in the worked examples.

WORKED EXAMPLE (1): Find the time and height of H.W. at Le Pellerin on the morning of March 1, 1973.

(i) Find Le Pellerin in the alphabetical General Index at the back of the *Admiralty Tide Tables* and note the reference number —1666.

(ii) Look for reference No. 1666 in Part II of the Tables—(page 386) Le Pellerin. Higher up the page is the name of the selected Standard Port, printed in heavy capitals—**BREST.** Note that predictions are on page 339 of the Tables.

(iii) Turn to page 339 and write down the time and height of H.W. at Brest on the morning of March 1st.

Brest, March 1st. H.W. 0242. 19·0 ft.

(iv) H.W. at Brest (0242) lies roughly halfway between 2300 and 0500 (the times for which the differences are given), so we could take a time difference halfway between the times given (say +0110).

To interpolate rigorously, proceed as follows. The required difference lies between +1 hr. 25 min. and +0 hr. 55 min., exactly 2 hr. 18 min. from 0500 (0500 −0242), or 2·3 hr. The correction to the Time Difference at the Secondary Port equals:

$$\frac{\text{Change in Time Difference at the Secondary Port}}{6h} \times \begin{array}{c}\text{Difference between nearest H.W.}\\ \text{and time of H.W. at the Standard}\\ \text{Port}\end{array}$$

$$\frac{30}{6} \times 2{\cdot}3 = 11{\cdot}5\text{m}$$

This correction is applied to the nearest Time Difference:

$$11\tfrac{1}{2} + 55 = +66\tfrac{1}{2}\text{m}$$

so that the time difference is 1 hr. 6½ min.

(v) The height of H.W. at Brest is 19·0 ft., which lies between M.H.W.S. 24·3 ft. and M.H.W.N. 18·9 ft. The required difference, then, lies between −5·6 and −4·4.

In this case, the height of H.W. on March 1st (19 ft.) is almost exactly the height of M.H.W.N. (18·9 ft.), so the M.H.W.N. difference is used (−4·4 ft.).

Should we wish to work out the correction to the Height Difference at the Secondary Port, the procedure is:

$$\frac{\text{Change in the Ht. Difference at the Secondary Port}}{\text{Change in Heights of H.W. at the Standard Port}} \times \begin{array}{c}\text{Difference between nearest Ht.}\\ \text{of H.W. and Ht. of H.W. for the}\\ \text{day at the Standard Port}\end{array}$$

$$\frac{1{\cdot}2}{5{\cdot}4} \times {\cdot}1 = {\cdot}2$$

This correction is applied to the nearest Height of H.W.:

$${\cdot}2 - 4{\cdot}4 = -4{\cdot}6$$

(vi) Apply the Time and Height Differences, according to their signs, to the Standard Port predictions and the result will be the predictions for the Secondary Port.

Le Pellerin 1666. Standard Port *Brest* (page 214).

Brest High Water	0242	Height	19 ft.
Differences	+0107		− 4·6 ft.
Le Pellerin	0349		14·4 ft.

(2) Find the times and heights of Low Water at Paimbeouf on March 1st, 1973.

March 1st. Times of L.W. at Brest 0903 hr., 9·3 ft.; 2127 hr. 8·8 ft.

Looking at the differences for Paimbeouf, we might take a time difference of +1 hr. and a height difference of −4 ft. for the day.

L.W. Brest	0903	9·3 ft.	2127	8·8 ft.
Differences	+ 100	−4·0 ft.	+ 100	−4·0 ft.
L.W. Paimbeouf	1003	5·3 ft.	2227	4·8 ft.

If we wish to work more precisely:
Times of L.W. at Brest 0500 and 1700, give time diffs. at Paimbeouf +0040
Times of L.W. at Brest 2300 and 1100, give time diffs. at Paimbeouf +0115

Change in time of L.W. Brest for 6h, give change in difference at Paimbeouf of 35m.

a.m. Tide

Change in time of 4h (0903−0500) $= \frac{35}{6} \times 4 = 23$m

Time of L.W. at Brest	0903
Difference for Paimbeouf (0040+23)	+0103
Time L.W. Paimbeouf	1006

p.m. Tide

Change in time of 1h 33m (2300−2127) $= \frac{35}{6} \times 1{\cdot}5 = 9$m

Time of L.W. at Brest	2127
Difference for Paimbeouf (0115−9)	+0106
Time L.W. for Paimbeouf	2233

Height of L.W. at Brest M.L.W.N.	9·9 ft.
Height of L.W. at Brest M.L.W.S.	4·4 ft.
Change in height of L.W. at Brest	5·5 ft.
Change in height of L.W. at Paimboeuf	2·7 ft.

$$\frac{\text{Change in height difference at Paimboeuf}}{\text{Change in height difference at Brest}} \times \frac{\text{Difference in height}}{\text{at Brest}}$$

a.m. tide

$$\frac{2{\cdot}7}{5{\cdot}5} \times {\cdot}6 = {\cdot}2$$

p.m. tide

$$\frac{2{\cdot}7}{5{\cdot}5} \times 1{\cdot}1 = {\cdot}5$$

M.L.W.S. height difference Paimboeuf	−4·7	−4·7
change	+ ·2	+ ·5
Height difference Paimboeuf	−4·5	−4·2
Height L.W. Brest	9·3	8·8
Height L.W. Paimboeuf	4·8	4·6

To find the Height of the Tide at times between High and Low Water.

The height of the tide at times between high and low water can be found, though too much reliance must not be placed on the accuracy of these predictions. Referring to Fig. 5.2, it can be seen that the tide does not fall (or rise) at a uniform rate. The shape of the tidal curve varies in each port, and a diagram, showing the mean curve at Spring and Neap tides, accompanies each set of predictions for Standard Ports (except Venice) in the *Admiralty Tide Tables*, Vol. I.

Inside the front cover of Volume I is an Index to the Standard Ports.

(i) Turn to the predictions for the Port and the date required.

(ii) Select the High Water and Low Water between whose times the given time (B.S.T.) lies, and write down the heights.

(iii) Write down the predicted time of High Water and the B.S.T. for which the height is required. The difference between these times is called the *interval* and should be labelled − or +, according to whether it is before or after High Water respectively.

(iv) Now turn to the Tidal Diagram. The vertical lines show the interval before and after High Water for every 20 min. The horizontal lines show the amount the tide rises above the height of Low Water, the 'factor' being the *proportion* of the range which has been risen.

Compare the predicted range with the mean ranges given in the diagram. Select the appropriate range curve and read off the factor where the *interval* line cuts the curve. When the predicted range lies between those given on the diagram, take the factor for each curve and interpolate. For ranges greater than the mean spring range use the spring curve and for those less than the mean neap range use the neap curve.

(v) Multiply the predicted range by the factor. (This can be done by Table II on page 260 or a slide rule.) The product is the *correction* to Low Water.

(vi) Add this correction to the predicted height of Low Water to obtain the height of the tide for the given time.

WORKED EXAMPLE: Find the predicted *height* of the tide at 0400 B.S.T. on Jan. 8th, 1973 at Brest. (*See* extract from *Admiralty Tide Tables*, Vol. I, on page 257.)

From Tables:

Height of L.W.	6·2 ft.	Time given	0400 hr.
Height of H.W.	23·4 ft.	Time of H.W.	0709 hr.
Predicted Range	17·2 ft.	Interval from H.W.	−0309 hr.

From Diagram

Mean Range		*Factor*
Spring	20 ft.	·49
Neap	9·2 ft.	·48
	10·8 ft.	Interpolated factor taken as ·48

Correction = Range × Factor (from Table II).

17·2 × ·48 = 8·3
Height of L.W. 6·2

Predicted height at 0400 14·5

To Find the Time at which the Tide will be at a Given Height.

(i) Turn to the Standard Port predictions for the day in question.

(ii) Write down the heights of the appropriate high and low waters and the time of H.W. Subtract height of L.W. to obtain *predicted range*.

(iii) Write down the height for which the time is required and subtract from it the height of L.W. to obtain *correction to L.W.*

(iv) Find the factor (correction to L.W. ÷ predicted range) from Table II or by slide rule.

(v) Select the appropriate curve (spring or neap) and read off the *interval* from the factor. If the predicted range lies between those given in the diagram, take the interval from both curves and interpolate.

(vi) Apply the interval to the time of H.W. to obtain the required G.M.T.

WORKED EXAMPLE: Find the *time* after the first H.W. at Brest on Jan. 8th at which the tide will have fallen to a height of 12·5 ft.

From Tables:

Height of H.W.	23·4 ft.	Required height	12·5 ft.
Height of L.W.	5·9 ft.	L.W. height	5·9 ft.
Predicted Range	17·5 ft.	Correction to L.W.	6·6 ft.

Factor = Correction ÷ Range   4·8 ÷ 17·5 = ·29
(Extracted from Table II.)

From Diagram:

Mean Range		*Interval*
Spring	20·0 ft.	−3h 50m
Neap	9·2 ft.	3h 50m
Diff.	10·8 ft.	0h 0m

Interval	3h 50m
Time H.W.	07h 9m
Required Time	1059

Predictions at Secondary Ports at Times between High and Low Water.

'The Tidal Curve' (says the *Admiralty Tide Tables*, Vol. I) 'for a Secondary Port will seldom be exactly the same as for the equivalent Standard Port; nevertheless, except for certain areas, such as "the Water between Swanage and the Nab Tower", the curves for the Standard Port will generally be similar in shape to those of the Secondary Ports, and they can be used to obtain predicted heights and times with a fair degree of accuracy.'

To obtain the predicted height at some time between High and Low Water at a Secondary Port, proceed as follows, using the predicted heights, times and ranges *at the Secondary Port*.

WORKED EXAMPLE: It is required to find the predicted height of the tide at 1000 on 9th January at Le Palais. In the General Index Le Palais is No. 1654. The corresponding Standard Port is Brest.

Place	*H.W. height*	*L.W. height*	*H.W. time*	*L.W. time*
Brest	23·1 ft.	6·2 ft	0745	1405
Le Palais differences	−6·9 ft.	−3·3 ft.	−18	−27
Le Palais predictions	16·2 ft	2·9 ft.	0727	1338

H.W. Le Palais	16·2 ft.	Time at H.W. Le Palais	0727
L.W. Le Palais	2·9 ft.	Time required	1000
Range at Le Palais	13·3 ft.	Interval from H.W.	+0233

From Spring Curve for Brest		*From Neap Curve for Brest*	
Factor for interval +0226	·66	Factor for interval +0226	·64
Predicted range at Le Palais	13·3	Predicted range at LePalais	13·3
Ht. of tide above L.W. Table II)	8·8	Ht. of tide above L.W.	8·5
Predicted height of L.W.	2·9	Predicted height of L.W.	2·9
Height at Le Palais at 1000	11·7	Height at Le Palais at 1000	11·4

Caution. This method should not be used for places in the Solent area between Swanage and the Nab Tower, including the whole of the Isle of Wight. For this area a specia table, Table III of Vol. I (not included in this book) is used.

Reed's Nautical Almanac

(*a*) *Reed's Nautical Almanac* gives only the Times and Heights of High Water.

To find the time of Low Water, subtract the Duration of Mean Rise from the previous High Water.

To find the height of Low Water, subtract the height of High Water from twice the Mean Rise. (The Duration of Mean Rise and the Mean Level are tabulated on the Tidal Difference pages.) Time and Height of H.W. at the Standard Port:

1st Mar. Brest H.W.	1508		Ht. 5·8m	2nd Mar. H.W. 0330	6·2m
Duration of Rise	−603	M.L.×2	8·8m	−603	8·8m
L.W. Brest	0905		3·0m	2127	2·6m

(Converting metres to feet by the Conversion Table in Reed's; Height of Low Water at Brest 9·84 ft. and 8·5 ft.)

Secondary Ports.

The *average* tidal differences for Spring and Neap tides are given as being generally sufficient. These are applied to the predictions for the Standard Port.

EXAMPLE: Find the times and heights of High and Low Waters at Paimboeuf on March 1st.

Brest H.W.	1508	5·8m		(2nd Mar.)	0330	6·2m
Differences	+07	−1·8			+7	−1·8
Paimboeuf H.W.	1615	4·0			0337	4·4
Duration of rise	526	M.L.×2	6·0		−526	6·0
Paimboeuf L.W.	1049		2·0		2211	1·6
			(6·6 ft.)			(5·2 ft.)

It will be seen that the calculation of the times and heights of H.W. at Secondary Ports by means of *average* Time and Height Differences are necessarily approximate.

(*b*) *Heights and Times between High and Low Waters.*

To find the Height of the Tide at times between High and Low Water, or the Times at which the tide will reach a certain height, 'Reed's' provides a Table. 'The Height and Times are necessarily approximations and therefore too much trouble should not be taken in trying to obtain exact interpolations,' because all tides are assumed to follow the shape outlined in Fig. 5.2. whereas, as we have seen, it differs in different ports.

To use the Table we need the Duration of Mean Rise and the Height of Mean Level, both of which are listed on the Tidal Difference pages.

(i) To find the *Height* of the tide at a certain time.

Enter the upper part of the Table with the Duration of Mean Rise (in the left hand column for intervals *before* H.W., in the right-hand column for intervals *after* H.W.) and the Interval from H.W.

Follow the column in which the Interval is found down to the lower part of the Table to the level of the H.W. Ht.—M.L.

Subtract the figure found here from the height of H.W. This is the Height of Tide at the time required.

EXAMPLE: Find the height of the tide at 1000. H.W. at Le Palais 0727, height 16·2. Duration of Rise 6h. 7m. M.L. 9·5 ft.

H.W.	0727	Ht.	16·2
Time required	1000	M.L.	9·5
Time after H.W.	2 33	H.W.–M.L.	6·7

Interval from H.W.	*Duration of Mean Rise*
	h. m. 6 20 6 00
2·32 ←	← 6 00
Correction to be subtracted from H.W. Ht.	H.W. Ht.—M.L.
4·1 ←	← 6

Correction to be subtracted from H.W.	4·7
Ht. of H.W.	16·2
Height of the tide at 1000 about	11·5 ft.

It may be easier to tabulate the work thus: (If the port required is a Standard Port, simply omit the Tidal Differences.)

Height of Tide Form	*Time*	*Feet or Metres*
Standard Port High Water		
Tidal Differences		
Secondary Port High Water		
Mean Level	//////	
Duration		//////
Interval Before/After H.W.		//////
High Water less Mean Level		
Correction from High Water		
Height of Tide		

(ii) To find the *Time* at which the tide will reach a certain height.

Enter the lower part of the Table with the H.W. Ht.—M.L. and follow this line horizontally to find the Correction to H.W. Ht.

Follow this column vertically upwards to the upper part of the Table until level with the Duration of Rise. If the figures in the left-hand column are used the Interval is *before* H.W., but if the right-hand column is used the Interval is after H.W.

EXAMPLE: Find the time at which the tide has fallen to 11.5. Work the previous example in reverse.

The Tidal Rise and Fall.

'The Twelfths' Rule'.

A rough and ready, but useful, method of finding the amount that the tide will rise or fall in a certain time is given by 'The Twelfths Rule'. The Rule is given in '*Reed's*,' but is easily memorised and the rise and fall can be reckoned mentally.

1st hour's rise or fall	= one-twelfth	of Range
2nd ,, ,, ,, ,,	= two-twelfths	,,
3rd ,, ,, ,, ,,	= three-twelfths	,,
4th ,, ,, ,, ,,	= three-twelfths	,,
5th ,, ,, ,, ,,	= two-twelfths	,,
6th ,, ,, ,, ,,	= one-twelfth	,,

Aid to memory: '1,2,3—3,2,1.'

EXAMPLE: H.W. at Le Palais is at 0727, Ht. 16·2, Range 13·3 ft. What will be the height at 1000?

1st hour fall	one-twelfth of 13·3	= 1·1		
2nd ,, ,,	two-twelfths ,,	= 2·2	H.W.	16·2 ft.
$\frac{1}{2}$ of 3rd hour	$\frac{1}{2}\times$ three-twelfths ,,	= 1·6	—	4·9
		4·9	Ht. 1000	11·3

Accuracy of Predictions.

The accuracy of all tidal predictions both for time and height, may be affected by the meteorological conditions prevailing at the time.

The pressure of the atmosphere, as shown by the barometer, if higher than normal will lower the sea level, while if lower than normal the sea level will rise. A variation of 1 inch (34 mb.) causes a change of about 1 ft. in the sea level.

An on-shore wind tends to pile up the water and so raise the sea level, while an offshore wind does the reverse.

Rain and melting snow in rivers and estuaries may raise the level of the water.

Storm surges or seiches. The deep depressions which travel east-wards to the north of the North Sea cause a raising and lowering of the sea level as they pass. The oscillations thus produced may cause tides in the southern part of the North Sea to rise and fall up to 1 or 2 feet above and below their predicted heights. Exceptionally, an even greater affect upon the tide level can occur, especially in the Thames Estuary, on the cost of Holland and in the Heligoland Bight.

Exercises in Finding the Predicted Heights and Times of High and Low Water at Secondary Ports

1. Find the predicted height of high water p.m. at St. Nazaire on 2nd January 1973.

2. Find the predicted height of low water a.m. at Le Palais on 2nd February.

3. Find the height of low water a.m. at Le Pellerin on 17th March.

4. Find the time of high water at La Trinité on 8th January, a.m.

5. Find the time of high water at Les Sables-d'Oloune on 31st January, p.m.

6. Find the time of low water at Le Pouliguen on 23rd February, p.m.

Exercises in finding the Heights and Times at a Standard Port other than High and Low Waters

1. What will be the approximate height of the tide at 1630 G.M.T., 20th January, 1973, at Brest?

2. What correction should be applied to a sounding of 5 fathoms taken off Brest on 29th March, at 1630 G.M.T., so that it may be compared with soundings on the chart?

3. Anchored off Brest at 1900 G.M.T. on 3rd February, in charted depth of $4\frac{1}{2}$ fathoms, what water will there be under the keel at 1900 G.M.T. and at low water, if the ship's draft is 9 ft?

4. How much water will there be on a shoal charted as 2 fathoms outside Brest at 1715 G.M.T. on 28th March?

5. When could you first pass over a shoal which dries 2 ft., if your draft is 7 ft. and you allow 1 ft. of water below the keel, near Brest, during the afternoon on 7th February?

6. A bridge near Brest has a clearance of 6 metres at M.H.W.S. Your mast height is 30 ft. above water level. Allow 2 ft. clearance. At what time will the tide permit you to pass under the bridge on the afternoon of 13th March?

Exercises in Finding Heights and Times between High and Low Water at Secondary Ports

1. Find the predicted height of the tide at 0945 Z.T., 6th January at Benodet.

2. What is the height of the tide at Lorient at 1615 B.S.T., 27th January?

3. At what time will the height of the tide be 10 ft. about mid-day on 31st January at Concarneau?

4. How much water should be beneath your keel draft 8ft. at 1045 and at L.W. on 27th February at Les Sables-d'Olonne?

5. You have anchored at 2300 on 10th March off Vannes, where the chart shows 4 ft. Your draft is 5 ft. 6 in. What water, if any, will be below your keel at Low Water?

6. You have grounded on a falling tide early on 13th March near Nantes, where the chart shows 4 ft. Your draft is 12 ft. At what time can you expect to float?

CHAPTER 6

TIME

'This (generally speaking) hazy subject of *Time*.'

Captain S. T. S. Lecky.

Required for this chapter: *Nautical Almanac* (extracts on pages 184-196); Burton's.

Apparent and Mean Time. The sun is our timekeeper. Once a day, at 1200 Local Apparent Time (L.A.T.), the sun crosses the meridian (when, in our latitude, it always bears south). The Apparent Solar Day is the time interval between two successive transits of the sun across the meridian. The Apparent Solar Day varies in length and, since clocks cannot be made to measure variable time intervals, the average interval is divided into 24 hours of Mean Time.

Mean Time and Apparent Time exactly coincide four times during the year. Between these dates the difference between Mean and Apparent Time (known as the Equation of Time) fluctuates, being nearly 15 minutes ahead of Apparent Time in February and nearly 15 minutes behind in November.

Relationship between Longitude and Time. The earth rotates through 360°, with reference to the Mean Sun, in 24 hours. The sun,

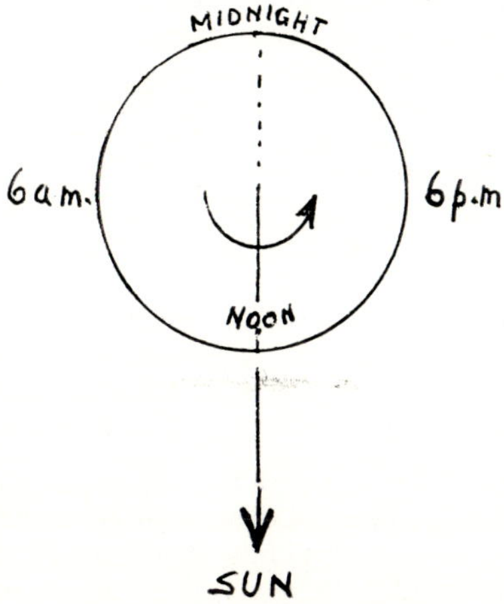

FIG. 6.1—Longitude and time

therefore, appears to travel 15° of longitude per hour. Thus if our time is 1200 hours, the time for an observer 90° to the west of us is only 0600 hours, because the sun will not cross his meridian for another six hours. To another observer 45° to the east of us, the time is 1500 hours, the sun having passed his meridian three hours ago. On the meridian 180° from our own the time is 0000 hours or midnight.

15° per hour is 1° in four minutes. Convenient tables for converting arc into time and time into arc are on page 134, Burton's, and in the *Nautical Almanac* and *Reed's Almanac*.

Standard Time (S.T.) It would be inconvenient if each town in the British Isles were to keep the time of its own meridian. The clocks in Dover would be about 22 minutes ahead of those in Devonport. It is therefore customary for every country to keep its own Standard Time, which is the Mean Time of one particular meridian most convenient for that country. In Great Britain our Standard Time is Greenwich Mean Time (G.M.T.). The Greenwich meridian passes through the Royal Observatory which was founded there in 1676.

In the *Nautical Almanac* there is a complete list of the Standard Times kept in each country.

British Summer Time (B.S.T.). From April to October we keep our clocks one hour in advance of G.M.T. This is known as British Summer Time.

Zone Time (Z.T.). Nearly all countries and many ships prefer to keep their clocks a whole number of hours ahead or behind G.M.T. For this purpose the world is divided into 24 zones, 12 east and 12 west of Greenwich, Greenwich being situated in the centre of Zone O. Each zone is one hour, or 15° wide, zones to the east of Greenwich being named minus, those to the west plus. For example, zone +1 begins at 7½° W. and extends to 22½° W. The 12th zone is divided by the Date Line into two half zones, +12 and −12. The zone number indicates the number of hours by which the zone time must be decreased or increased to obtain G.M.T. To find the G.M.T., given zone time, add or subtract the hours as indicated by the sign. It may occasionally be necessary to find in what zone a certain place lies. To do this convert the longitude of the place into time and the nearest whole hour is the zone number. Cyprus, for example, is in Long. 33° E. 33° is 2 hrs. 12 min.; therefore Cyprus lies in Zone −2 and keeps mean time of 30° E. or 2 hrs. in advance of G.M.T.

The Date Line. When crossing the 180th meridian (the date line is varied in places to include all islands in an archipelago) the date must be changed. If the ship has sailed westward (in the same direction as the sun appears to travel) the clocks have been periodically put back making the days longer. We must now allow for this by advancing the date by one day, for instance the day after Sunday June 16th becomes Tuesday June 18th. When sailing eastward, days have been shortened, so we borrow another day having, for instance two Sundays June 16th.

Local Mean Time (L.M.T.). The mean time of any locality, or ship, not keeping standard time, i.e. the time of its meridian. It can be found by adding or subtracting its longitude (in time) from Greenwich Mean Time using the rhyme:

'Longitude East; Greenwich is least;
Longitude West; Greenwich is best.' (i.e. 'most').

EXAMPLE: What is the L.M.T. of a ship in Long. $75^\circ\ 15'$ E. when the G.M.T. is 2240 (i.e. 22 hours 30 minutes or 10.40 p.m.?)

Long. $75^\circ\ 15' =$ 5h 1m (Burton's, p. 134).
G.M.T. 2240+5h 1m = 27h 41m, i.e. 0341 on the following day.

Chronometers. The G.M.T., (which is essential for position-finding out of sight of land) is kept on board ship by chronometers, which are accurately constructed clocks specially compensated for changes in temperature. They are kept slung in gimbals, like a compass, in a box with a glass lid which is opened only when the chronometer is wound. Chronometers are wound daily, at about the same time each day.

The hands of a chronometer are never disturbed, the day's error of the chronometer being noted in the Chronometer Ratebook at the time of winding and this error applied to the chronometer time to get the G.M.T.

To wind a chronometer. Turn the instrument upside down on its gimbals, push the dust cap to one side and insert the key. About seven anti-clockwise turns are necessary, a small hand on a dial on the face showing if it is fully wound.

Now that time signals can be frequently obtained by radio, a 'chronometer watch,' a good 'deck watch' or wristwatch may be used instead of a chronometer to keep G.M.T. A chronograph, a watch combined with a stop watch mechanism, is very convenient for taking the Greenwich time of sights. Shortly before a time signal is to be received the hands are set to this time and the chronograph re-started at the exact instant. The G.M.T. can then be read directly from the chronograph—provided that the interval between the time signal and sight is not too long.

Whenever the term 'chronometer' occurs in this book it is meant to include chronographs and watches used to keep Greenwich time.

The Greenwich Date. It should be noted that a chronometer face only shows 12 hours—not 24, and if there is any doubt about the correct reading the 'Greenwich Date,' the approximate G.M.T. must be found. To do this apply to the time at ship the longitude in time, according to the rhyme:

'Longitude East; Greenwich time least,
Longitude West, Greenwich time best.'

EXAMPLE: In Long. 75° 15′ E., on April 3rd at 1917, L.M.T. at ship the chronometer showed 2h 14m 50s; its error was 2m 15s slow. What is the correct G.M.T.?

April 3rd L.M.T.	19 17	Chron,	2h 14m 50s
Long. in Time	5 1E.	Error +	2 15
Corrected Date April 3rd and approx. G.M.T.	14 16		2h 17m 5s
			+12
		G.M.T.	14h 17m 5s

Finding the Chronometer Error. The error of the chronometer is found by comparing its reading with a time signal. Radio time signals can now be received all over the world and details of their transmission are published in Vol. 5 of the *Admiralty List of Radio Signals*. Remember that it is the last dot of the 6 which indicates the exact hour of B.B.C. time signals.

Since it is not always convenient to get time signals every day it is usual to take them periodically and calculate the Daily Rate at which the chronometer is losing or graining. The Daily Rate, multiplied by the number of days since the last time signal and then added or subtracted to the last error obtained will give the present error of the chronometer.

EXAMPLE: (i) On Sept. 29th the chronometer was 4m 6s slow of G.M.T. The daily rate is 2·5s gaining. What is the error on Oct. 21st?

Error Sept. 29th	4m 6s slow	Daily rate	2·5s gaining
Accumulated rate	— 55	Days	22
Error Oct. 21st	3m 11s slow	Acc. rate	55s gaining

EXAMPLE: (ii) The error of a chronometer on May 19th was 0m 5s fast of G.M.T. and on June 16th 3m 1s slow. What is the error on July 24th?

When intervals extend over a month it is convenient to make use of the days of the year listed on the calendar in the *Nautical Almanac* (page 5).

May 19th	139 days	Error	0m 5s fast	Daily rate	6·65 losing
June 16th	167 days	Error	3m 1s slow	days	38
Days	28	Lost	3m 6s	acc. rate	252·7s
					= 4m 12·7s losing

July 24th	205	
June 16th	167	error 3m 1s slow
	—	+4m 12·7
Days	38	
July 24th		7m 13·7s slow

It is desirable to get a time signal as shortly before sights are taken as possible, as even with good chronometers the daily rate may vary slightly.

Exercise in Chronometer Rating and Correction

1. May 16th 1900 a radio time signal showed the chronometer to be 1 min. 35 sec. slow. On May 9th, 1900, the chronometer was 57 sec. slow. What is the chronometer error on May 18th?

2. On 4th March at 2400 G.M.T., the chronometer was 4 min. 40 sec. fast on G.M.T. On 30th May at 1200 G.M.T. it was 2 min. 20 sec. fast. Find the chronometer's daily rate and the G.M.T. on 25th June when it showed 12 hrs. 01 min. 40 sec.

3. On 5th August at 1200 G.M.T., the chronometer was 2 min. 10 sec. slow on G.M.T. and, on 14th October, at 1200 G.M.T. it was found to be 2 min. 38 sec. fast. Find the daily rate and the G.M.T. on 6th November when the chronometer showed 18 hrs. 04 min. 00 sec.

4. On 2nd April at 1200 G.M.T., the chronometer showed 11 hrs. 56 min. 04 sec. On 18th April at 0600 G.M.T. it showed 05 hrs. 56 min. 52 sec. Find the G.M.T. on 4th May when the chronometer showed 10 hrs. 00 min. 27 sec.

To find the times of Sunrise and Sunset (and the beginning and end of twilight). The times of sunrise and sunset tabulated in the *Nautical Almanac* are for the *visible* phenomena, when the observer sees the sun's upper limb appear or disappear over the sea horizon. All times in the *Nautical Almanac* are G.M.T. and these times are G.M.T. for the Greenwich meridian and are given for every third day (the middle day on the page) and may be regarded without serious error as the L.M.T. on any meridian for one of the three days appearing on the page. This is accurate enough for practical purposes but interpolation may be made for the odd day.

If the G.M.T. of sunrise or sunset is required apply the longitude in time to the time given in the *Nautical Almanac*. If Standard or Zone Time is required, find the G.M.T. first and add or subtract the appropriate number of hours.

EXAMPLE: What is the Standard Time of sunset in the Azores (Lat. 36° N., Long. 26° W.) on 4th January, 1973?

From page 13 of the *Almanac*, change for 5° Lat. is 13 min., for 1° = 2·6
Change for 3 days is 2 min., for 1 day = 0·67.

Sunset Jan. 5th Lat. 35° N.		1702
		− $2\frac{1}{2}$
	Lat. 36	$1659\frac{1}{2}$
		− $\frac{1}{2}$
Sunset Jan. 4th Lat. 36°	L.M.T.	1659
Long. in time	W.	144
	G.M.T.	1843
Standard Time kept —		2 0
Sunset	S.T.	1643

To find the Time of Moonrise and Moonset. The times tabulated in the *Nautical Almanac* are for the *visible* phenomena when the observer sees the Moon's *upper* limb on the sea horizon.

These times are given in G.M.T. for the Greenwich meridian. To obtain the time of moonrise or moonset for any other meridian a correction for longitude must be applied to the tabulated time. (See Table II on page xxxii of the *Nautical Almanac*). Except for rare occasions in very high latitudes, this correction is additive for west longitudes and subtractive for east. (*See* explanation at bottom of Table II), page 254 in this book.

EXAMPLE: Find G.M.T. of moonrise on January 4th in Lat. 41° 46′ N., Long. 48° 10′ W.

Write down the time of moonrise for the nearest *lower* latitude (40° N.) for the day required and also for the following day (in W. longitude). If the longitude was E., used the preceding day.

Moonrise in Lat. 40° N., Jan. 4th, 0715. Jan. 5th 0753. Note the next latitude tabulated (45° N.); it is 5° higher and moonrise there is 19 minutes later.

(i) Interpolating for latitude. Turn to Table I for Latitude (page xxxii) in the *Nautical Almanac*. (This table can also be used for interpolating the latitude for sunrise and sunset).

Enter the 5° column of Tabular Interval and find the amount to add to Lat. 40° to get the ship's latitude (1° 36′). 1° 30′ is the nearest entry. Move horizontally to the right, to the column 'Difference between the times for consecutive latitudes'; 20 minutes is the nearest to the wanted 19 minutes. The correction is 6 minutes later.

(ii) Interpolating for longitude. Use Table II for longitude. (This table is exclusively for the moon). Enter the longitude column with the longitude nearest to the ship's longitude (50°). More horizontally to the column for 'Difference between the times of given date and following date (for W. longitude)': 40 minutes is the nearest to the wanted 38 minutes. The correction is 6 minutes later.

Moonrise Lat. 40° N. L.M.T.	0753	
Cor. for Lat. Table I	+6	
Cor. for Long. Table II	+6	
Correct L.M.T.	0805	
Long. in time W.	3 3	('Greenwich is best')
G.M.T.	1108	

(2) WORKED EXAMPLE: Find the Z.T. at which moonrise occurs in Lat. 41° 36′ N., Long. 140° 10′ W., on 19th September (at ship). *Procedure:*

(1) Write down the time of moonrise for the nearest *lower* latitude (40° N.) tabulated on the day required and also for the day

following (in west longitude). If the longitude had been east we would have used the preceding day.

(2) *Note the following*:

(i) The interval at which latitudes are tabulated (5° in this case);

(ii) The difference between the times for consecutive latitudes (18 minutes *earlier* for the higher latitude).

(iii) The difference between the times for the given date and the following date (69 minutes).

(3) Turn to Table I (page xxxii in the *Nautical Almanac*) to interpolate for latitude. Enter the 5° column of Tabular Interval and find the difference between the ship's latitude and lat. 40° (1° 36′). 1° 30′ is the nearest entry. Move horizontally to the right, to the column 'Difference between the times for consecutive latitudes'; 20 minutes is the nearest to the required 18. The correction is 6 minutes to subtract.

(4) Obtain the longitude correction from Table II on the same page. Enter with argument 'Long.' on the left and 'Difference between times for the given date and the following date' along the top. 70 minutes is the nearest to our 69. The correction is 27 minutes to add, for west longitude.

(5) To the L.M.T. at ship apply the longitude-in-time to obtain G.M.T.

		Moonrise		*Diff.*
Lat. 40° N., long. 0°	19th Sept.	2241	20th Sept. 2350	69
Table I, Cor. for 1° 36′ of lat.		−6		
Lat. 41° 36′ N., long. 0°		2235		
Cor. for long. 140° West		+27		
L.M.T. at ship	19th Sept.	2302		
Long. 140° 10′ W.		+921		
G.M.T.	20th Sept.	0823		
Zone+9		−9		
Z.T.	19th Sept.	2323		

The following symbols are used in the *Nautical Almanac*:

 sun or moon continuously above the horizon.

 " " " " below " "

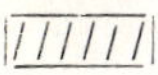 Twilight lasts all night.

Exercises in Finding the Times of Rising and Setting of the Sun and Moon and of Twilight

1. Find the L.M.T. of civil twilight and sunrise on 3rd Jan. 1973, in Lat. $16\frac{1}{2}°$ N.

2. What is the Z.T. of sunrise and sunset on 5th Jan. in Lat. 48° S., Long. 17° 45′ W.?

3. Find the Standard Time and Summer Time of sunset in Newfoundland on 21st Sept. (Lat. 48° N., Long. 53° W.).

4. Find the times of sunrise and sunset and moonrise and moonset for Standard Time at Blacksod Bay (Lat. 54° N., Long. 10° W.) on 4th January.

5. Find the Standard Time of sunset, evening civil twilight and moonrise in the Cook Islands (Lat. 20° S., Long. 160° W.) on 20th September.

6. Find the Zone Time of sunrise and sunset and moonrise and moonset in the Pacific Ocean in Lat. 7° N., Long. 158° E., on 2nd January.

CHAPTER 7

THE SEXTANT

'It is fine to bounce the sun along the skyline and chase the moon in its swift flight.' *From a cadet's essay.*

The sextant is used for accurately measuring the angle at the observer's eye subtended by objects at a distance.*

In off-shore work these are:

(*a*) The angle between the top of some object and the water line, when the sextant is held vertically; and

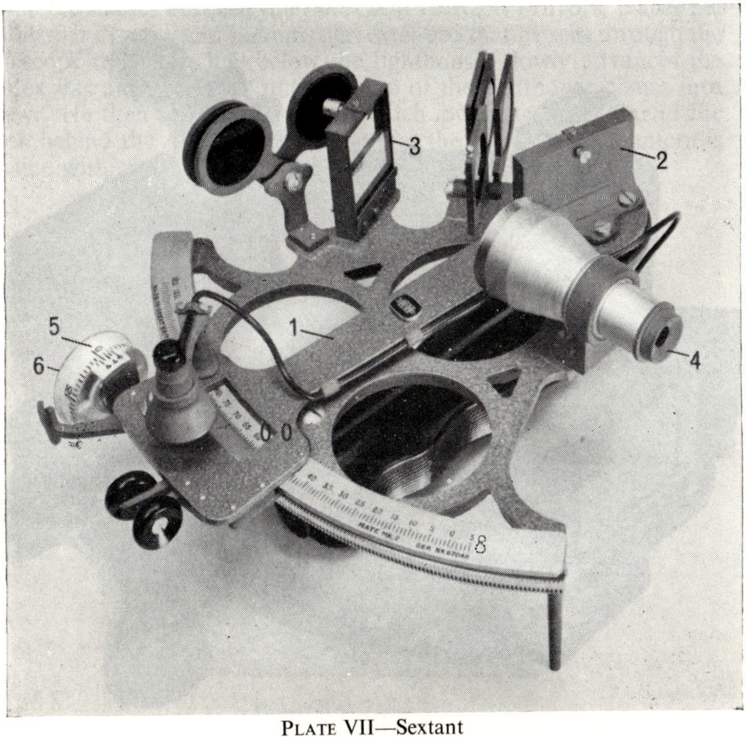

PLATE VII—Sextant

The principle parts of the sextant, shown on Plate VII, above are:

1. Index Bar. 2. Index Mirror. 3. Horizon Mirror. 4. Telescope. 5. Vernier. 6. Tangent Screw. 8. Graduated Arc.

* The sextant is a versatile instrument. Francis Galton, on one occasion, used his to measure unobtrusively the vital statistics of an exceptionally well-developed Hottentot woman.

Figure 7.2 shows the method of reading more clearly. The index arrow points to rather more than 41°. The vernier arrow points to rather more than 27′. 41° 27′ would usually be sufficiently accurate

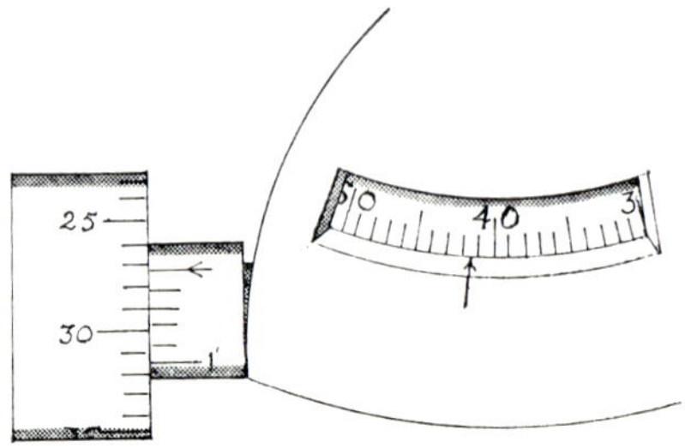

FIG. 7.2

reading, but if greater precision is needed the vernier, shown for this sextant for every ·2′, is read. Here the second division down (·4) is in line with a division of the micrometer; so the complete reading is 41° 27·4′.

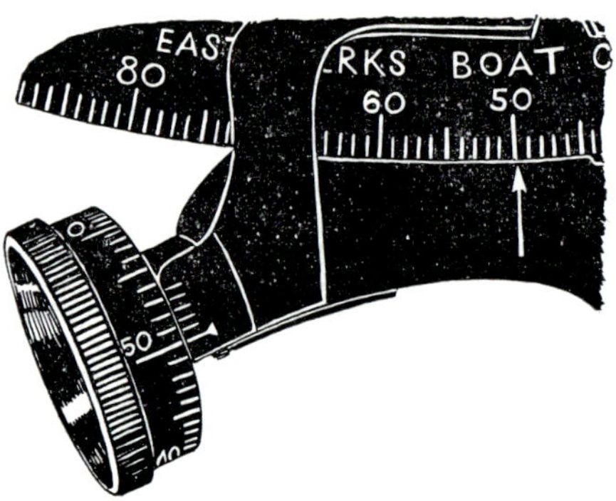

PLATE IX—Micrometer sextant reading 49° 50′.

Older types of sextants have a vernier which slides along the arc when the tangent screw is turned. The arc of the sextant is divided into degrees and tens of minutes, the unit minutes and decimals (most sextants read to ·2 of one minute) being read from the vernier. With this type the angle is read as follows. First, read the number of degrees and tens of minutes on the arc indicated by the Index—the arrow head on the right of the vernier.

Suppose that it is to the left of the first division after 2° on the arc, that is 2° 10′. Then note which division of the vernier is in line with a division on the arc, suppose it to be the first division after 3′, i.e. 3·2′. The angle will therefore be 2° 10′+3·2′ = 2° 13·2′. The sextant is provided with a magnifier but it may be difficult to tell exactly which divisions are in line, however, a reading to the nearest half minute is accurate enough for the purposes mentioned.

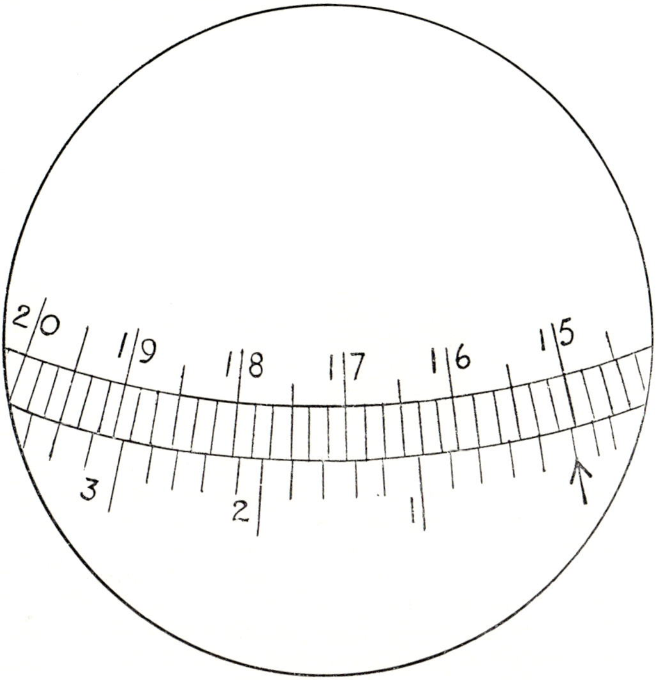

FIG. 7.3

Part of sextant arc and vernier as seen through the magnifier. The upper row of divisions are on the arc: the lower on the vernier. The sextant reads 15° 3·4′.

Measuring a Horizontal Angle. To measure horizontal angles the sextant is held horizontally with the handle below. Push the shades aside, place the index at zero, look through the telescope at the right-hand object. Keeping this object in view in the lower half of the telescope, turn slowly to the left, advancing the index bar while so doing, until the left-hand object appears in the upper half of the telescope. Lock the index bar by releasing the pressure of thumb and finger and bring the two objects into coincidence with the tangent screw. Read the sextant to the nearest half degree.

Measuring the Altitude of a Heavenly Body. (*a*) To observe the sun, first select a shade or shades pushing aside the others, and place the index about zero. Look through the telescope at the horizon below the sun and advance the index bar along the arc until the reflected sun appears in the horizon glass, with its lower limb (or edge) close to the line of the horizon. (It is customary to observe the lower limb, but if clouds obscure it the upper limb may be used.) Lock the index bar and, by turning the micrometer or tangent screw, bring the lower limb of the sun in contact with the horizon. Rock the sextant gently from side to side so that the sun skims the horizon touching it only at the point directly beneath it. Note the exact time of your observation (Fig. 7.4).

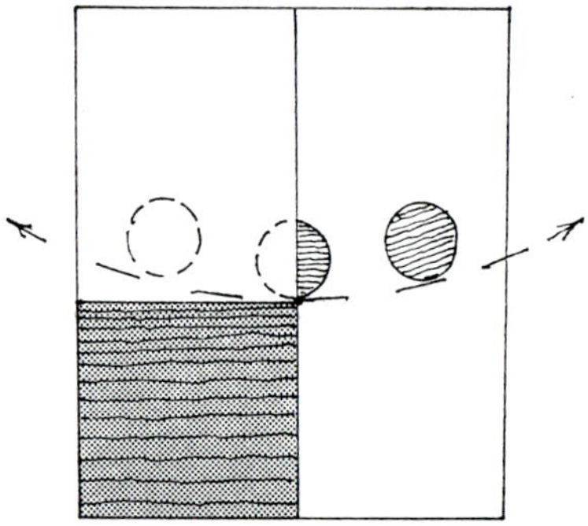

FIG. 7.4

(*b*) To observe a star, set the approximate altitude on the arc, push the shades aside and look through the telescope at the horizon in the direction of the star's bearing. Its reflected image should appear in the horizon glass somewhere near the horizon. Make the final contact as with the sun.

(*c*) To observe the moon, planet or star whose altitude has not been computed, place the index at zero and look through the telescope directly at the heavenly body. Without moving the index bar —so that the reflected image will remain visible—slowly dip the sextant until the horizon can be seen through the unsilvered part of the horizon glass. When the heavenly body has thus reached the horizon, clamp the index bar and complete the observation as before.

Sights in Small Craft. When taking sights in a small craft one should be as high up as possible to ensure getting the heavenly body in contact with the horizon and not on an intervening wavetop, yet at the same time, reasonably sheltered so that the sextant can be held steady. A towel may be useful to protect the sextant. Several 'shots' should be taken and the average altitudes and times used to work the sight. Standing on deck with the shoulders braced against the shrouds is often a good position to adopt.

Adjustments to the Sextant. If a sextant is to measure angles correctly the index and horizon mirror must be perpendicular to the plane of the sextant and, when the index is at zero, the two mirrors must be parallel to each other. To make sure that this is so, the following procedure should be carried out:

The First Adjustment. To ensure that the Index Mirror is perpendicular to the plane of the sextant. Unship the telescope which will otherwise be in the way, and place the index bar about half way along the arc. Now, holding the sextant horizontally with the arc away from you, look into the index mirror and see whether the reflected image of the arc is in line with the arc itself. If the true and reflected arcs are not continuous the mirror can be tilted until it is perpendicular by turning the screw situated at the top and back of the index mirror frame. If the index mirror is not set correctly the sextant has an 'error of perpendicularity.'

The Second Adjustment. To ensure that the horizon mirror is perpendicular to the plane of the sextant, ship the telescope and place the index at zero. Hold the sextant nearly horizontally and look through the telescope at the horizon. The reflected image of the horizon should be in line with the horizon itself. If it is not, the mirror can be tilted by turning the screw at the top of the horizon mirror frame. If the horizon mirror is not set correctly the sextant has 'side error.'

The Third Adjustment. To ensure that the index and horizon mirrors are parallel when the index is at zero. Hold the sextant vertically and look at the horizon, through the telescope. As the index is at zero the reflected and real horizon should be in line. If not, the horizon mirror can be slewed by turning the screw at the bottom of the horizon mirror frame. If the third adjustment is necessary the sextant has 'index error.'

The Fourth Adjustment. Another error, collimation, will be introduced if the axis of the telescope is not parallel to the plane of the sextant. (For this reason, it is important when observing to make sure that the heavenly body is in the middle of the telescope's field when contact is made.) In a few sextants the collar which holds the telescope is split into two parts allowing the axis of the sextant to be tilted slightly to correct this error.

It should be very seldom necessary to make *any* adjustments, but the *Index Error* should be frequently checked.

The Index Error of the Sextant. It may not be possible to make the third adjustment correctly without putting out the second adjustment. In any case, every sextant may have a slight instrumental error and this residual error, when the adjustments have been made, is known as the Index Error. It is found as follows:

Place the index near zero and, holding the sextant vertically, look through the telescope at the horizon (which should be clearly defined). Turn the tangent screw until the reflected and real horizon are in line. The sextant should, of course, read exactly zero; if it does not, the angle, now measured (the index error) will have to be added or subtracted to all angles measured. If the index lies to the left of zero, 'on the arc,' all observed angles will be too large, by the amount of the index error, so the index error must be subtracted, but if the index lies to the right of zero, 'off the arc,' it must be added to observed angles (Fig. 7.5).

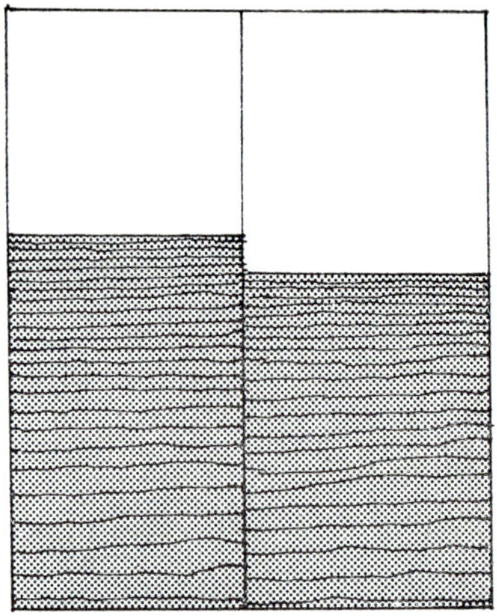

FIG. 7.5

It will be noticed that to permit angles 'off the arc,' i.e. below zero, to be measured, the divisions on the arc are continued for some few degrees to the right of zero. When reading an angle off the arc we are moving from zero in the opposite direction and must therefore use the *vernier* in reverse, taking the index to be on the left at 10, 9 now being read as 1, 8 as 2 and so on.

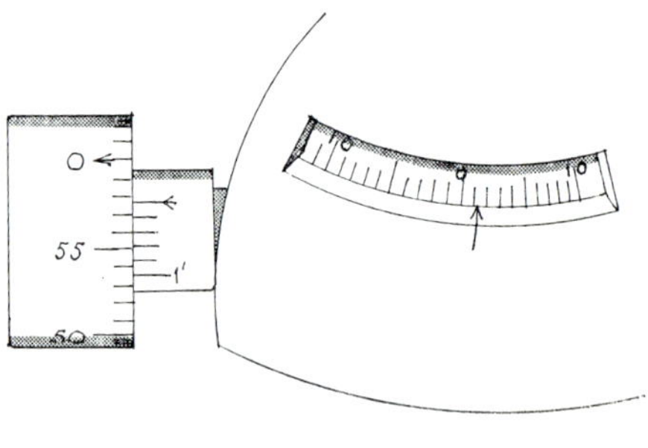

FIG. 7.6

Reading Off the Arc. In Fig. 7.6 the index arrow points just beyond -1°. Reading the micrometer in reverse from 0 gives $2'\cdot4$: so the angle is $-1^\circ\ 2'\cdot4$. One can equally say the index arrow shows -2°, but the micrometer shows $+57\cdot'6$: which is also $-1^\circ\ 2'\cdot4$.

CHAPTER 8

THE TRAVERSE TABLE: DEAD RECKONING AND DAY'S WORK

'A good D.R.—the basis of sound navigation.'

Problems involving course and distance, or Estimated and Dead Reckoning positions, hitherto worked graphically upon a chart, may sometimes be more conveniently solved by calculation. During an ocean passage, for instance, the scale of the chart may be too small for an accurate result to be had by measurement, or, at other times, the starting point and destination may lie on different charts.

For distance under 600 miles the curvature of the earth's surfaces may be disregarded and the problems all require the solving of a plane right-angled triangle, whose remaining sides and angles can be found when two parts are known.

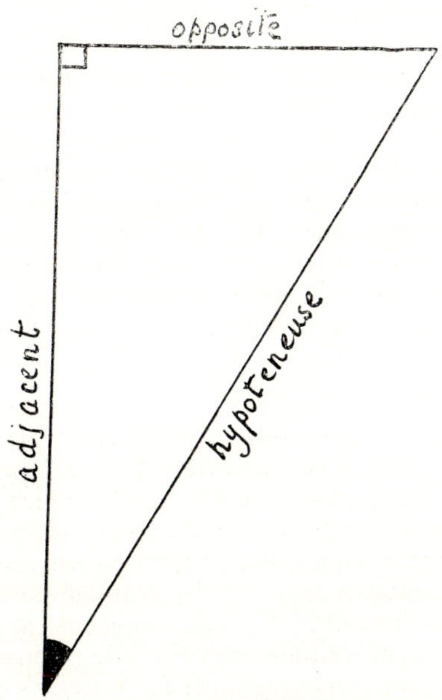

FIG. 8.1

The 'navigation triangles': (1) Angle = course, hypotenuse = distance, adjacent = difference of latitude, opposite = departure.

(2) Angle = mean latitude, hypotenuse = difference of longitude, adjacent = departure.

These various parts need not be worked out by trigonometry because every right angled triangle with angles between 1° and 89° and with hypotenuses of up to, at least, 100 units have been calculated and tabulated in the Traverse Tables, and these solutions will generally be sufficiently precise for our purposes.

The Traverse Tables enables us to solve the following problems very simply:

(1) To find the position arrived at after steering a certain course and distance.

(2) To find the course and distance to steer between two known positions.

(1) During an ocean passage we need to find our D.R. position at the time of taking observations of, for instance, the sun.

(2) When sights are obtained and the ship's position fixed, the course and distance made good from the last fix is worked out and the course and distance to the destination found. The difference in the D.R. and Observed Positions is the amount of current experienced. This can also be found by using the Traverse Table.

During ocean passages the course and distance made good is usually worked out from 1200 on one day to 1200 to the next, and when this is done the Problem is known as a '*Day's Work*.'

Before giving examples of the use of the Traverse Tables it may be as well to make a rather obvious point about calculations in general. Calculations may be easy: but they must be correct! A wrong answer will be misleading and may be dangerous. To err is human; so great care must always be taken to avoid mistakes in our working. For example; when extracting figures from the tables, read the figures, write them down, and then read them again. When adding, add the figures twice; after subtraction, add the 2nd and 3rd line to see if their sum agrees with the top line. Work methodically and write clearly.

The Traverse Table (Burton's p. 113) contains the tabulated lengths of the sides of right-angled triangles for every degree from 1° to 89°. (Refer to Fig. 8.1). In order to avoid printing twice the number of pages, the angles from 46° to 89° (which are the complements of the angles 44° to 1°) are tabulated at the *bottom* of the pages and, when using these angles, the column names must be read from the *bottom* of the columns also. In the problem under consideration the hypotenuse = distance, the side adjacent to the angle (the course) = difference of latitude, and the side opposite = departure.

Departure is a new term and must be defined. Departure is the number of sea miles (or minutes of latitude) made in an east or west direction. The name probably comes from 'Departure from the Meridian' (just as Dead Reckoning is supposed to derive from 'Deduced Reckoning.') Note that departure is not the same as difference of longitude. Refer to Fig. 1.1. Assume an observer at *A* moved east to *B* on the meridian *NGS*. His difference of longitude would be *QG*, but his departure would be *AB*. At the equator 1′ of departure equals 1′ of difference of longitude, but as the poles are approached and the meridians come closer together the length of one minute of longitude gets less. In our latitude 10′ departure equals about 16′ difference of longitude. The relationship between departure and difference of longitude thus depends upon the latitude. Note that 'd. long.' of two places is always larger than the 'dep.' between them, except at the equator when they are the same.

Departure must always be converted to difference of longitude in order to find the new longitude; and the difference of longitude must be changed to departure before the course and distance can be found. This can easily be done by the Traverse Table (or the Table 'Departure into D. Longitude' in *Reed's Nautical Almanac*.).

Examples in the Use of the Traverse Tables.

1. Leaving a position off the Needles, course is set 227° and the log is set to 0. What will be the latitude and longitude of the ship's position when the log reads 71 miles?

Opening the Traverse Table at course 227° (angle 47°) and entering distance 71 we extract the d. lat. (48·4) and dep. (51·9). We can apply the d. lat. directly to the latitude 'from.'

Position off Needles Lat.	50° 38′ N.	Long. 1° 40′ W.
d. lat.	48·4 S.	
Lat. in	49 49·6 N.	

The departure must be changed into d. long. before we can apply it to the longitude 'from.' The mean latitude is 50°, so using this angle and the italic column headings, we enter dep. 51·4 (the nearest entry to 51·9) and get a d. long. of 80, so our d. long. is about 80·4′.

Long. from	1° 40′ W.
d. long.	1 20·4 W.
Long. in	3 0·4 W.

2. What is the course and distance between positions off Portland Bill (50° 26′ N., 2° 28′ W.) and the Lizard (49° 55′ N., 5° 12′ W.)?

Find the d. lat. and the d. long. between the two positions, also (to the nearest half degree) the mean latitude.

Off Portland	50° 26′ N.		2° 28′ W.
Off Lizard	49 55 N.		5 12 W.
d. lat.	31′ S.	d. long.	2° 44′
			60°
			164° W.

Find the departure corresponding to d. long. 164 in latitude 50°. This Traverse Table stops listing the hypotenuse for every unit after 120 so we cannot find 164. However, if we divide 164 by 2 and look up the departure for 82 (which is 52·7), 52·7×2 will be the correct departure.

Now look for 105·4 (dep.) and 31 (d. lat.) in the columns of the Traverse Table. The nearest entries are 105·7 and 30·3, which gives angle 74° and distance 110. D. lat. is south and dep. is west, so the course is 254° and the distance 110 miles to the nearest degree and mile.

3. To find the course and distance to be sailed along a rhumb line joining points on a Great Circle track (*see* Chapter 3, page 68).

Although the distance may be rather large for much precision to be obtained the method is sufficiently accurate. In the example of a projected passage from the Bishop Rock to the Bahama Islands, the position of Bishop Rock is 49° 42′ N., 6° 27′ W. The next point on the Great Circle track has been found to be 48° 24′ N., 20° W. What course should be steered and what is the distance to this point?

Lat.	49° 42′ N.	Long.	6° 27′ W.
	48° 24′ N.		20°
	1° 18′		13° 33′
	60		60
diff.	78′		813′

813 is too large to be found in the Traverse Table, but we may divide it by 10 and multiply the corresponding departure by the same amount. This gives a departure of about 534. Entering the Table with dep. 53·4 and d. lat. 7·8—or the nearest figures we can get to them—we find the course is 262° and the distance 540′.

Traverse Table Exercises.

From the given positions find the new positions:

1. In the Atlantic, from Latitude 1° 17′ S., Longitude 11° 42′ W., Course steered 333° for 102 miles. Note that the equator is crossed.

2. From a position south of the Royal Sovereign LtH. Lat. 50° 36′ N., Long. 0° 26′ E. Course 261°, distance 55 miles. Note that the Greenwich meridian is crossed.

3. In the Bass Strait, Lat. 38° 12′ S., 144° 52′ E. Course 132°, distance 115 miles. Find the course and distance between the given positions. Note that if a side of the triangle is too large for exact entry in the Table, sides may be divided by 2 (or 10), and the extracted sides doubled (or multiplied by 10) to get the correct answers. The *angle* must *not* be altered in any way.

4. Off Shetland in Lat. 59° 55′ N., Long. 1° 35′ W. Off Bergen in Lat. 60° 15′ N., Long. 5° 7′ E.

5. Off Flamborough Head in Lat. 54° 8′ N., Long. 0° Off Dowsing LtV. in Lat. 53° 34′ N., Long. 0° 50′ E.

6. Off Bishop Rock in Lat. 49° 52′ N., Long. 6° 25′ W. Off Fastnet Rock in Lat. 51° 22′ N., Long. 9° 37′ W.

The Day's Work. A Day's Work will be rather more complicated if more than one course has been steered and perhaps a current allowed for. When there is more than one course it is best to tabulate the problem as shown below. The data is obtained from the log book.

Work the following example of a partial Day's Work, being careful (*a*) to use the correct column headings, i.e. those closest to the angle (course), (*b*) to place the d. lat. and dep. in the right columns.

0900 Compass Course 050°. Dev. 4° E. Var. 8° W. Wind S.E. leeway 3° log 37.

1130 Altered Course 105° C. Dev. 2° E. Var. 8° W. Wind S.E. leeway 1°. log 52. It was estimated that the tidal stream from 0900 to 1000, set 300° at 2 knots, and from 1000 to 1200, 320° at 1 knot. What was the course and speed made good between 0900 and 1200 when the log read 56 miles?

Compass Course	*Compass Error*	*True Course*	*Lee-way*	*Course made good*	*Dist.*	*D. Lat.*		*Dep.*	
						N.	*S.*	*E.*	*W.*
050°	4° W.	046°	3°	043°	15	11·0	—	10·2	—
105°	6° W.	099°	1°	098°	4	—	0·5	4·0	—
		Set		300°	2	1·7	—	—	1·0
				320°	2	1·5	—	—	1·3
						14·2	0·5	14·2	2·3
						−0·5	—	−2·3	—
					d. lat.	13·7 N.	*dep.*	11·9 E.	

We now enter 13·7 in a d. lat. column and 11·9 in a dep. column and find that the nearest entry gives distance 18 M, angle 41°. Since the d. lat. is N. and the dep. E. the course made good is 041°.

We may also wish to know the latitude and longitude of our new position, assuming we knew it at 0900. The d. lat. may be applied directly to the 0900 latitude to give the 1200 latitude, but we cannot apply departure to longitude and the dep. must first be changed into d. long.

Latitude 50°, dep. 11·9 gives d. long. 18·5.

0900 Lat. 50° 0′ N.		Long. 61° 13′ W.	
d. lat.	13·7 *N.*	*d. long.*	18·5 *E.*
1200 *Lat.* 50	13·7 *N.*	*Long.* 60	56·5 *W.*

To find the speed when the time taken to travel a certain distance is known.

The above method is reversed. The number of minutes taken in hyp. column must equal the distance run in another column. The ship's speed is read aginst 60 in the hyp. column.

EXAMPLE: In 72 minutes I have travelled 12·5 M. What is the ship's speed? Under angle 10° we find 72 in the hyp. column = 12·5 in the opp. column. Speed, found against 60in in the hyp. column is 10·4 Kn.

Approximate Conversions

Kilometres to Sea Miles	angle	57°	hyp. col	Km.	adj. col.	Miles.
Metres to Fathoms.	,,	57°	,,	Metres	,,	Fathoms.
Metres to Feet.	,,	72°	,,	Feet	,,	Metres.
Pounds to Kilograms.	,,	63°	,,	Pounds	,,	Kg.
Litres to Gallons.	,,	77°	,,	Litres	,,	Gallons.

Fahrenheit to Celsius (Centigrade).
When converting °F to °C, first subtract 32° from Fahrenheit.
,, ,, °C ,, °F, add 32° after conversion.
Angle 56° hyp. column Fahrenheit adj. column Celsius.

EXAMPLE: What is the equivalent Fahrenheit reading to 19° C? 34°+32° = 66° F.

Currency

Find the angle where the larger number in the hyp. column equals the lower number in another column.

EXAMPLE: When £1 = 10·5 francs, what is the sterling value of 45 francs. Angle 6° gives 100 (pence) in the hyp. column and 10·5 (francs) in the opp. column (or angle 84° gives 100 and 10·5 in the adj. column). Dividing by 10, 4·5 gives 43 pence, so 45 francs = £4·30.

PART II

METEOROLOGY

'The winds and the waves are always on the side of the ablest navigator.'

Edward Gibbon.

CHAPTER 9

THE ATMOSPHERE, WATER VAPOUR AND CLOUDS

The great aerial ocean which envelopes the earth and at whose bottom we dwell, is densest at the sea surface. About half the mass of the atmosphere lies below the summit of Mt. Blanc (15,800 ft.) and about two-thirds below the top of Everest (29,000 ft.). Pressure and the resulting winds are described in the next chapter.

Water Vapour in the Atmosphere. Evaporation. The escape of water vapour from the surface of water, snow and ice into the atmosphere is continually taking place, so that the lower layers of the atmosphere always contain a certain amount of invisible water vapour, especially over the warmer parts of the ocean. This has important meteorological results.

Saturation. The amount of water vapour which the atmosphere can hold in this way is governed by its temperature; the warmer the air the greater its capacity for holding water vapour. When the air contains as much water vapour as its present temperature permits, it is said to be saturated.

Condensation and Precipitation. Any excess moisture beyond the saturation point is condensed either into water droplets or ice crystals and becomes visible, first as fog (at ground level) or cloud (above ground level) and later, perhaps, as precipitation. *Precipitation* is the name given to any form of deposit or fall of water drops, whether frozen or otherwise; i.e. dew, hoar-frost, rime, glazed frost, drizzle, rain, sleet, snow or hail.

Relative Humidity is the proportion, expressed as per cent, of the amount of water vapour actually present in the air compared with the quantity it would hold when saturated at the same temperature. When the air is saturated the relative humidity is 100 per cent, but when holding only half the water vapour it is able to contain *at the same temperature* the relative humidity is 50 per cent. When the relative humidity is high a general feeling of dampness, and sometimes oppression, is noticeable.

Dew Point is the temperature at which cooling unsaturated air becomes saturated. If the air temperature falls below the dew point condensation of the *excess* water vapour will occur. If the air is cooled below dew point temperature by contact with a ground surface dew, mist or fog, or hoar frost will form. Hoar frost can form only when the temperature is below freezing point in addition to being below dew point.

Heating of the Atmosphere. The sun's rays heat the earth's surface but not the atmosphere which is 'transparent' to this short-wave radiation. The atmosphere, in turn, is warmed mainly by direct contact with the earth's surface. The heat is spread upwards by conduction, turbulence and convection. It follows that, since the atmosphere receives its heat from the earth, temperature decreases with height.

The Formation of Clouds. Cloud formation results from the rising of unsaturated air which, as pressure decreases with height, undergoes expansional cooling (adiabatic cooling). If the ascent continues the air will eventually be cooled below its dew point and the *excess* water vapour condensed into visible water droplets, i.e. cloud.

There are four main modes in which a rise of air may occur:

(i) *By a general ascent or upsliding.* When a warm air mass travelling along the surface encounters a colder air mass, it tends to override and climb over the colder air. A moving mass of cold air when overtaking a warm air mass tends to undercut and force the warm air up.

(ii) *By turbulence* (especially in strong winds). Surface friction causes eddy currents which mix the surface layers and clouds form at the condensation level (cloud base) which is often low.

(iii) *By convection* when a portion of air is heated more than the surrounding atmosphere by contact with a warm surface, e.g. a small island heated by the sun, it will expand, become lighter and float upwards.

(iv) *By orographic uplift.* When winds encounter hillsides the air is forced upwards and if it rises above the condensation level clouds will form. This can be below or above the summit of the hills.

The Principal Cloud Types. Most clouds can be placed in one of two classes:

(*a*) Clouds in a *layer* or sheet (Stratisform cloud).

(*b*) Clouds in more or less separate *heaps* or masses. (Cumuliform cloud).

They can also be classified by their approximate heights; High, Middle or Low Clouds.

The basic types are named:

(i) *Cirrus* (Ci) feathery, fibrous, detached clouds of most varied forms, some as straight or curved lines, wisps and tufts. Commonly called 'mares' tails.' High clouds.

(ii) *Cumulus* (Cu) clouds with rounded tops above a flat base, often brilliantly white with contrasting shadow. Low based clouds.

(iii) *Stratus* (St). More or less continuous sheets of cloud, like fog, though not resting on the ground. Low clouds.

These can be modified to describe the varying appearances, as in the Table below, which shows in addition their approximate heights.

High Clouds	Middle Clouds	Low Clouds	Low Clouds of Marked Vertical Extent
45,000 ft. to 18,000 ft. 13 700m to 5 500m	18,000 ft. to 8,000 ft. 5 500m to 2 500m	8,000 ft. to close to ground	45,000 ft. to close to ground
Cirrus (Ci) Cirrocumulus (Cc) Cirrostratus (Cs)	Altocumulus (Ac) Altostratus (As)	Nimbostratus (Ns) Stratus (St) Stratocumulus (Sc)	Cumulus (Cu) Cumulonimbus (Cb)

Cirrocumulus (Cc). Very small separate rounded cloudlets, without shadows, arranged in groups or lines, often resembling the ripples in the sand of the seashore.

Altocumulus (Ac). Lower and rather larger cloudlets, of ovoid shape. Blue sky usually shows between them, but their edges may join. Arranged more or less regularly in groups or lines. Both cirrocumulus and altocumulus are often described as mackerel sky.

Cirrostratus (Cs). A thin veil, sometimes fibrous and at others so diffused as to give a milky appearance to a blue sky. Never thick enough to obscure the sun or moon.

Altostratus (As). More fibrous and often thicker than cirrostratus. Blurs the outlines of sun and moon and in its thicker form, may completely hide them. May be described as a watery sky.

Nimbostratus (Ns). A low layer of completely opaque rain cloud, of nearly uniform structure and colour, which may or may not give rain or snow. The dark and sometimes threatening appearance distinguishes it from stratus.

Cumulonimbus (Cb). A mass of cloud of great vertical height, flat-bottomed but with fibrous tops of ice crystals often brilliantly white in colour in contrast to the dark base. It may bring thunderstorms and showers, especially when the tops are anvil shaped. Commonly called thunder cloud. Note: Nimbostratus gives continuous rain; Cumulonimbus gives showers).

Occasionally other terms are used to describe clouds more precisely:

Anvil Cirrus. A well developed cumulonimbus may spread out to an anvil-shaped top with a fringe of fibrous cloud called anvil cirrus. At times the anvil cirrus may be left after the anvil cumulonimbus has dispersed.

Lenticular Cloud. Hard-edged ovoid forms, or 'whale-back' clouds. Usually seen in small numbers in a bright blue sky, with a strong, gusty wind and an unsteady barometer.

Castellatus. Altocumulus Castellatus (Ac Cast). Here the individual clouds extend vertically upwards in 'castles'. This cannot be distinguished overhead, but can be seen at low elevations. Often precedes thundery weather.

Mammatus Cloud. The opposite to castellatus. The lower surface of the cloud hangs down in udder-like bulges. Usually occurs beneath cumulonimbus in thundery conditions and is then called mammato cumulus.

Fog and Mist may be described as cloud on the earth's surface. Fog is denser than mist and when visibility is reduced below 1100 yards fog is said to occur. (*Note*: The essential difference between cloud and fog is in the cooling process. Cloud is formed through adiabatic cooling whereas fog is formed by cooling by *contact* with the underlying surface, the temperature of which is below the dew point of the air).

Audibility of Sound Signals in Fog. Sound travels in a very uncertain manner in fog; not only is it impossible to judge distance by the loudness of a fog signal but the direction from which it appears to come can be very misleading. Navigators must be on their guard against this.

The Formation of Fog. There are two main types to distinguish:

(i) *Advection Fog.* This fog occurs when moist warm air moves over a colder surface of sea, e.g. the frequent fogs of the Grand Banks of Newfoundland which are formed when the warm damp air over the Gulf Stream is blown over the adjacent cold Labrador Current. Over the land it is likely to form in winter after a cold spell when a milder wind blows in from the sea, but this type seldom persists and is usually confined to the time of the onset of the warm air. The air is cooled by the land but if the wind is more than moderate, turbulence will diffuse the surface cooling through a greater depth of air and low cloud will form instead of fog.

The most widespread type of advection fog is that caused by a warm maritime air mass moving over a surface which becomes only gradually colder, e.g. an air mass moving from the Azores towards the British Isles which may cause fog to extend over many thousands of square miles as it approaches our cooler latitudes.

(ii) *Radiation Fog* (ground or land fog), forms almost entirely at night over land and is due to the air being cooled by contact with the ground which has cooled by radiation. Radiation fog may drift seawards and is often encountered in rivers and harbours. Sometimes it spreads from England across the Straits of Dover to the French coast. Conditions favourable to its formation are (*a*) much moisture in the lowest layers of the atmosphere; (*b*) a clear sky at night; (*c*) a light surface breeze; (*d*) surface of the ground cold and wet.

The lowest temperature occurs just before dawn and the slight increase of turbulence aroused by the first warmth of the sun means that the fog is likely to be thickest one hour after sunrise. After the sun has risen higher the increased heat tends to clear the fog. In the British Isles radiation fog seldom lasts more than a few hours in summer, but is more prevalent in the longer nights of autumn and winter and may persist for some days in an anticyclone. Ground fog rarely extends to a great height and occasionally it is possible to see over it from the masthead.

Two less common types of fog are:

(iii) *Mixing Fog* which may occur when a cold air current meets a warm air current and both air masses are nearly saturated. This can happen near a warm front or occlusion in temperate or high latitude (*see* Chap. 11) and for this reason the fog is also known as *Frontal Fog*.

(iv) *Sea Smoke* is mist or fog forming on the sea's surface when the temperature of the air is very low relative to the sea surface. It is only common in polar regions and the east coasts of continents when strong off-shore winds are blowing in winter. The cold air close to the surface rapidly becomes saturated with the moisture evaporating from the relatively warm water and this moisture quickly condenses and becomes visible as the warmed air rises through its colder lower layers. Generally sea smoke is shallow and patchy, having the appearance of columns of steam or smoke, and does not much hamper navigation.

CHAPTER 10

ATMOSPHERIC PRESSURE AND WIND SYSTEMS

'From east to north the transient breezes play . . .'
Wm. Falconer.

'And the stormy winds do blow.' *Thos. Campbell.*

Atmospheric Pressure. Air, like any other substance, has weight (we can feel it exerted horizontally when standing in a wind). The pressure of the atmosphere at sea level is usually about 15 lbs./sq. in., equivalent to about 29·5 inches of mercury, or about 1 bar or 1,000 millibars (mb.), the *millibar* being a unit denoting pressure and defined as 1,000 dynes/sq. cm. (1 dyne is the unit of force, in the centimetre-gram-second system, which accelerates a unit mass (1 gram) by 1 centimetre per second.*

Lines joining places with equal pressure on weather maps are known as *isobars* (isos. = equal) (Fig. 10.1).

Pressure Variation over the Earth's Surface. Pressure is not constant over the earth's surface, but varies almost continually and often rapidly in the temperate zones, though in the tropics it remains fairly constant for long periods apart from a small daily range which is primarily due to atmospheric tides.

Atmospheric Pressure and Temperature are closely linked because air expands when heated, becoming less dense in consequence and lighter. Lines joining places with equal temperatures on weather maps are called *isotherms* (Fig. 10.2).

The General Pattern of Atmospheric Pressure. In the tropic ocean area the warm expanding air produces a belt of low pressure between two belts of higher pressure on the polar sides. These belts 'follow the sun' moving northwards when the sun has north declination and southwards when its declination is south. This general arrangement is altered by the huge Europe-Asian land-mass which, becoming very hot in the summer produces a low-pressure area centered over northern India, while, in winter, being excessively cold, a high pressure area exists centred over Siberia. (*See* Figs. 10.2 and 10.3).

At the polar caps the earth loses heat by radiation throughout the greater part of the year, thus the air is very cold at all levels. This in turn causes permanent high pressure in each of these regions and the consequent outflow of air at surface level is replaced by subsidence from higher levels. In the Arctic, therefore, north-east and easterly winds are common and in the ocean round Antarctica south-easterly and easterly winds are frequent.

---

* In the SI system 1 mbar$=10^2$N/m^2, that is 100 newtons per square metre.

Wind. Winds are named from the compass direction *from* which they blow. Surface winds blow along the isobars but with an inclination towards the low-pressure side. Winds are said to *veer* if they change their direction clockwise; or *back* if they change their direction anti-clockwise.

Relationship between Wind and Isobars. Wind is caused by pressure difference in adjacent areas. A *pressure force* is exerted in a direction perpendicular to the isobars tending to drive the air particles from high pressure towards low pressure. The rotational effect of the earth, however, deflects the moving air to the right of its original path in the northern hemisphere and to the left in the southern hemisphere. This deflecting force is called the *geostrophic force.* The result of these two forces, plus surface friction, is that at the surface the wind blows at an angle of indraft of from 15° to 30° across the isobars towards the low pressure side. The rotational effect of the earth is not effective between latitudes 5° N. and 5° S., and here the wind blows at 90° to the isobars.

Circulation Round Areas of High and Low Pressure. It follows that, in the northern hemisphere as a wind blowing from an area of high pressure is deflected to the right, it will circulate clockwise, while air will circulate anti-clockwise round an area of low pressure. In the southern hemisphere the reverse is the case and winds circulate clockwise round low pressure areas and anti-clockwise round areas of high pressure.

Buys Ballot's Law. It also follows, that if we *face the wind in the nothern hemisphere, low pressure is on our right.* The reverse is true of the southern hemisphere. This is known as Buys Ballot's Law.

The Permanent Winds. The Trade Winds and Westerlies. In the tropics the air is heated, expands and rises and is replaced by colder, denser, heavier air which flows from higher latitudes. As this cool air moves towards the equator it passes over areas which are rotating faster and thus the original wind from the north in the northern hemisphere becomes north-east, while in the southern hemisphere the south wind becomes south-easterly. These north-east and south-east winds are known as the *Trade Winds* and blow more or less steadily all the year round from about latitude 30° towards the equator in the North and South Atlantic and Pacific Oceans and in the South Indian Ocean. In the North Indian Ocean the North-east Trade Wind is replaced by the North East and South West Monsoon (*see* below). The average speed of the Trades is 15 knots in the Atlantic and 18 knots in the South Indian Ocean. They blow most strongly towards the end of the winter and their limits move up to 10° latitude northwards and southwards following the sun's declination.

The low pressure areas of calms, light airs and occasional heavy showers between the Trades is known as the *Doldrums.*

On the polar side of the high pressure areas of light variable winds (known as the *Horse Latitudes*) there are belts of *Westerly*

Winds. The British Isles lie in the northerly belt and this profoundly affects our climate. In the southern hemisphere these winds blow almost uninterrupted by land and are known as the *Roaring Forties.*

Seasonal Winds. The Monsoons. The permanent wind systems are modified chiefly by the continents which heat and cool more rapidly than the oceans and this upsets the balance of temperatures. The best example is the enormous land-mass of Europe-Asia. The low pressure over northern India (due to summer heat) causes an anti-clockwise circulation producing the *South West Monsoon,* strong to gale force south-westerly winds in the North Indian Ocean, Arabian Sea and Bay of Bengal and moderate south-westerly winds in the China Sea. In winter the now cold, high pressure area over Siberia causes a clockwise movement and the *North East Monsoon* with fresh to strong north-east winds in the China Sea and moderate to light north-east winds in the Bay of Bengal, Arabian Sea and North Indian Ocean.

Daily Winds. Land and Sea Breezes. The diurnal range of sea surface temperature is not more than 1° C. in any latitude, whereas the land temperature may vary by 20° C. in some places. During the day the warm air rising from the land is replaced by a cool breeze blowing from seaward, which may set in by noon though often not until afternoon. At night the reverse may occur, though the land breeze is usually weaker than the sea breeze. They are experienced most often in summer and in the tropics where the land is strongly heated by day. There the sea breeze can extend 20 miles, and the land breeze several miles off shore.

Katabatic Winds (*kata* = down). During the night, particularly with a clear sky, when the land is cold and consequently cools the air immediately above it, if the ground is sloping the cool, denser air flows down the slope, and is known as a *katabatic wind.* Off mountainous coasts this wind can blow strongly when combined with a steep pressure gradient (e.g. the Bora of the East Adriatic coast).

A wind blowing up the slope of a mountain is known as *anabatic wind.* It occurs in calm, sunny weather when the land's surface is greatly heated.

Tornadoes and Waterspouts are miniature cyclones, (*see* Chap. 12), miniature, that is, in area though not in force. *Tornadoes* are violent whirlwinds a few hundred feet in diameter, with wind speeds of 200 knots or more near the centre. When these form over the sea they are called *waterspouts* which, though less violent than tornadoes ashore are a hazard to shipping and a danger to small craft. Waterspouts form below a heavy cumulonimbus cloud. A funnel-shaped cloud stretches downwards towards the sea where a cloud of spray forms. Waterspouts travel quite slowly and may last for half an hour; they are more frequent in the tropics than in the temperate zones.

CHAPTER 11

WEATHER SYSTEMS: DEPRESSIONS AND ANTI-CYCLONES

'A depression situated south of Iceland . . .' *Weather Report.*

A look at any weather map will show closed isobaric systems, like the contour lines denoting hills on topographical maps, some enclosing *Lows* or *Depressions*, others, probably, *Highs* or *Anticyclones.* It will further be observed that in the northern hemisphere the wind circulation is anti-clockwise round the centres of low pressures and clockwise round the centres of high pressure, while in the southern hemisphere the opposite is the case.

These systems are enormous eddies which often form along what is known as the polar front and which travel in a general easterly direction.

The Polar Front. The cold surface air from the polar regions is separated from the warmer air of lower latitudes and this well-defined surface is known as a polar front. (*See* Fig. 11.1). The polar front, though not continuous, extends round the world. In the North Atlantic it extends, in winter, from Florida to Cornwall, in summer, generally, from Newfoundland to Scotland. When the Azores high is well developed the polar front is pushed northwards and the depressions which are formed along it pass to the north of the British Isles. While this happens Britain gets better weather than usual.

Depressions. The formations of depressions is aided by large temperature differences between warm and cold air masses. This occurs in the western North Atlantic and western North Pacific and in the Southern Oceans, particularly in winter.

For a depression to form on the polar front it is necessary that the warm air should be moving to the eastward at a greater speed than the cold air. When this happens a small wave-like disturbance may form on the frontal surface. It develops (in the northern hemisphere into an anti-clockwise circulation and moves away to the east or north east. The warm air overrides the cold air at what is called the *warm front*; the cold air undercuts the warm air at what is called the *cold front*. Pressure is reduced in the centre and the depression deepens. The whole system meanwhile drifts along roughly with the speed of the warm air.

As the depression proceeds the cold front slowly gains on the warm front, gradually diminishing the warm air at the surface until it finally lifts it clear of the ground. This shutting-off of the warm air from the surface is called an *occlusion*. Most of the depressions which reach north west Europe from the Atlantic are occluded. Once a depression is occluded it usually fills up within a few days.

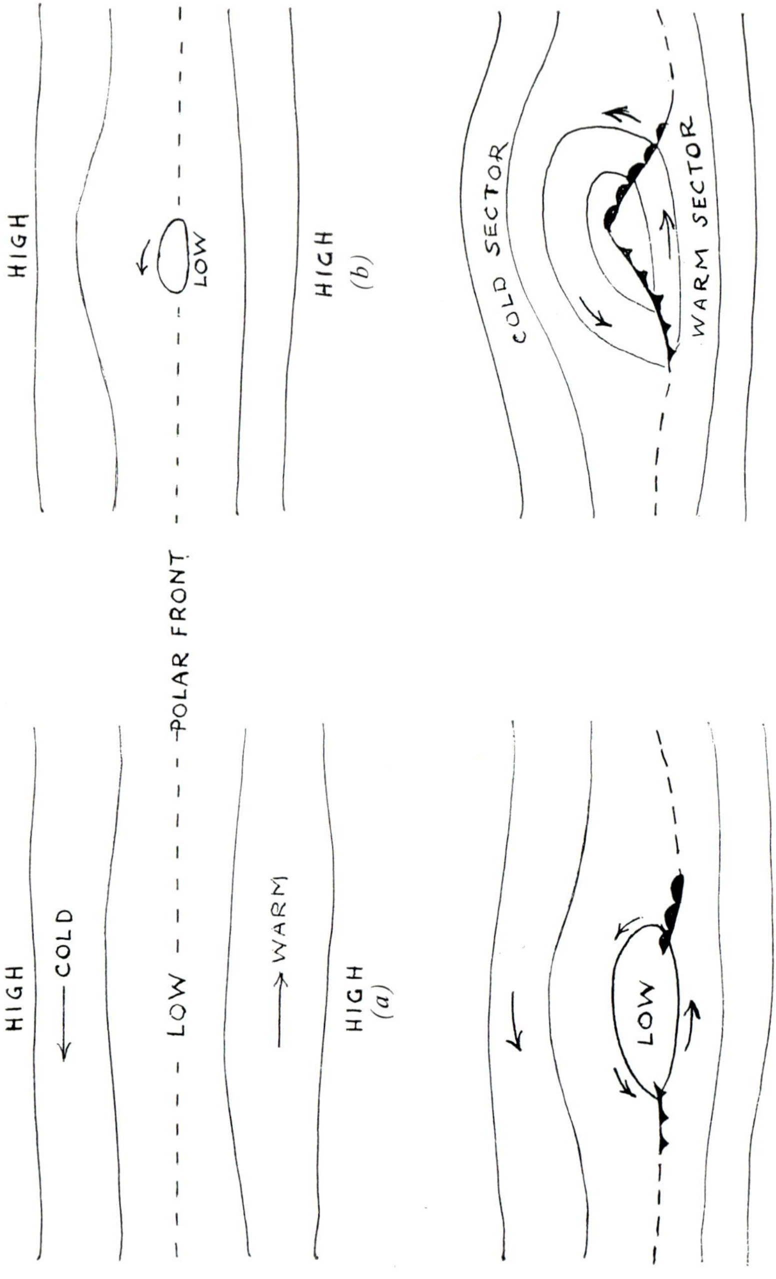

FIG. 11.1—Stages in the Development of a Depression (Northern Hemisphere)

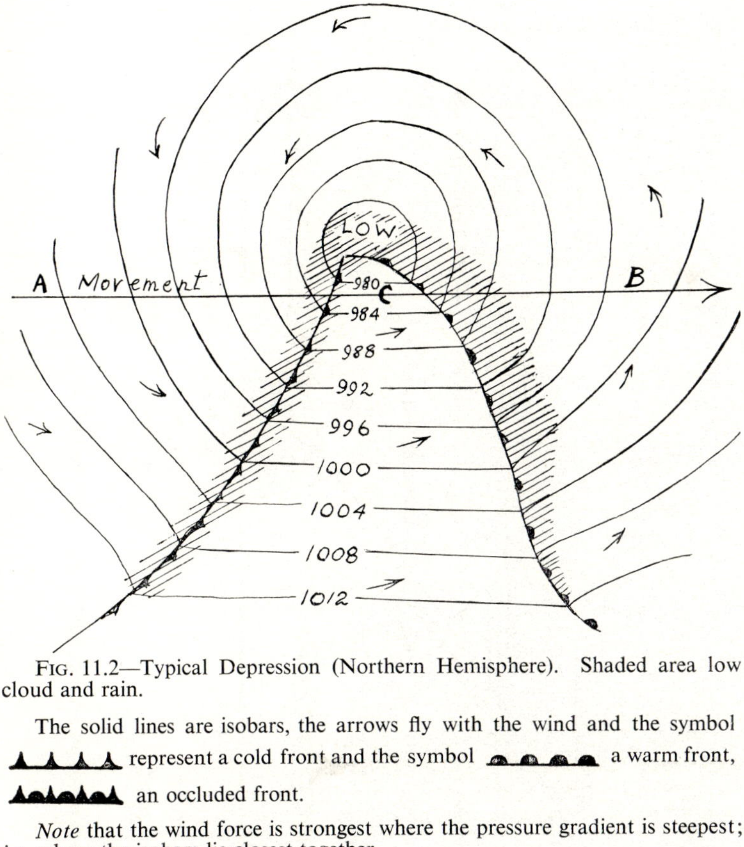

FIG. 11.2—Typical Depression (Northern Hemisphere). Shaded area low cloud and rain.

The solid lines are isobars, the arrows fly with the wind and the symbol represent a cold front and the symbol a warm front, an occluded front.

Note that the wind force is strongest where the pressure gradient is steepest; *i.e.* where the isobars lie closest together.

Families of Depressions. Conditions which favour the formation of one depression usually permit the formation of others. The average number is about four and the depressions are usually separated by ridges of high pressure which give brief intervals of fine weather.

Secondary Depressions. A 'Secondary' is a depression which becomes embedded in the circulation of a larger 'primary' depression. Sometimes one depression in a family, moving faster, gets captured by a more vigorous one preceding it. Usually the secondary tends to move round the primary in a cyclonic direction and sometimes especially when the primary is filling up, the secondary absorbs the primary.

L

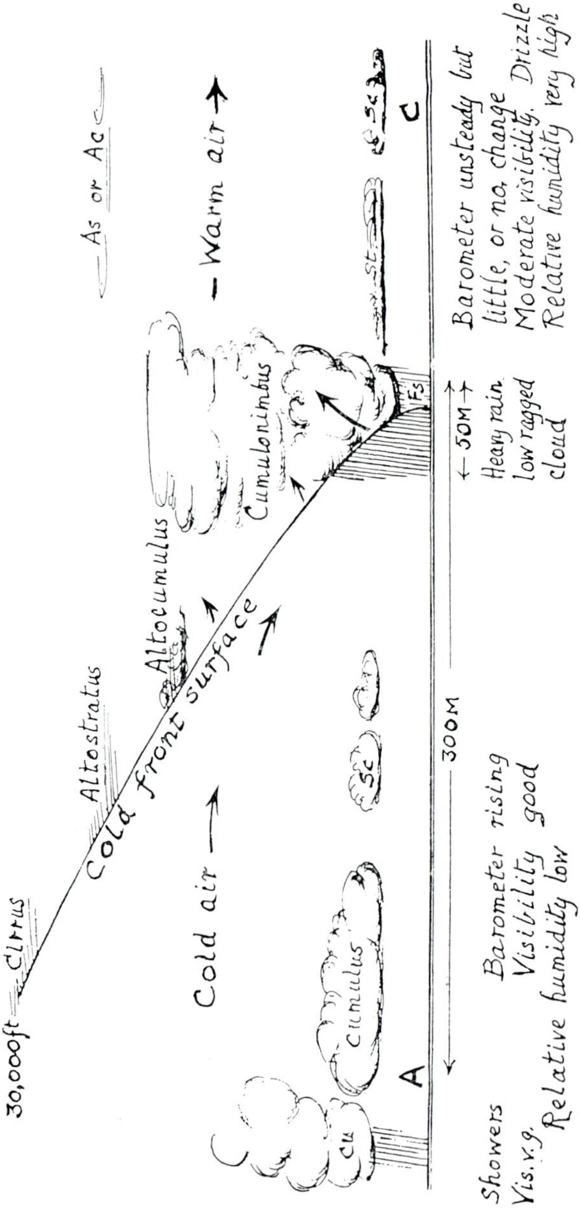

FIG. 11.3—Elevation of an average cold front along the line *AC* in Figure 11.2

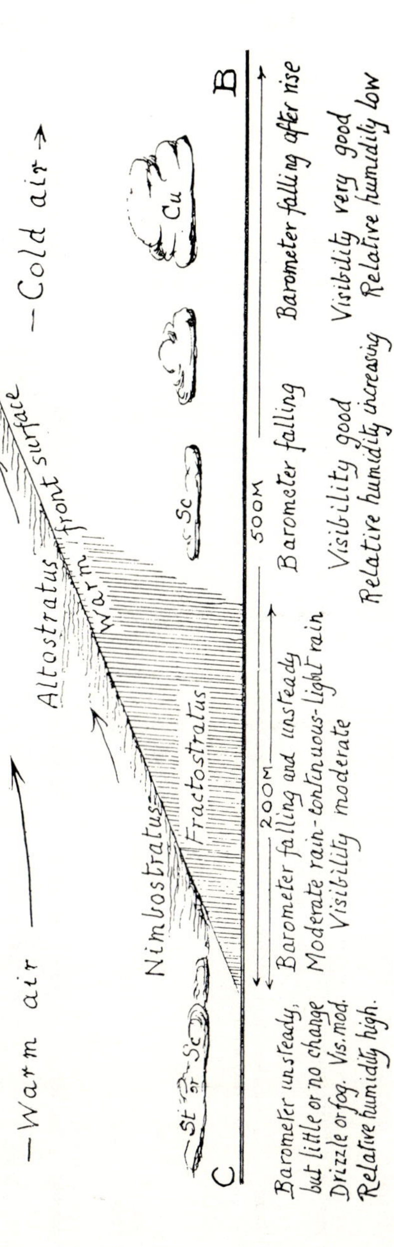

FIG. 11.4—Elevation of an average warm front, along the line *CB* in Figure 11.2

Weather experienced in Depressions. Since the British Isles lie directly in the course often taken by depressions the changes of weather they produce are important to us; they can be summarised as in the following table.

	Barometer	*Wind*	*Temperature*	*Cloud*	*Weather*	*Visibility*
Approach of Warm Front	Fallings unsteady	Increasing, sometimes backs a little	Steady, or slow rise	Ci, Cs, As, Ns in succession ; scud below As and Ns	Continuous rain or snow	Very good or good except in precipitation
At the passage	Fall ceases	Veers and sometimes decreases	Rise	Ns and scud	Rain gives way to drizzle	Poor, sometimes mist or fog
In the warm sector	Little change, but unsteady	Steady direction	Little change	St and Sc	Mainly cloudy, or drizzle or intermittent slight rain	Usually moderate; perhaps mist or fog
At near approach of Cold Front	Fall	Backing a little, increasing and often squally	Steady but starts to fall on near approach of cold front	Ac and As then heavy Cu or Cb	Usually some heavy rain, perhaps thunder	Usually moderate
At the passage	Sudden rise	Sudden veer and sometimes heavy squall	Sudden fall	Cb with low scud	Rain, often heavy, perhaps thunder and hail	Rapid improvement when front has gone by
In the rear	Steady rise	Backing a little after squalls, then often strengthens and may steady or veer further in a later squall	Little change, or perhaps steady fall	Lifting rapidly. Then As or Ac; later Cu or Cb, becoming more scattered	Heavy rain gives way to showers which become less frequent as cold front retreats	Very good, except during showers

Weather experienced in Occlusions.

Weather at the approach of an occluded front.	Similar to the approach of a warm front.
At the passage of an occluded front.	The wind veers and may strengthen. No rapid clearance of the sky. Heavy rain with cb and, perhaps, squalls.

Anticyclones are systems of isobars enclosing a region of high pressure. Winds circulate in the opposite manner to a cyclone; clockwise in the northern hemisphere and anti-clockwise in the southern hemisphere. The deficiency of air in the centre, caused by the outflow of air across the isobars (divergence) is made up by subsidence from upper levels. Thus the relatively moist surface air is replaced by warm dry air from above.

Weather is usually quiet, fine and settled with light winds near the centre. There may be much cloud and some precipitation in the outer part of the system. In summer it may be bright and sunny; in winter there may be radiation fog, clear sky and hard frost.

Troughs of Low Pressure. The isobars resemble contours round a valley. 'V' troughs often contain a front where the isobars are sharply bent. 'U' troughs do not usually contain a front but often

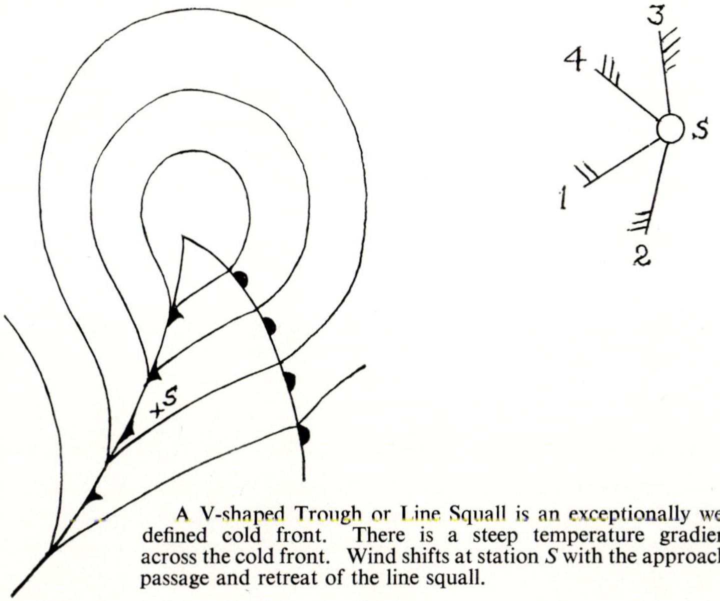

A V-shaped Trough or Line Squall is an exceptionally well defined cold front. There is a steep temperature gradient across the cold front. Wind shifts at station *S* with the approach, passage and retreat of the line squall.

FIG. 11.5

occur in the cold air mass behind an occluded depression. Strong winds may blow as the pressure gradient increases and bad weather is usual on the advancing side of troughs though the weather clears when the bottom of the trough has passed.

Ridges (or Wedges) of High Pressure. Outward extensions of high pressure from anti-cyclones protruding between two areas of low pressure. Weather will be anti-cyclonic in character but, if the ridge is moving fast, fine weather will not last long.

The Col. A neutral region of indeterminate pressure between two highs and two lows located diametrically opposite each other. The weather is usually changeable with light variable winds. Fog may form in autumn or winter. In summer the weather may be brilliantly fine, but over the land, thunderstorms may occur.

CHAPTER 12

TROPICAL REVOLVING STORMS

'. . . the stormy wind ariseth which lifteth up the waves thereof. They are carried up to the heaven, and down again to the deep.'
Psalm 107.

Within the tropics, particularly in the western part of the oceans and in the neighbourhood of archipelagoes, destructive storms originate and the seaman must avoid them if he can. They are of cyclonic form, the wind blowing in an inward spiral towards the low-pressure storm centre. Though less extensive than the depressions of higher latitudes they are usually more violent.

Path. Originating generally between 7° and 15° of latitude, the storms move westward, later tending to turn away from the equator and, when between latitudes 15° and 25°, often recurve to the north east (in the northern hemisphere) and south east (in the southern hemisphere), though some do not recurve at all.

Extent. The most dangerous part of the storm is the centre, for there, although comparatively windless, the sea is extremely confused and capable of damaging or destroying any vessel. The centre, perhaps 10 miles across, is surrounded by furious winds blowing along the isobars, the area of gale-force winds extending some hundred miles. As the storm proceeds it grows in extent and by latitude 35° may be twice as wide, but the intensity will be less.

Speed of Advance. Storms usually move slowly at first, gradually increasing to about 12 knots before recurving. While recurving they slow down and afterwards increase speed to 20 or 25 knots as a rule, though they have been known to exceed 40.

When the path is erratic, as is sometimes the case, the speed of advance is low.

Indications of Approach.

(i) Long swell coming from the direction of the centre, sometimes extending for 1,000 miles.

(ii) Usually thick bands of cirrus cloud converging towards the centre.

(iii) Unsteady barometer, followed by a definite fall.

(iv) Increase in wind force and gusting.

(v) Ugly threatening appearance of the sky.

(vi) Barometer falling more rapidly.

Apart from radio warnings, barometer readings are most valuable, particularly within 30° of the equator. If the diurnal variation in pressure ceases; if the (corrected) barometer reading is 3 millibars below the mean for the time of year, a storm may be approaching, and if the pressure has fallen 5 mb. it will be some 200 miles away and time that avoiding action be taken. In an area likely to be visited by storms hourly readings of the barometer are desirable.

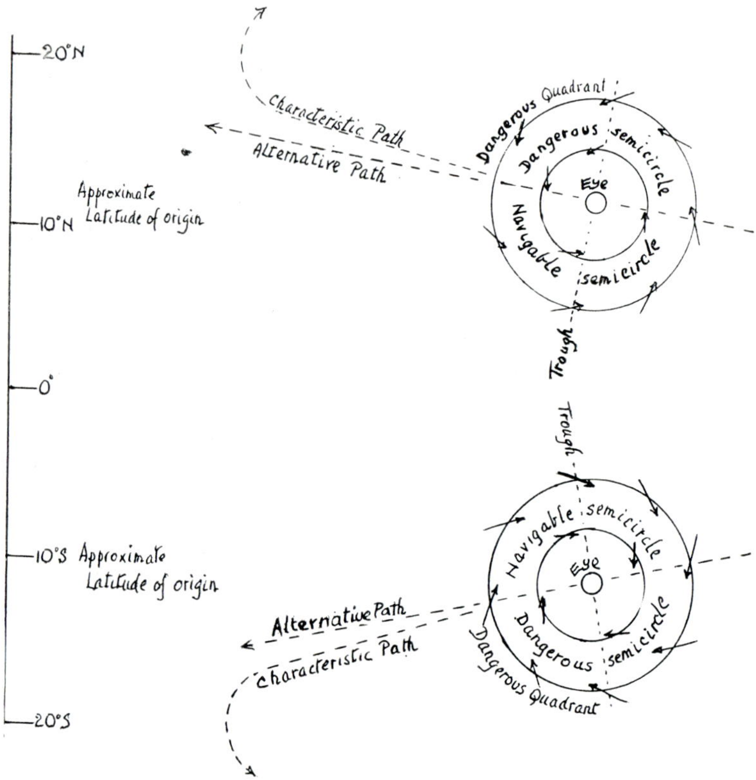

FIG. 12.1—Tropical Revolving Storms

Note the terms used. The *eye of the storm* is the small, cloudless area at the centre, almost windless but highly dangerous.

The Dangerous Semicircle, to the right of the path in the northern, but to the left of the path in the southern hemisphere, that is; the side towards the usual direction of recurvature, and so called because a vessel caught in it may be blown towards the path of the storm and also because the storm may recurve and the centre pass over her.

The Dangerous Quadrant is the leading quadrant of the Dangerous semicircle.

The Navigable Semicircle lies on the other side of the path and a vessel within it will tend to be blown away from the centre, and any recurvature will increase her distance from it.

The Vertex is the most westerly point of the path.

Radio stations ashore broadcast warnings of the position, intensity and probable movement of storms. Ships also have the duty of reporting storms. Because of the slow rate of advance it is often possible to avoid the worst of the storm if warning is received in time.

The information which a seaman requires before he can take avoiding action is: (*a*) the bearing of the storm centre; (*b*) the semi-circle in which he is situated; (*c*) the path of the storm.

Avoidance of the Storm Centre. If, in the absence of sufficient information by radio, the approach of a tropical revolving storm is expected:

(i) *Stop the ship* for about 3 hours. This is essential to determine the true wind behaviour and barometric tendency, both of which can be masked by the ship's progress.

(ii) *Determine the bearing of the centre* by Buys Ballot's Law (page 135). The centre will bear 11 or 12 points from the wind when the storm commences, 10 points when the barometer has fallen 10 millibars and 8 points when the barometer has fallen 20 mb. or more.

(iii) *Determine in which Semi-circle the ship is located.* Wind shifts to the right in the right-hand semi-circle and to the left in the left-hand semi-circle in both hemispheres.

If the wind remains steady in direction and increases in force, the ship is in the direct path of the storm.

(iv) *The Quadrant* is determined by the behaviour of the barometer. If the pressure is falling, the ship is ahead of the trough; if rising, behind it.

Action. *In the Dangerous Semi-circle.* If in a powered ship, steam to windward away from, and at right angles to, the path of the storm. If in a sailing vessel, bring the wind on the starboard bow in the northern hemisphere (port bow in the southern hemisphere), as indicated in the figure. If impossible to make much headway, heave-to with the wind in the same position, head away from the centre.

In the Navigable Semi-circle. Run with the wind on the starboard quarter in the northern hemisphere (port quarter in the southern) until the barometer begins to rise. Keep as much as possible at a right angle to the estimated path.

In the direct path. Run as if in the navigable semi-circle, but if there is insufficient sea-room in that direction, endeavour to get into the dangerous semi-circle which is at least preferable to remaining in the direct path of the storm.

It is of vital importance to avoid passing within 50 miles or so of the centre and as well to keep outside a radius of 200 miles where the wind force is unlikely to be more than force 7.

A harbour must be a very sheltered one with a good holding ground before it can be considered safe. It may be better to put to sea if there is sufficient time to get out of the path of the storm.

All tropical oceans, except the South Atlantic, experience a season of cyclones. Some Atlantic hurricanes curve before reaching the Carribean and continue up north-eastwards towards Europe.

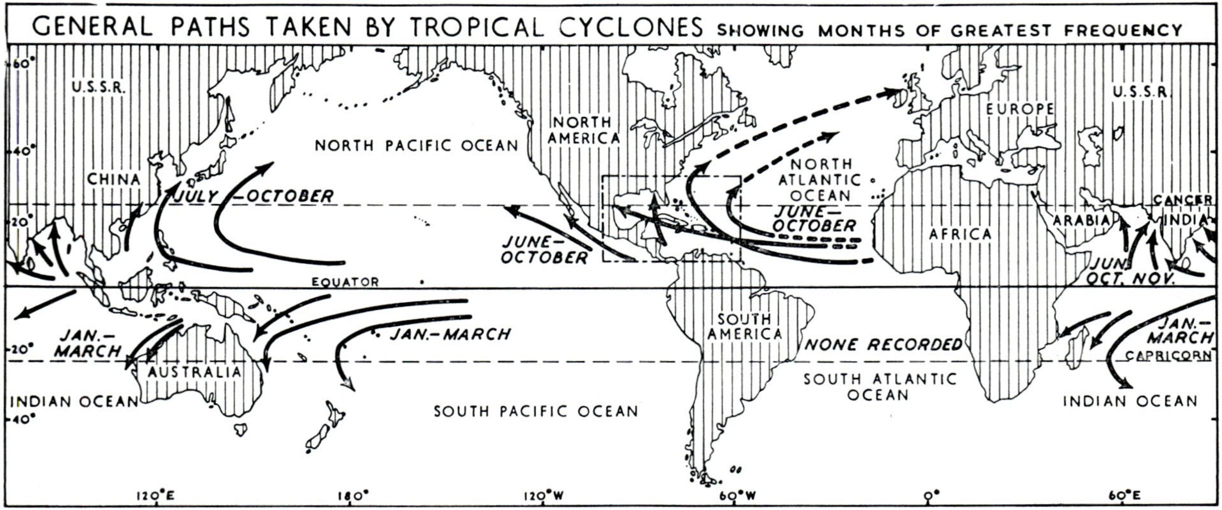

FIG. 12.2

Seasons and Localities. Storms are most frequent during the late summer and early autumn of their hemispheres, but in the Arabian Sea they usually occur at the change of the monsoons, that is, in October—November and May—June.

In the West Indies:

'June; too soon. July; stand-by.
August; look out you must.
September; remember. October; all over.'

Occasionally out of season storms occur, particularly in the western North Pacific.

Local Names. Tropical revolving storms are known as *hurricanes* in the western North Atlantic, eastern North Pacific and South Pacific; *typhoons* in the western North Pacific; *cyclones* in the Indian Ocean, Bay of Bengal and Arabian Sea; *willy-willies* in North-west Australia.

CHAPTER 13

METEOROLOGICAL INSTRUMENTS

THE THERMOMETER, HYGROMETER, BAROMETER AND WIND SPEED INDICATORS

Said Anna Royd to Millie Barr: 'These science fellows go too far.
'Inches,' 'degrees' and 'bars' and such, of differing kinds—it is too much!'
Said Millie Barr to Anna Royd: 'There's not much cause to be annoyed.
For Fahrenheit's no longer made, and Celsius *is* Centigrade.'

It is clear from Chapters 9 and 10 that the constituents of the atmosphere which need to be measured for analysis or prediction of the weather are: temperature, relative humidity and pressure. The instruments used for this are: the thermometer, hygrometer, barometer.

The Thermometer (Thermos = warm). Mercury, enclosed in glass, expands as heat increases and, travelling up a tube of very fine bore, permits graduations of its length to indicate change of temperature.

Thermometer Scales: Celsius (Centigrade), Absolute and Fahrenheit.

Scale	*Freezing Point*	*Boiling Point*
Celsius	0° C.	100° C.
Absolute	273° A.	373° A.
Fahrenheit	32° F.	212° F.

0° Absolute has been calculated as the temperature at which a substance contains no heat calories. The degree divisions are the same as in the Celsius scale.

CONVERSION OF SCALES:

(*a*) *Celsius to Fahrenheit* $100°\ C. = 180°\ F.$

$\therefore 5°\ C. = 9°\ F.$

$°C. \times 9/5 + 32° = °F.$

(*b*) *Fahrenheit to Celsius* $9°\ F. = 5°\ C.$

$(°F. - 32°) \times 5/9 = °C.$

(*c*) *Celsius to Absolute and vice versa*

$°C. + 273° = °A.$

$°A. - 273° = °C.$

(*d*) *Absolute to Fahrenheit.* First convert °A. to °C. and then to F. as shown above. Conversion tables are given in meteorological text books and some books of nautical tables. For conversion by Traverse Table, *see* page 128.

The Hygrometer (hugros = wet). An instrument for determining the humidity of the atmosphere. It consists of two similar thermometers in a slatted wooden screen (a Stevenson screen) which is designed to protect them from the direct rays of the sun but to allow a free passage of air past them.

One thermometer indicates the shade temperature and is known as the *dry bulb*. The other, called the *wet bulb*, has a piece of muslin or cambric tied round the bulb by a few strands of cotton wick whose free ends are immersed in a small jar of distilled water. Water passes up the wick by capillary action and moistens the muslin but is evaporated by the air current flowing through the screen.

Evaporation causes loss of heat from the wet bulb. The greater the evaporation the more the loss of heat, thus the temperature of the wet bulb is generally lower than that of the dry bulb. If, however, the relative humidity is 100 per cent no evaporation can take place and both thermometers will indicate the same temperature.

The difference in readings between the two thermometers is called the *depression of the wet bulb*. This rarely reaches 5° C. at sea and sometimes 10° C. in England, but in hot dry climates can be as much as 20° C.

During, just before, and after heavy rain or when fog or dew is forming, there will be little or no evaporation and the depression of the wet bulb will be zero or very small.

The Barometer (baros = weight), used for measuring the pressure of the atmosphere, is of two distinct types, (1) the mercurial barometer and (2) the aneroid (a-neros = not wet) barometer.

The Aneroid Barometer. The compact aneroid is the type used in yachts. It consists of a flat, corrugated, metal box from which most of the air has been extracted. When the atmospheric pressure increases one wall of the box is pressed inwards; when the pressure decreases it expands. This movement, amplified by a spring and lever is transmitted to a pionter on a dial graduated in millibars and/or inches.

Corrections: *Index Error*. This can be determined by comparison with the correct barometer reading, which can be obtained by telephoning to a nearby weather centre. An adjustment for Index Error can be made by a screw in the back of the instrument.

Height above Sea Level. The higher the barometer is above sea-level the less atmosphere there is above it. Thus pressure falls with height. The standard height for comparison is sea-level, so the height correction is always additive. Near the earth's surface pressure falls with height at the average rate of 1 mb. per 30 ft. or ·03 in. per 30 ft.

Temperature. Compensation for changes in temperature is made either by using a bimetallic link or leaving a calculated small amount of air in the box. Aneroids therefore do not require correcting for temperature.

Wind Speed Indicators. The *anemometer* has a number of cups, fixed at the mast head or held in the hand, which are rotated by the wind: the stronger the wind the faster the rotation. The cups drive a little electric generator and the voltage is shown on a dial graduated in wind speeds.

The *ventimeter* is a glass tube marked with a scale, containing a disc which is free to rise and fall. When it is held vertically with the small opening at the bottom pointing to the wind, the air pressure causes the disc to rise up the scale. The air escapes at the top.

Wind speed is measured either by numbers in the Beaufort Scale (See Chapter 15) or in knots. The speed of 1 metre per second equalls almost 2 knots, more precisely 1·944 knots.

CHAPTER 14

CURRENTS OF THE WORLD

'The Gulf Stream' (alone) 'carries more than 25 times as much water as all the rivers in the world added together.'

T. F. Gaskell.

Causes and Types. The principal cause of surface currents, the only ones of importance to most navigators, is wind. *Drift Currents.* Wind sweeps the surface water to leeward, and this type of current is known as a drift current. The strength of drift currents is about 2 per cent of the prevailing wind speed in high latitudes and 4 per cent in low latitudes; that is, in low latitudes, the currents speed in miles per day is about the same as the wind speed in knots.

Gradient Currents. Where there is a slope in the water level water will run down the slope. This can happen when water banks up on the lee side of an ocean (the slope may be 3 inches in 1,000 miles), or it may be caused by masses of water of different densities. Differences in temperature or salinity causes differences in density. A gradient current may run in a direction contrary to the wind.

Warm and Cold Currents. Currents may be classified also as warm or cold. Warm currents come from warmer regions; cold currents may either flow from higher latitudes or have their origin in the cold dense water which wells up on the eastern side of the Atlantic and Pacific oceans to replace the surface water which is blown westward by the Trade Winds. Cold currents from high latitudes have especial significance for navigators since they cause fogs and may bring ice to lower latitudes.

General Circulation. The surface water is kept in motion by the wind, but its direction is affected by the earth's rotation in the same way as the air is affected. The current is deflected to the right in the northern hemisphere and to the left in the southern hemisphere. In the open ocean the deflection is about 45° so that both the North East and South East Trades produce a westerly current. The Westerlies produce a general drift to the east, so that a circulation, clockwise in the North Atlantic and North Pacific Oceans and anticlockwise in the South Atlantic, South Pacific and Indian Oceans, is set up. Outside these enormous central eddies there may be other currents, usually flowing in a contrary direction.

Current Variation. All currents, however regular, may at times vary considerably both in direction and strength, mainly from change of wind. In the Arabian Sea, Bay of Bengal and China Sea, they are completely reversed by the monsoons, though it should be noted that a gradient current flows north east along the African and Arabian coasts and east coast of India against the wind for a part of the year.

The Principal Currents. (Fig. 14.1). Currents which have been known to exceed 3 knots are so marked, with the season at which this can be expected.

NORTH ATLANTIC.

The North Equatorial Current flows westwards propelled by the North East Trades, south of about latitude 20° N. It is bounded by the less constant *North Subtropical Current* flowing in the same direction between 20° W. and 32° W. Off the north coast of Brazil the North Equatorial Current is joined by part of the South Equatorial Current and both currents sweep through the Caribbean Sea, the Yucatan Strait and out through the Florida Strait as the *Gulf Stream.* (3 Kn. Florida Str. to 58° W.). This warm, well-defined stream runs strongest along the 100 fathom line until north of Cape Hatteras where it turns gradually eastwards. East of 46° W. it weakens and fans out to form the north easterly and easterly *North Atlantic Current.* The North Atlantic Current splits in the eastern Atlantic, the southerly part turning clockwise into the *Azores Current* and *Portugal Current*, these merging into the *Canary Current* and the North Subtropical Current. The northern part of the North Atlantic Current continues past the British Isles to the Norwegian coast, which it keeps free of ice, before turning anti-clockwise and merging with the *East Greenland Current* which flows south west to the south of Greenland and then north west as the *West Greenland Current.* The cold *Labrador Current*, caused chiefly by the northerly wind piling water upon the Labrador coast, flows southwards past Newfoundland and then south west inshore of the Gulf Stream as far south as the vicinity of Cape Hatteras. It brings icebergs down to the Grand Banks of Newfoundland in the summer months, especially in May, and its cold water is the cause of the frequent fogs of this region.

South of the North Equatorial Current, in a narrow belt just north of the equator, the gradient *Equatorial Counter Current* flows eastwards. Its eastern part flows along the African coast as the *Guinea Current* (3 Kn. May to July).

SOUTH ATLANTIC

The *South Equatorial Current* (extending to about 4° N.) flows westerly (3 knots between 34° W. and 70° W. except from February to April) bordered about Lat. 6° S. by the weaker *South Subtropical Current.* The South Equatorial Current divides off Cape Recife. Most of it flowing along the North Brazil Coast and joining with the North Equatorial Current, but some of it turning southward and joining the South Subtropical Current which curves anti-clockwise south of Cape San Roque and flows south west as the warm *Brazil Current*, this at length turns eastward and joins the *Southern Ocean Current* which flows round the southern part of the world.

The *Benguella Current*, a cold current flowing northwards along the western side of South Africa, completes the circulation.

The *Falkland Islands Current* is not part of the main circulation, but flows northwards between the Falkland Isles and Argentina continuing along the coast and extending from May to November as the *Brazil Inshore Counter Current.* Sometimes as far as Rio de Janeiro.

NORTH PACIFIC

The main circulation is clockwise and resembles the North Atlantic. The *North Equatorial Current* flows westwards north of about Lat. 8° N., its northern boundary being undefined. In the western North Pacific the current is deflected clockwise and flows northwards past the Philippine Islands and east coast of Formosa, forming the warm *Kuro Shio* (3 knots), the Pacific equivalent of the Gulf Stream. South of Japan the Kuro Shio fans out and becomes the *North Pacific Current*, which, with the cold *Aleutian Current* flows westward until it approaches the North American coast when it divides in about the latitude of Vancouver, the north division becomes the *Alaskan Current* flowing north and west, the southern becoming the *Californian Current* which eventually joins the North Equatorial Current. The Californian Current does not reach the coast, and from November to February the cold *Davidson Current* runs northward close inshore.

The counterpart of the Labrador Current is a current flowing from the Behring Sea, first called the *Kamchatka Current* and, later, the *Oya Shio*, which runs down the east coast of Japan until it meets the Kuro Shio in about 36° N.

The *Equatorial Counter Current* (3 knots east of Mindinao, June to August) is formed by the recurving of the North Equatorial Current and (except in December to February) the South Equatorial Current.

SOUTH PACIFIC

The main circulation is, as in the South Atlantic, anti-clockwise. The west flowing *South Equatorial Current* (3 knots from 92° to 112° W. between August to October) extends from about 4° N. and merges in about 6° S. with the less pronounced *South Subtropical Current.* In June to August when the sun is furthest north the South Equatorial Current flows north west along the north New Guinea coast to recurve northwards and join the Equatorial Counter Current. From December to February, however, when the sun is furthest south, the current recurves south west and southwards and flows in the opposite direction along the Guinea coast.

The west side of the circulation is not well marked. Some water sets through the Torres Strait and the *East Australian Current* (3 knots, north of 34° S. from October to April) sets southwards along the coast of New South Wales. This is not very constant and northerly sets may at times be met.

The south side of the circulation is, of course, the *Southern Ocean Current* which appears to be weaker and more variable than in the South Atlantic and Indian Ocean and with occasional north sets. The Southern Ocean Current encounters the southern part of the Chilean coast, where it divides, part turning south to continue past Cape Horn, and part northwards where it becomes the cool *Peru Current* (earlier known as the *Humbolt Current*) which follows the coastline to the equator.

SOUTH INDIAN OCEAN.

The general circulation is anti-clockwise. The *Equatorial Current* (3 knots near Madagascar, May to October) flows westward, its north boundary between 6° and 10° S., much further south than in the Atlantic and Pacific. Passing north of Madagascar it encounters the African coast near Cape Delgado in Latitude 10° S. and there divides, part turning north and the rest south to form the *Mozambique Current* (3 knots occasionally) flowing southwards between Madagascar and the mainland. This stream continues as the *Agulhas Current* (3 knots) to Cape Agulhas, the southern point of Africa, some of it rounding the cape and joining the Benguella Current, the residue curving anti-clockwise to join the *Southern Ocean Current.* The circulation is completed by a weak off-shoot of the Southern Ocean Current to the north west which becomes the *West Australian Current.*

The *Equatorial Countercurrent* (3 knots, occasionally), flows mainly south of the equator, though it extends northward of it during the North East Monsoon.

NORTH INDIAN OCEAN, ARABIAN SEA AND BAY OF BENGAL.

The currents are reversed by the monsoons; westward and counter-clockwise along the coasts during the North East Monsoon from November to January. The coastal currents, however, reverse before the end of the North East Monsoon and flow clockwise in March and April. During the South West Monsoon, May to September, the coastal circulation still flows clockwise, and, in the open waters, east.

The *East African Current* (3 knots) flowing southwards along the coast from Ras Hafun at the north east corner of Africa since November, reverses in March. From May to September during the height of the South West Monsoon it is strong and north easterly. Some now flows into the Gulf of Aden but most continues eastwards, south of Socotra where, from July to September, it is the strongest ocean current known in the world, attaining up to 7 knots. October is the month of change.

CHINA AND JAVA SEAS.

The currents are controlled by the monsoons, flowing north east in the China Sea from May to August and south west from October to March, the current persisting longer in the southern part. In the

Java Sea an easterly current flows from November to March and a westerly current from June to September.

MEDITERRANEAN SEA

There is a general anti-clockwise circulation most marked along the North African coast. Since the evaporation of sea-water is not equalled by the inflow from rivers, there is a continuous surface inflow from the Atlantic, part of the Portugal Current entering the Strait of Gibraltar. There is also a nearly constant surface current from the Black Sea, through the Bosphorus, Sea of Marmora and Dardenelles.

BLACK SEA

The general circulation is anti-clockwise.

BAY OF BISCAY.

There is a slight clockwise current south east and south down the coast of France and westward along the nrth coast of Spain.

Current Atlases, with detailed current charts for the different oceans are issued by the Meteorological Office.

CHAPTER 15

RECORDING AND REPORTING THE WEATHER

'Here is today's weather forecast' . . .

B.B.C. Weather Report.

Wind. The *direction* of the wind, from which it blows, is usually logged in points of the compass, always true (i.e. with reference to the true meridian).

The *force* of the wind is logged by using the appropriate number in the Beaufort Code, reproduced below. In the International Code, however, it is expressed in knots. If expressed in metres per second (Système Internationale), 1 metre/sec = 1·944 knots.

Beaufort Scale No.	Mean Wind Speed in knots (Measured at a height of 33 ft. above sea level)	Limits of Wind Speed in knots (Measured at a height of 33 ft. above sea level)	Descriptive Terms	Sea Criterion	Probable Wave Height in ft.
0	0	Less than 1	Calm	Sea like a mirror	0
1	2	1—3 0·5	Light air	Ripples with the appearance of scales	¼
2	5	4—6 2·1	Light breeze	Small wavelets, crests have glossy appearance and do not break	½
3	9	7—10 3·6	Gentle breeze	Large wavelets. Crests begin to break. Foam of glassy appearance. Perhaps scattered white horses	2
4	13	11—16 5·7	Moderate breeze	Small waves; fairly frequent white horses	3½
5	18	17—21 8·7	Fresh breeze	Moderate waves of more pronounced long form; many white horses. Chance of some spray	6
6	24	22—27 11·3	Strong breeze	Large waves begin to form; white crests more extensive. Probably some spray	9½

Lower Figures in 'Limits of Wind Speed' column are in metres per second.

Beaufort Scale No.	Mean Wind Speed in knots	Limits of Wind Speed in knots	Descriptive Terms	Sea Criterion	Probable Wave Height in ft.
7	30	28—33 14·4	Near Gale	Sea heaps up and white foam from breaking waves begins to blow in streaks along the direction of the wind	13½
8	37	34—40 17·5	Gale	Moderately high waves of greater length; edges of crests begin to break into spindrift. Foam blown in well-marked streaks along the direction of the wind	18
9	44	41—47 21·1	Strong Gale	High waves. Dense streaks of foam along the direction of the wind. Crests of waves begin to topple, Spray may affect visibility	23
10	52	48—55 24·7	Storm	Very high waves with long overhanging crests. Foam in great patches is blown in dense white streaks along the direction of the wind. The whole surface of the sea takes a white appearance. The tumbling of the sea becomes heavy and shocklike. Visibility affected	29
11	60	56—63 28·8	Violent storm	Exceptionally high waves. Medium sized ships might be for a time lost to view behind the waves. The sea is completely covered with long white patches of foam lying along the direction of the wind. Everywhere the edges of the wave crests are blown into froth. Visibility affected	37
12	—	64+ 32·9	Hurricane	The air is filled with foam and spray. Sea completely covered with driving spray. Visibility very seriously affected	45

Lower Figures in 'Limits of Wind Speed' column are in metres per second.

The apparent wind felt by the observer is the resultant of the wind caused by the ship's motion and the wind actually blowing. The feel of the wind is thus not a true guide and the direction of the true wind can best be found by noting the direction from which the waves are approaching.

In estimating the direction of the wind look to windward from behind the compass and note the direction of the crests of the waves or ripples which lie perpendicularly to the direction from which the wind blows. Swell (waves caused by wind *not* blowing at the ship at the time of observation) must be ignored, if present. The force of the wind is then estimated by appearance of the sea. The Beaufort Scale may be modified by local conditions: (*a*) a sudden increase of wind strength will take time to affect the height of waves. (*b*) An off-shore wind does not produce its appropriate sea close inshore but requires a certain distance, or fetch, before its full effect is produced. (Fetch is the distance of sea over which the wind has blown). (*c*) Tides and strong currents can affect the sea surface; a weather tide (one moving against the wind) will produce more disturbance than a lee tide. (*d*) Heavy rain produces a smoothing effect.

There are excellent photographs of the state of the sea for different Beaufort Scale numbers in the *Marine Observers' Handbook* and *Meteorology* by C. W. Roberts.

The State of the Weather can be succinctly recorded in Beaufort notation.

The Beaufort Notation:

b = Blue sky or sky not more than one-quarter clouded, whether with clear or hazy atmosphere.

bc = Sky between one-quarter and three-quarters clouded.

c = Mainly cloudy (not less than three-quarters covered).

d = Drizzle or fine rain.

e = Wet air (without precipitation).

f = Fog (Visibility less than ½ mile).

g = Gale (force 8 or 9). *G* = Storm and above.

h = Hail.

jp = Precipitation in sight of ship.

kq = Line squall.

l = Lightning.

m = Mist. (Visibility not less than ½ mile).

o = Overcast sky. (Sky completely covered with a uniform layer of thick or heavy cloud).

q = Squalls. (*Squalls* may last for minutes; *gusts* only momentarily. There is no letter allocated for gusts).

r = Rain.

rs = Sleet (i.e. rain and snow together).
s = Snow.
t = Thunder.
tl = Thunderstorm.
u = Ugly, threatening sky.
v = Unusual visibility.
w = Dew.
x = Hoar frost.
y = Dry air.
z = Dust haze; the turbid atmosphere of dry weather.

The system can be extended by using:

Capital letters to indicate increased intensity, e.g. R = heavy rain.

A small suffix 'o' indicates slighter intensity, e.g. q_0 = slight squalls.

A prefix 'p' indicates 'shower of', e.g. pR = shower of heavy rain.

A prefix 'i' indicates 'intermittent', e.g. if = fog patches.

A solidus '/' is used after present weather to distinguish weather in the past hour, e.g. c/r_0 = cloudy after slight rain during past hour.

Continuity is indicated by repeating the letter, thus: *rr* = continuous moderate rain.

In past weather the letters are used in the same way but their order indicates sequence in time, e.g. *b,bc,cpr* = cloudless, became cloudy, followed by clouds with showers of rain.

WEATHER REPORTS AND FORECASTS.

A weather *report* is a statement of weather conditions at the time the report is made: a *forecast* is a prediction of what the weather will become, and thus of a lesser degree of accuracy. A Weather Bulletin may include Gale Warnings, Weather Reports and Forecasts.

Gale Warnings (i) Visual signals are exhibited at certain places round the coasts of Great Britain and indicate that a gale is expected within 12 hours, or is already in progress in the neighbouring coastal sea area. The signal is a cone, 3 ft. high and 3 ft. wide at base, which has the appearance of a triangle when hoisted. It is usually black, but when local conditions make it desirable, white or yellow. In some places, three white or three red lights in the form of a triangle are exhibited at night in place of the cone. The signal is lowered when the wind is below gale force and a renewal of gale force winds is not expected within 6 hours, so that the gale warning is left flying during a lull.

The gale warning signals are:

The North Cone. The North Cone (point upwards) is hoisted for gales from any point north of the east-west line.

The South Cone. The South Cone (point downwards) is hoisted for gales from any point south of the east-west line.

When the direction of the gale is expected to change from the northern side to the southern side of the east-west line, the North Cone is lowered and the South Cone hoisted. Conversely, when the direction of the gale is expected to change from the southern to the northern side of the east-west line, the Cone signal is changed accordingly.

(ii) Radio Gale Warnings are broadcast on Radio Two of the B.B.C. at the first programme juncture. If the first broadcast of any warning does not occur at half-past the hour, it will be broadcast again at the first programme juncture that does fall at half-past the hour. Where the news summary is scheduled at half-past the hour, the gale warnings will be given immediately after it.

(iii) Gale Warnings by R/T and W/T are also included in Bulletins from G.P.O. Coastal Stations and The Atlantic Weather Bulletin. Full details of the Meteorological Office Services are given in Met.O. Leaflet No. 3 1973, *Weather Bulletins and Gale Warnings for Shipping including Fishing Vessels.*

Weather Bulletins for Shipping—B.B.C. Radio 2. Weather bulletins for shipping, originated by the Meteorological Office, are broadcast on 200 kc./s. (1500 m.) at the times shown below:

Monday—Saturday	*Sunday*
(clock time)	(clock time)
0033—0038	0033—0038
0633—0638	0633—0638
1355—1400	1155—1200
1757—1800	1757—1800

Contents of broadcast: (1) Gale warnings in force. (2) Coastal sea area forecasts for the next 24 hours, (wind direction and force, weather and visibility). (3) Station reports (wind direction and force, present weather, visibility, barometric pressure and tendency).

B.B.C. Radio 4. Forecasts for coastal waters (50 miles seawards) off Great Britain and Northern Ireland.

Contents of broadcast: Forecast of wind and visibility until 1800 the next day.

The times of the various bulletins may be amended occasionally.

The words 'wind,' 'force,' 'millibar' and 'visibility' are omitted from B.B.C. bulletins. This is to enable the forecast to be given in a more concise form.

By Telephone. Weather reports, constantly up-dated, are recorded on tape and can be had at any time simply by dialling a local number. *See* Preface of the P.O. Telephone Directory.

FIG. 15.1

It need hardly be added that full use should be made by yachtsmen of this excellent service, but its limitations should be recognised. Some sea areas, for instance, are very large. A gale in one part of the area may be quite consistent with Force 4 or 5 in another part of the same area at the same time. A gale warning for an area does not mean that a gale is expected over the whole of that area. Neither does the wind necessarily blow at a uniform strength for a given pressure gradient, but tends to rise and fall in strength in a fairly regular time pattern.

Reading Weather Bulletins. It is of great practical importance that the weather bulletins broadcast on 1,500 metres in the B.B.C. Light Programme should be recorded correctly. Some kind of prepared form is necessary, for they are read at above ordinary

SHIPPING WEATHER FORECAST AREAS

FIG. 15.2

dictation speed. The Royal Meteorological Society publish a map and form in pads of 24 on which the bulletins may be taken down,[1] or a form such as the one shown in reduced size on the opposite page may be ruled in ink on the back of a chart, each bulletin being recorded in pencil.[2]

[1] 'Weather,' Royal Meteorological Society, 49 Cromwell Road, London, S.W.7.
[2] Reproduced by permission from an article in *The Marine Observer* and *The Nautical Magazine*, by Lt.-Commander L. B. Philpott, D.S.C., R.N.R.

SHIPPING FORECAST (B.B.C. 1500m.) AT .... GMT 1964

GALE WARNINGS FOR AREAS

GENERAL SYNOPSIS:-

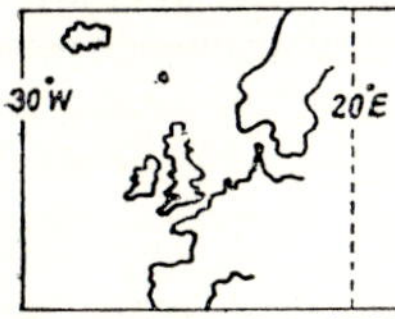

FORECAST FOR SEA AREAS FOR NEXT 24HRS.:-

AREA		GALE ?	WIND		WEATHER	VISIBILITY	REMARKS
			FORECAST	BECOMING			
1	VIKING						
2	FORTIES						
3	CROMARTY						
4	FORTH						
5	TYNE						
6	DOGGER						
7	FISHER						
8	GERMAN BIGHT						
9	HUMBER						
10	THAMES						
11	SMITH'S KNOLL						
12	DOVER						
13	WIGHT						
14	PORTLAND						
15	PLYMOUTH						
16	BISCAY						
17	FINISTERRE						
18	SOLE						
19	LUNDY						
20	IRISH SEA						
21	FASTNET						
22	SHANNON						
23	ROCKALL						
24	MALIN						
25	HEBRIDES						
26	MINCHES						
27	FAIR ISLE						
28	BAILEY						
29	FAEROES						
30	S.E. ICELAND						

REPORTS FROM COASTAL STATIONS AT ....GMT:-

STATION	WIND		WEATHER	VIS.	BARO.	TENDENCY
	DIR.	FORCE				
WICK						
BELL ROCK						
DOWSING						
GALLOPER						
R. SOVEREIGN						
PORTLAND BILL						
SCILLY						
VALENTIA						
RONALDS WAY						
PRESTWICK						
TIREE						

Suggested form on which the weather bulletins for shipping can be taken down.

FIG. 15.3

It should be used as follows:

Gale Warnings. The bulletin always opens with any gale warnings in force, and a tick can be placed against the area named in the column headed 'Gale?' and the names entered in the top panel later.

The words 'imminent' means that the gale is expected within 6 hours; 'soon,' approximately after 6 hours and before 12 hours; 'later' approximately 12 hours after the time of origin of the bulletin.

General Synopsis. 'Low,' 'Ridge,' 'Trough,' etc., can be written in the appropriate place on the map; the position of a depression shown by a circle containing the barometer reading; the direction of travel indicated by an arrow.

The speed of travel of depressions and fronts are indicated in bulletins as follows:

'Slowly,' less than 15 Kn. 'Steadily'; from 15 Kn. to 25 Kn. 'Rather quickly'; from 25 Kn. to 35 Kn. 'Rapidly'; from 35 Kn. to 45 Kn. 'Very rapidly'; above 45 Kn.

Forecasts for Sea Areas. The numbers are useful when areas are bracketed together. For example, if the forecast for Tyne is the same as for Humber, 9 (i.e. Humber) is written against Tyne. The shaded squares are a reminder that these two areas are not always named. The weather is best recorded in Beaufort notation.

Weather Charts and Their Use. Weather charts covering Home Waters and the eastern North Atlantic can be obtained at the Weather Centres at many British ports. They consist of a weather chart for the previous midnight, a forecast chart for midnight of the day of issue, a statement of the general synoptic ('affording a general view at a given time') situation, forecasts for the next 24 hours, gale warnings in operation at the time of issue, and a local forecast. Weather charts are also published in newspapers, and shown on television.

A synoptic chart shows the wind directions and forces, atmospheric pressures and temperatures, and the weather at a number of stations, all observations being taken at the same G.M.T. Isobars, fronts and weather systems are drawn on the synoptic chart by the forecaster to complete the weather chart.

Wind Speed and Pressure Gradient.

When the Isobars are close together the Pressure Gradient (the change in pressure with distance) is described as 'steep'; when far apart as 'slack', 'slight' or 'flat'. The closer together the Isobars are, the stronger will be the wind. In our latitude the rough rule that a change of 1 mb in 30 M gives a surface speed of about 24 knots giving results tabulated:-

Probable Wind Speeds Between Isobars

Isobars plotted 4 Isobars apart		Isobars plotted 5 Isobars apart	
Perpendicular distance between Isobars in Miles	Probable Wind Speed in Knots	Perpendicular distance between Isobars in Miles	Probable Wind Speed in Knots
50	58	50	72
100	29	100	36
150	19	150	24
200	14	200	18
250	12	250	14
300	10	300	12
350	8	350	11
400	7	400	9

The weather is plotted closely round a small ring representing the station. The directions of the wind is shown by an arrow flying with the wind, the speed being indicated by the number of feathers or pennants on the arrow. Wind speed is indicated to the nearest 5 knots; each full feather represents 10 knots, half feathers represent 5 knots, while a pennant means 50 knots. Symbols or Beaufort, letters are used to show the cloud amount, present weather (ww) and past weather (W). Pressure (PPP), temperature (TT), and visibility (VV) are plotted as reported. A + denotes an increase and – denotes a decrease, e.g. +f = increasing fog.

It must be remembered that a weather chart is bound to be some hours out of date, so that a knowledge of the probable movement and development of pressure systems is important.

Movement of Depressions.

(*a*) Depressions move from where pressure is rising to where pressure is falling.

(*b*) A depression with a warm sector moves in a direction parallel to the isobars in the warm sector. The closer the isobars the higher the wind speed and the faster the progress of the depression.

(*c*) When a depression has nearly occluded it becomes stationary or slow moving but it tends to follow the pre-existing flow of isobars.

(*d*) A small depression caught up in the circulation of a larger will tend to move round it in a cyclonic direction. If the two are not greatly different in size they tend to rotate anti-clockwise about a point midway between them.

(*e*) Depressions tend to move round large well established anti-cyclones in the direction of the wind round their boundaries.

(*f*) Frontal depressions tend to occur in families, each depression following approximately in the track of its predecessor, though nearer to the equator.

(*g*) Non-frontal depressions tend to move in the direction of the strongest wind circulating round it.

Movement of Anti-cyclones

(*a*) Anti-cyclones move from areas of falling pressure to areas of rising pressure.

(*b*) Anti-cyclones move more slowly than depressions.

(*c*) Anti-cyclones and ridges of high pressure between successive depressions of a family move with the depressions.

(*d*) Anti-cyclones forming behind the cold front of a frontal depression tend to move with the mass of cold air, usually towards the equator.

Movements of Fronts.

(*a*) The speed of advance of a warm front is usually the same as the surface wind speed.

(*b*) The speed of advance of a cold front is usually a little more than the surface wind speed.

Forecasting Fog.

(*a*) *Temperature.* If the sea temperature is below the air temperature and dew point, fog may form. The sea may cool the air to below dew point and the excess water vapour will condense into fog.

Smoke can be a guide. If it lies horizontally it indicates an inversion (a layer of atmosphere through which the temperature increases, instead of falling with height), caused by the cooling of the lower layers of the air. This helps the formation of shallow fog.

(*b*) *Wind.* Wind from a warmer oceanic region may produce fog. Cold wind replacing warm moist air will thin or clear fog. In the northern hemisphere a wind veering or backing to the north will reduce the chance of fog. On the Grand Banks of Newfoundland this is especially noticeable.

Force of the wind: Force 3 is about the most likely strength to produce fog. Fogs form in calms and lighter breezes, but with force 4, or above, the chance of fog is slight, except on the Grand Banks where a southerly wind of any force can cause fog.

SELF-EXAMINATION PAPERS

The numbers in brackets refer to the pages containing the answers to the questions set.

Paper I

1. What is (*a*) a millibar, and (*b*) an isobar? (134)

2. State Buys Ballot's Law. (135)

3. What is the direction of rotation of cyclonic winds in the northern hemisphere? (135)

4. 'A cyclone circulates about an area of low/high pressure.' Which is correct? Is your answer the same for both hemispheres? (137)

5. Distinguish between 'sea' and 'swell.' (160)

6. How can you judge the direction of the wind? (160)

Paper II

1. How is fog caused? (132)

2. Describe the different types of fog. (132)

3. For what clouds do the following abbreviations stand? State whether they are high, medium or low clouds and their general appearance. Ac, Ci, Cu, Cb, St. (131)

4. What are (*a*) the Westerlies, (*b*) the Doldrums, (*c*) the Roaring Forties? (135)

5. Distinguish between the Trade Winds and the Monsoons. (136)

6. What is the 'Beaufort Scale?' (158)

Paper III

1. At what time of day would you expect a sea breeze to blow in a tropical port? (136)

2. Off what sort of coast are you likely to encounter katabatic winds? (136)

3. Distinguish between a Revolving Storm in the tropics and a Depression in temperate zones. (145)

4. In which direction do Tropical Revolving Storms revolve? (146)

5. Why is one semi-circle of a tropical revolving storm designated 'navigable' and the other one 'dangerous'? (146)

6. Distinguish between fog, mist and haze. (132)

Paper IV

1. How could you tell of the approach of a tropical revolving storm? (145)

2. The month is August in the West Indies; the wind is E., force 5: the barometer has fallen 6 mb. below its mean; the swell is southerly; there is much cirrus cloud. Is a tropical storm brewing and, if so, what is its approximate bearing and distance? (*See* Paper VI, end). (179)

3. What action would you take on the information given in question 2, and why would you take it? (147)

4. Draw a rough map of the world and show the various permanent winds. (135)

5. Distinguish between a current and a tidal stream. (20)

6. How is the direction of a drift current affected by the earth's rotation? (153)

Paper V

1. Distinguish between the Polar Front and a Cold Front. (137)

2. Draw (*a*) plan and (*b*) a vertical section of a depression. (139)

3. What is an occlusion? (137)

4. Draw the symbols used on weather maps for (*a*) cold fronts, (*b*) warm fronts and (*c*) occluded fronts. (139)

5. Distinguish between an anti-cyclone and a depression. (143)

6. In what direction, and why, does a current flow in the Straits of Gibralter? (157)

Paper VI

1. Describe (*a*) a Col, (*b*) a Wedge or Ridge. (144)

2. In what direction does the wind blow relative to the isobars? (135)

3. How soon would you expect the onset of a gale described in a British Storm Warning as (*a*) imminent, (*b*) soon, (*c*) later? (166)

4. What visual storm signals are used on the British coasts? (161)

5. What do the following letters of the Beaufort Notation stand for? *bc*, *g*, *kq*, *opr*, *is*, *V*. (160)

6. Draw on a map of the world the principal currents and name them. (154)

(*Answers* to Paper IV—2. Yes. S.S.W., 200 M.)

PART III

ASTRONOMICAL NAVIGATION

CHAPTER 16

ASTRONOMICAL NAVIGATION: PRINCIPLE

'. . . there is nothing in the realm of navigation that lifts one's heart up more in adoration.'

Captain Joshua Slocum, the first solo circumnavigator.

The Celestial Sphere. All the heavenly bodies—the sun, moon, planets and stars—are imagined to lie on the surface of an immense sphere of infinite radius, the earth being at the centre point (Fig. 16.1).

To describe the position of the heavenly bodies on the sphere it is provided with a graticule in the same way as the earth (*see* Chap. 1). The axis of the earth is extended, and the points at which it meet the celestial sphere are called the north and south *Celestial Poles*. Half-way between these poles lies the *Celestial Equator* or *Equinoctial* as it is occasionally called. Circles parallel to the celestial equator, equivalent to parallels of latitude on the earth, are known as *Parallels of Declination*. The *Declination* of a heavenly body corresponds to terrestial latitude and is measured north or south of the celestial equator. (All measurements are in angular measurement, made in degrees and minutes.) Semi-great circles joining the celestial poles are called *Celestial Meridians*. The prime meridian, from which all others are measured is that which passes through the point where the sun crosses the equinoctial in the spring. This was originally in the constellation of Aries, 'The Ram,' with ram's horns symbol (♈) and is generally known as 'Aries.' The angle between the meridian passing through Aries and a heavenly body, or the arc of the celestial equator between the two, is known as the *Sidereal Hour Angle* (siderus = a star), abbreviated to S.H.A. The position of a heavenly body is thus described by its declination and S.H.A.

Navigators on the earth's surface however measure their distance from the Greenwich meridian and not from Aries, so the brightest of the heavenly bodies—those most used by navigators—the sun, moon, the four planets Venus, Mars, Jupiter and Saturn, and also Aries, have their Greenwich Hour Angle (G.H.A.) calculated and recorded. *Greenwich Hour Angle* is the angular distance a meridian is *west* (never east) of the Greenwich meridian. Should the G.H.A. of a star be required it is found simply by adding its S.H.A. to the G.H.A. of Aries. Subtract 360° should the sum come to more than that.

EXAMPLE:	Canopus S.H.A.	264°	9′·4
	+G.H.A.	193°	17′·6
		457°	27′
	−	360°	
	Canopus G.H.A.	97°	37′

These quantities of G.H.A., S.H.A. and Declination are listed in the *Nautical Almanac*. The position of the stars changes slowly and the S.H.A. and Declination of 57 of the brightest are listed for every month. The G.H.A. and Declination of the sun, moon, planets and Aries are tabulated for every hour of each day, and a table of Increments, at the end of the Almanac is given, so that the G.H.A. can be found to the nearest second of G.M.T. and the declination to the nearest decimal of a minute of arc.

The whole of this vast sphere appears to an observer on the earth

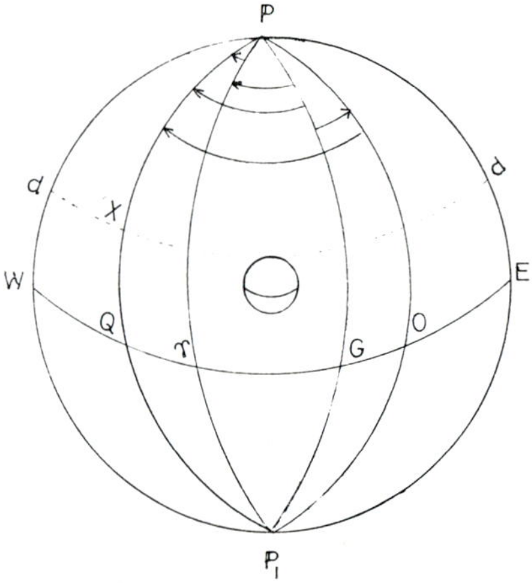

FIG. 16.1—The Celestial Sphere. (Compare with FIG. 1.1)

P and P_1 represent the north and south celestial poles
$WQ♈GE$ represents the celestial equator
dXd is the parallel of declination of X, the letter always used to denote a heavenly body
XQ is the heavenly body's declination
PX is the polar distance (the complement of the declination).
The semi-great circles joining the poles are all meridians
PXP_1 is the meridian passing through the heavenly body
$P♈P_1$,, ,, ,, ,, ,, Aries
PGP_1 ,, ,, ,, ,, ,, Greenwich
POP_1 ,, ,, ,, ,, ,, the observer's position
The angle $♈PX$ and arc $♈Q$ = the S.H.A. of the heavenly body
,, ,, GPX ,, ,, GQ = ,, G.H.A. ,, ,, ,, ,,
,, ,, $GP♈$,, ,, $G♈$ = ,, G.H.A. Aries
,, ,, GPO ,, ,, GO = ,, Longitude of the observer (in this case East)
,, ,, OPX ,, ,, OQ = ,, L.H.A. of the heavenly body.

to rotate round him, making one complete revolution each day, the increase of the sun's G.H.A. being 15° per hour. (Extracts from the *Nautical Almanac* are given on pages 241-263).

Fixing the Ship's Position. The essential principle of fixing a ship's position by celestial or astronomical navigation; 'astro'; is to

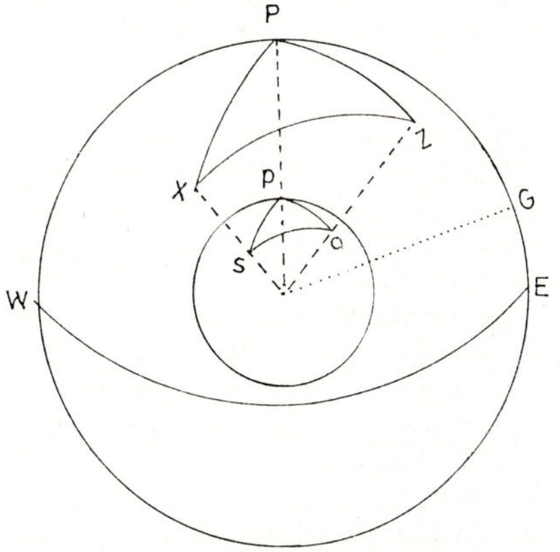

FIG. 16.2—The Celestial and Terrestial Spheres and triangle *PZX*

(*s* = the sub-stellar spot or geographical position of the heavenly body, the point on the earth's surface directly beneath the heavenly body.

Z, the zenith, is the point on the celestia lsphere vertically above the head of the observer.)

In the figure, drawn on the plane of the Greenwich meridian, with the earth to a much enlarged scale, shown at the centre, it can be seen that there are two similar triangles; *PZX* on the celestial, and *pos* on the terrestial sphere.

The position of *X* is determined by (1) *PX*, the polar distance (which is 90°−Dec., or when the heavenly body is in the other hemisphere and the latitude and declination have therefore different names, 90°+Dec.), and (2) the angle *GPX*, the G.H.A. of the heavenly body. The position of *Z* (and *o*) is determined by (1) *PZ*, the co-latitude (90°−Lat.) and (2) the angle *GPZ*, or angle *gpo*, the longitude.

In the triangle pos

p = the north terrestial pole
o = the observer's position
s = the sub-stellar spot or geographical position of the heavenly body

In the triangle PZX

P = the north celestial pole
Z = the observer's zenith
X = the heavenly body
PZ = the co-latitude
PX = the polar distance
ZX = the zenith distance
angle *P* = the Local Hour Angle (found by applying the lontitude, angles *GPZ* and *gpo*, to the Greenwich Hour Angle, *GPX*).
angle *Z* = the azimuth of the heavenly body.

The solution of the triangle *PZX* forms part of almost all problems in astro navigation.

observe the angular distance of a heavenly body whose position is precisely known at the time when it is measured. The distance from it then gives us a (circular) position line in the same way as the vertical angle of a lighthouse, whose height above sea level is known, provides one (*see* page 24).

When the heavenly body is very close to the zenith—that is when the zenith distance is very short—the position line can be drawn on the chart. The geographical position of the heavenly body is plotted on the chart using the declination as the latitude and the G.H.A. (subtracted from 360° for east longitude) on the longitude scale. The zenith distance is then taken on compasses from the latitude scale and an arc swept with the Geographical Position of the heavenly body as the centre of the circle.

It should be noted that one heavenly body provides one position line. To get a fix more than one position line is necessary.

Because the geographical position is often hundreds of miles away from the zenith the position line can rarely be plotted in this manner and a different method has to be employed. It is first assumed that the ship is in a certain position and the (angular) distance of the zenith distance is then found and compared with the length actually measured. If the computed and measured lengths are the same the ship is on a position line passing through the assumed position; if less it is moved closer to the heavenly body. The distance that the position line is moved as known as the *intercept*, named 'towards,' if it is moved towards the heavenly body, or 'away' if it is moved away from it.

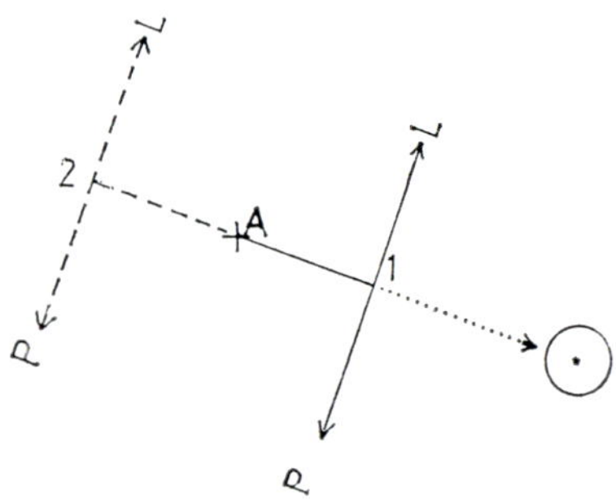

FIG. 16.3—Intercepts and Position Lines

A = Observer's assumed position
A_1 = Intercept drawn towards the heavenly body
A_2 = Intercept drawn away from the heavenly body
PL = Position lines drawn at right angles at the end of the intercepts.

CHAPTER 17

ASTRONOMICAL NAVIGATION: PRACTICE

'It is fine to bounce the sun along the skyline and chase the moon in its swift flight.'

From a cadet's essay.

Working a 'Sight'. The stages of finding (*a*) the length of the intercept and (*b*) the bearing or azimuth of the heavenly body, necessary for drawing the position line are, briefly, as follows:

(1) The heavenly body is observed with the sextant and the time noted and the log reading recorded.

(2) The assumed position of the ship is calculated.

(3) The position of the heavenly body observed is extracted from the *Nautical Almanac.*

(4) The distance and bearing from the assumed position is found.

(5) The assumed zenith distance and the measured zenith distance are compared and the intercept found.

(6) The intercept is drawn from the assumed position, either on the bearing or its reciprocal depending on whether the intercept is towards or away, and the position line is drawn from the end of the intercept and at right angles to it.

Each stage is first described in detail, the complete process then shown as worked.

(1*a*) **The Observation of the Heavenly Body.** It has been shown in the *PZX* triangle that what is required is the zenith distance. This cannot be measured directly as there is no spot in the sky marking the zenith. The *altitude* however can be measured—the height that a heavenly body is above the horizon—and this subtracted from 90° equals the zenith distance, the altitude and zenith distance being complementary (Fig. 18.1). In theory, to obtain this true altitude, the observer should be placed at the centre of the earth and not at some little distance above its surface; Furthermore, the light ray from the heavenly body is refracted (bent) in its passage through the atmosphere, and in the case of the sun and moon, it is difficult to judge the centre of the heavenly body and so, in practice, the lower or upper limb (edge) of the heavenly body is brought into contact with the horizon. These discrepancies must be allowed for to get the true altitude.

Altitude Correction. In Figure 17.1 the height of the observer above the earth's surface and the diameter of the earth are both greatly exaggerated in comparison with the distance of the heavenly body in order to illustrate the angle observed (*BOV*) and the angle required (*HCR*).

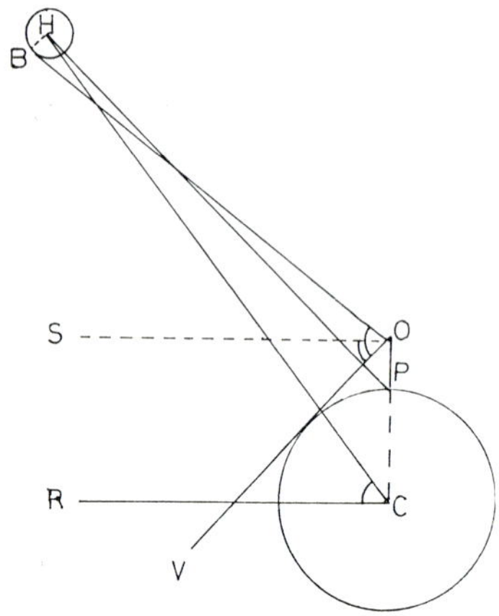

FIG. 17.1—Altitude Corrections

C represents the centre of the earth: the observer's theoretical position
P ,, ,, observer's position on the earth's surface
O ,, ,, observer's eye, above the earth's surface
OV ,, ,, the observer's visible horizon
H ,, ,, centre of the heavenly body observed.

The scale is grossly disproportionate in order to show the very small angles involved:

$\angle HCR$ = the True Altitude
$\angle BOV$ = ,, Observed Altitude
$\angle SOV$ = ,, dip
$\angle CHP$ = ,, parallax
$\angle HOB$ = ,, semi-diameter.

The method of observing a heavenly body with the sextant is described in Chapter 7. The Index Error (if the sextant has one) is usually constant and known to the observer. It is customary to apply the Index Error mentally to the sextant altitude and then to write down the resulting observed altitude.

The other corrections are, for convenience, combined and tabulated in 'Altitude Correction Tables.' The 'total' correction comprises:

(1) *Refraction.* The bending of the light ray is greatest at low altitudes. *Change* in refraction is then largest and, if there is any variation from the normal refraction due to difference in temperature and pressure it is then most marked. To avoid this uncertainty in the refraction altitudes of less than 10° should not normally be taken.

(2) *Dip.* This is the difference between the angle measured by the observer whose eye is at some height above sea level and the angle at sea level. The dip increases as the height of eye increases.

In bad weather the observer is rising and falling considerably with the heaving of the ship and height of eye, and thus the dip, is continually changing. To overcome this a succession of 'shots' may be taken and the average altitude used. The height of eye in still water should be known to the nearest foot or so. Except in misty weather the greater the height of eye, the more distinct the horizon will be.

(3) *Parallax.* The correction necessary because the observer is situated on the surface of the earth and not its centre is known as parallax. Its amount depends upon the distance of the heavenly body. The stars are too far away to have any parallax and the sun and planets have very little—not more than ·1′—but the moon is close enough to have a horizontal parallax (when the altitude is 0°) of about 1°. Her parallax decreases with her altitude.

(4) *Semi-diameter* is the term used for the correction made when either the upper or lower limb of the sun or moon is brought into contact with the horizon. The distance of the sun and moon from the earth varies throughout the year, so the semi-diameter does also.

Altitude Total Correction Tables. *The Nautical Almanac* contain the most extended tables. They consist of a Dip table arranged for heights of eye (in metres and feet), used for *all* the heavenly bodies.

For the *sun* a table of combined Refraction, Parallax and Semi-diameter arranged for the months October-March and April-September, for both lower and upper limbs.

For the *stars* and planets a refraction table (with a supplementary table for the planets).

For the *moon* a table for upper and lower limb observations.

Nautical Tables and Brown's and Reed's *Nautical Almanac* contain shorter tables.

For the *sun* enter with the Observed Altitude of the Lower Limb and Height of Eye, with small monthly corrections. If the upper limb is observed, subtract twice the semi-diameter for the day (found in the *Nautical Almanac*) from the Observed Altitude and proceed as usual. Corrections for the lower limb are always plus. The *star* and planet corrections table is similarly arranged. This correction is always minus. The moon's corrections are given for both upper and lower limbs and entered with the Observed Altitude and Horizontal Parallax and there is a supplementary table for Height of Eye correction.

(1*b*) **Recording the Time of the Observation.** The G.M.T. is required because all quantities in the *Nautical Almanac* are computed for G.M.T. The instant at which the heavenly body is brought into contact with the horizon is noted on a chronometer or watch. This may be done by the observer calling out 'Time!' to an assistant who then reads the nearest second, minute and hour, (writing it down in that order), or by the observer himself using a stop-watch which he starts when contact is made and stops when he has reached the chronometer. An experienced observer may dispense with a stop-watch and count 10 seconds, reading the chronometer 10 seconds after making contact.

A chronometer is never moved from its stowed position, nor are its hands altered, but its error is recorded daily in the chronometer journal. (*See* page 107).

(1c) *The Clock Time and Log Reading is Recorded.* These are necessary for the next step—finding the ship's approximate position.

(2) *Calculating the Assumed Position of the Ship.* The D.R. position (or E.P. if a current is suspected) is calculated by traverse table, using the course or courses steered and distance run by log since the last fix. (*See* Chap. 8.)

(3) *Finding the Geographical Position of the Heavenly Body.* Opening the *Nautical Almanac* at the correct date (G.M.T.), write down the G.H.A. and Dec. for sun, moon or planets, or if a star has been observed, the G.H.A. Aries, for the hour of G.M.T. From the Table of Increments and Corrections add the amount for the minutes and seconds to the G.H.A., and if a star has been observed, add its S.H.A. to get the correct G.H.A. Apply the correction to the Declination, adding it if the declination is increasing, but subtracting it if the declination is decreasing. The correction for the sun's declination can generally be estimated with sufficient accuracy and a star's declination will need no correction. The Declination correction is found by entering the table with 'd,' obtained from the bottom of the daily page (d = 0·2 in this example) which is the amount that the declination changes in 1 hour.

EXAMPLE: January 1st G.M.T. 4h 5m 23s.

G.H.A. 4h	239° 7′·8	Dec. S	23° 01′·2
inc. 5m 23s	+1° 20′·8	cor.	0′·0
cor. G.H.A.	240° 28′·6	cor. Dec. S	23° 01′·2

(4) *Finding the Zenith Distance and Azimuth of the Heavenly Body*, that is; the side *ZX* and angle *Z* in the triangle *PZX*. This is basically done by trigonometry, using the known parts of the triangle *PZ* (co-lat.), *PX* (polar distance) and angle *P* (L.H.A.). We already have the G.H.A. and to find the Local Hour Angle (the hour angle of the heavenly body at the place where the ship is) add or subtract the longitude according to the rhyme

'Longitude west, Greenwich is best' (i.e. greatest)
'Longitude east, Greenwich is least.' (*See* Fig. 16.1)

In order to save the navigator from the necessity of calculating each sight special 'Sight Reduction Tables' have been computed.

(4.1) *Finding the Zenith Distance and Azimuth by the Use of Special Tables.*

Sight Reduction Tables for Air Navigation A.P. 3270 (H.O. No. 249 in the U.S.A.) are generally considered to be the best for yachtsmen, for they require the least figuring and give an accuracy commensurate with the data obtainable in a small craft. They are published in three volumes. Volume I gives the computed altitudes and azimuths of seven selected stars for every whole degree of latitude from 0° to 89° N. and S. and for whole degrees of L.H.A. Volumes II and III, for the other heavenly bodies, cover latitudes 0° to 39° and 40° to 89° respectively and declinations from 0° to 29°. Volume I is published at 5-year intervals, but Volumes II and III are permanent. Volumes I and III are used with this book. Examples: Volume I, page 189; Volume III, page 180.

Sight Reduction Tables for Marine Navigation N.P. 401(4) (H.O. 229) give a greater degree of accuracy, but are not so convenient for the working of star sights. They are published in six Volumes of latitude and are suitable for observations of all heavenly bodies at all times.

There are other 'short method' tables also. All such tables contain full instructions as to their use and the reader who has progressed thus far will have no difficulty in following them.

Sight Reduction Tables. In order to keep the size of sight reduction tables within bounds, all such tables are restricted to whole numbers of degrees of latitude, declination and local hour angle. To use them we therefore decide upon a *Chosen Position*, which is the nearest whole degree of latitude, and the degree and minutes of longitude which will give a whole degree of local hour angle.

EXAMPLES: What is the Chosen Position for D.R. 50° 52′ N., 27° 33′ W.? G.H.A. 115° 43′.

	G.H.A.	115° 43′
Chosen latitude 51° N.	Chosen long.	27° 43′ W.
	(L.H.A.	88° 0′)

What is the Chosen Position for D.R. 50° 2′ N., 14° 25′ E.?

	G.H.A.	115° 43′
Chosen latitude 50° N.	Chosen long.	14° 17′ E.
	(L.H.A.	130° 0′)

It will be further noticed that the tables give *altitudes* and not zenith distances. This is done to reduce the figuring, it being just as useful to compare the length of the computed altitude with the length of the true altitude as to compare the length of the computed and true zenith distances, since altitude and zenith distance are complementary to each other.

N. Lat. { LHA greater than 180° . . . Zn = Z
LHA less than 180° . . . Zn = 360 − Z }

DECLINATION (15°–29°) *Contrary* NAME TO LATITUDE

LAT. 46°

	15°	16°	17°	18°	19°	20°	21°	22°	23°	24°
LHA	Hc. d. Z.	Hc. d. Z.	Hc. d. Z.	Hc. d. Z.	Hc. d. Z.	Hc. d. Z.	Hc. d. Z.	Hc. d. Z.	Hc. d. Z.	Hc. d
	° ′ ′ °	° ′ ′ °	° ′ ′ °	° ′ ′ °	° ′ ′ °	° ′ ′ °	° ′ ′ °	° ′ ′ °	° ′ ′ °	° ′
34	21 43 54 144	20 49 55 145	19 54 54 145	19 0 55 146	18 5 55 146	17 10 55 147	16 15 55 147	15 20 55 148	14 25 55 148	13 30
33	22 07 55 145	21 12 56 146	20 18 55 146	19 23 55 147	18 28 55 147	17 33 55 148	16 38 56 148	15 42 55 148	14 47 55 149	13 52
32	22 31 55 146	21 36 56 147	20 40 55 147	19 45 55 148	18 50 55 148	17 55 56 148	16 59 55 149	16 04 55 149	15 09 56 150	14 13
31	22 53 55 147	21 58 55 148	21 03 56 148	20 07 55 149	19 12 56 149	18 16 55 149	17 21 56 150	16 25 56 150	15 29 55 151	14 34
30	23 16-56 148	22 20-56 149	21 24-55 149	20 29-56 150	19 33-56 150	18 37-56 150	17 41-56 151	16 45-55 151	15 50-56 151	14 54
29	23 37 56 149	22 41 55 150	21 46 56 150	20 50 56 150	19 54 56 151	18 58 56 151	18 02 57 152	17 05 56 152	16 09 56 152	15 13
28	23 58 56 150	23 02 56 151	22 06 56 151	21 10 56 151	20 14 57 152	19 17 56 152	18 21 56 153	17 25 57 153	16 28 56 153	15 32
27	24 19 57 151	23 22 56 152	22 26 56 152	21 30 57 152	20 33 56 153	19 37 57 153	18 40 57 153	17 43 56 154	16 47 57 154	15 50
26	24 38 56 152	23 42 57 153	22 45 56 153	21 49 57 153	20 52 57 154	19 55 57 154	18 58 56 154	18 02 57 155	17 05 57 155	16 08
25	24 57-56 153	24 01-57 154	23 04-57 154	22 07-57 154	21 10-57 155	20 13-57 155	19 16-57 155	18 19-57 156	17 22-57 156	16 25

Fig. 17.2—Sight Reduction Tables for Air Navigation (A.P. 3270) Volume 3. Part of page.

EXAMPLE: In D.R. 46° 13′ N., 17° 53′ W., G.H.A. 45° 38′, Declination S. 20° 41′. What is the computed altitude and azimuth of the heavenly body? (*See* Fig. 17.2).

	G.H.A.	45° 38′
Chosen latitude 46° N.	Chosen long.	17° 38′ W.
	(L.H.A.	28° 0′

Open Volume III at 'Lat. 46° Declination (15-29) *Contrary* name to Lat.'. The whole number of degrees of Dec. is found along the top of the page. The whole number of degrees of L.H.A. is found in the left or right hand columns (in this case the left hand). The entry reads:

Hc	d	Z
19° 17′	56′	152°

Hc. is the computed altitude ('Height-computed'). d. is the difference between Hc. for Dec. 20° and Dec. 21°. Z. is the azimuth angle, and the rule at the top of the page states 'N. Lat. L.H.A. less than 180° . . . Zn. = 360°−Z. Zn. ('azimuth from north') is the true bearing of the heavenly body.

d. is the difference in altitude for 1° of declination. At the end of the book, and also printed as a separate book-marker, Table 5 gives the Correction to Tabulated Altitude for Minutes of Declination. In this example we have 41′ of declination not yet allowed for. Entering Table 5 with d. in the left hand column and 41 along the top, we find 38′. Hc. is decreasing at each degree of declination, so the correction is minus:

Hc.	19° 17′	Z.	152°
−	38′		360°
	18° 39′	Bearing	208°

Notes: In selecting the Chosen Latitude the *nearest* whole number of degrees is taken. In entrying the Declination take the whole number of degrees given in the *Nautical Almanac*, not necessarily the nearest.

Be careful in opening the table to use 'Declination *same* name as Latitude' or 'Declination *contrary* name to Latitude' as the case may be. Be careful to apply the Declination correction the right way.

(4.2) *To Calculate the Zenith Distance by Trigonometry.* This method takes longer but does not require any special volumes of table; *all* standard nautical tables, and Reed's *Nautical Almanac* contain the necessary tables.

The formula used is:
Haversine (or Versine) Zenith distance = cosine latitude × cosine declination × haversine (or versine) L.H.A. + haversine (or versine) latitude ± declination.

Note that when the latitude and declination are of the same name the lesser is subtracted from the greater: when they are of contrary names the two are added.

The calculation can be arranged as follows (Burton's Versine table is used), the cosines being looked up first and the remainder of the problem being finished in the versine table.

The D.R. position is used, there being no need for a Chosen Position.

Lat. (N.)	46° 13′	log. cos.	9·8401
Dec. (S.)	20° 41′	log. cos.	9·9711
L.H.A.	27° 45′	log. vers.	9·0607
		log. vers.	8·8719
(The nat. vers. is found alongside the log. vers.)		nat. vers.	·0745
Lat. ± Dec. (in this case +)		nat. vers.	·6077
Zen. Dist.	71° 28′	nat. vers.	·6822
	90°		
Altitude	18° 32′		

The *Azimuth* is found by the use of the ABC Tables as described in Chapter 8.

Another worked example will be found in Burton's, page v.

A Further Note on Finding Calculated Altitudes and Azimuths.

There are, broadly, three methods available to the navigator (two of which have been described) any of which he may use as he prefers.

CALCULATION.

(1) The basic method of calculation by trigonometrical formula (for altitude or zenith distance and ABC Tables for azimuth). The necessary log and ABC Tables are found in all standard Nautical Tables and in Reed's *Nautical Almanac*, so no special tables are required. The subsequent plotting is simple.

SHORT METHOD.

(2) Using *Sight Reduction Tables for Air Navigation* (A.P. 3270), 3 Volumes, or *Sight Reduction Tables for Marine Navigation* (H.O. 229), 6 Volumes. The method is very quick and, more important, the few figures reduce the chance of arithmetical errors being made. The plotting is slightly more complicated.

MIXED METHOD.

(3) Less calculation is required than in (1) but more than in (2). Plotting as in (2). Special tables are needed (unless Inman's or

Norie's are used) but these are not so bulky or expensive as the '*Sight Reduction Tables*' and cover all navigable latitudes.

Examples of methods follow for comparison.

Required the altitude and azimuth of the sun in D.R. position 45° 35′ N., 39° 15′ W.

(1) CALCULATION: Using the Versine formula (Burton's, Reed's etc.)

	D.R. Lat.	45° 35′	l.cos.	9·8450	A+1·28
	Dec.	22° 56′·7	l.cos.	9·9642	B+0·68
(D.R. Long. used)	L.H.A.	320° 46′·4	l.ver.	9·3528	
					C+1·96
			l.ver.	9·1620	= 36·5
			= nat. ver.	·1452	Az. S36·5E.
	Lat.±Dec.	68° 31′·7	nat.ver.	·6339	
	Z.X.	77° 14′·2=	nat.ver.	·7791	
		90°			
	Cal. Alt.	12° 45′·8			Bearing 143·5

(2) SHORT METHOD: Using *Sight Reduction Tables for Air Navigation*. Chosen Latitude whole degree 46° N., Chosen Longitude to give tabulated H.A. for a whole degree 39°.

Hc.	13° 22′	−54	Z. 143°
	— 51′	(correction for minutes of declination)	
Computed Alt.	12° 31′		Bearing 143°

(3) 'MIXED' METHOD: Table from Norie's *Nautical Tables*. Chosen Latitude 46° and Chosen Longitude to give Tabulated H.A. for whole degree 39°.

A.	0·04606	K	53° 7′	A+1·28
		Dec.	22° 56′·7	B+0·68
	10·61815←	K+D	76° 3′·7	C+1·96
				Az. S .36·5° E.
	10·66421→	l.sec.	77° 29′·2	
			90°	
Calculated Alt.			12° 30′·8	Bearing 143·5°

(5) *Comparing the computed altitude tabulated in the Sight Reduction Tables or calculated by trigonometry, with the altitude (corrected) observed with the sextant.*

When the true altitude is greater than the computed altitude the position line must be moved from the Chosen (or D.R.) Position *towards* the heavenly body observed. The intercept is then *towards*.

When the true altitude is less than the computed altitude the position line must be moved away. The intercept is then *away*. This can easily be remembered if one recollects that the nearer one approaches a lighthouse the greater the altitude of the light becomes.

EXAMPLE:

Computed, corrected *tabulated* altitude, (Hc.)	18° 39′
True, corrected *observed* altitude, (Ho.)	18° 33′
Intercept	6 away

The whole working of a sight may now be shown as it appears in the work book, for the remainder of the problem has to be drawn on a plot, or chart.

Sights should always be laid out and worked in a systematic manner and, to begin with, it is best to write down the headings in

WORKED EXAMPLES:

	Sun Sight	*Star Sight*	*Moon Sight*
Day and Date	Wed. 17th January	Tues. 2nd January	Sun. 23rd September
Time Log	1412 23	1705 51	0450 77
Zone	+1	0	−1
Greenwich Date	17. 1512	2. 1705	23. 0350
Course and Dist.	235° 23 M.	062° 51 M.	005° 77 M.
from Lat. and Long.	46° 26′ N. 17° 26′ W.	48° 34′ N. 5° 14′ W.	38° 58′ N. 18° 40′ E.
d. lat. (dep.) d. long.	13·25′ (18·8) 27′ W.	24′ N.(45) 1 9′ W.	1° 17 N. (5·4) 7′ E.
D.R. at sight	46° 13′ N. 17° 53′ W.	48° 58′ N. 4° 5′ W.	40° 15′ N. 18° 47′ E.
Body: Obs. Alt.	☉ 18° 23′	Altair 29° 25′	− 31° 38′
cor.	− + 10	(H.E. 12) −5	☾ +23′
			(H.E. 15) +6·1′
Ho. True Alt.	18° 33′	29° 20′	32° 7·1′
Chronometer	3h 9m 30s	5h 2m 15s	3h 50m 50s
error	+ 3 15	+ 1 13	− 1 5
G.M.T.	15h 12m 45s	17h 3m 28s	03h 49m 45s
G.H.A. h.	42° 26·6′	357° 11·2′	270° 01·0′ v 9·6
m. s.	3° 11·3′	52·1′	11° 52·3′
S.H.A. or V. cor.	—	62° 38·1′	+ 7·9′
G.H.A.	45° 37·9′	420° 41·4′	282° 1·2′
±360° if required	—	60° 41·4′	—
Chosen Long.	W. 17° 38′	W. 5° 41·4′	E. 18° 59′
L.H.A.	28°	55°	301°
360°−L.H.A. if req'd.			
Tab. L.H.A.	28°	55°	301°
Dec. d.	S. 20° 41·3′	N. 8° 47·8′	N. 13° 02·2′ d 12·5 H.P. 59·1
d. cor.	—		− 10·3′
Dec.	S. 20° 41·3′		N. 12° 51·9′
Hc. d. Z.	19° 17′ −56′ 152°	28° 32′ +47′ 113°	31° 18′ +38′ 101°
d. cor.	− 38′ 360°	+ 38° 360°	+ 33′
Hc. Computed Alt.	18° 39′	29° 10′	31° 51′
Ho. True Alt.	18° 33′	29° 20′	32° 7′
Intercept. Bearing	6′ away 208°	10′ towards 247°	16′ towards 101°

FIG. 17.3

the sight book before working the calculation. They should always be placed in the same order. A good type of sight book is a stiff-backed foolscap or A4 size. The page is large enough for plotting and vertical columns may be drawn to allow more than one sight to be worked on one page.

Examples of three sights, laid out in a convenient way, are given in Figure 17.3 (Sight Reduction Forms N.P. 400(*a*) are printed and can be obtained from chart agents.)

The **Sun Sight** has already been described in detail. Exercises in Sight Reduction are given at the end of the chapter.

Working a Planet Sight. The G.H.A. and Dec. for the four planets useful to the navigator are tabulated in the *Nautical Almanac* in the same way as the Sun, but there is a small *v* correction to be applied to the G.H.A. (added unless preceded by a minus sign) in the same way as correction *d* is applied to the Declination. The Star Total Correction to Altitude table is used.

Working a Moon Sight. (See worked example on the Sight Form). (Fig. 17.3). The G.H.A. and Dec. are found as for the Sun, but the G.H.A. must have the *v* correction, and the Declination the *d* applied. Note and write down the H.P. (Horizontal Parallax) at the same time.

Altitude correction (Burton's pages 50 and 51) for observations of the Upper Limb are on the left-hand page; for the Lower Limb on the right-hand page. Enter with the observed altitude and the Horizontal Parallax. Note whether the correction is additive or subtractive. Add the correction for height of eye at the bottom of the page.

Working a Star Sight. Observations of stars are worked in the same way as sun sights except in two particulars.

(i) Finding the L.H.A. It is impractical to include in the *Nautical Almanac* the G.H.A. of every star used in navigation. The G.H.A. of the first point of Aries (the Vernal Equinox) is therefore tabulated for each hour, and the Sidereal Hour Angles (S.H.A.) of 57 selected stars are tabulated for each month. The Sidereal Hour Angle is the angular distance the star is west of Aries, so that G.H.A. Aries + S.H.A. Star = G.H.A. Star. No interpolation is needed for the star's S.H.A. or Dec. as they change very slowly.

(ii) Correcting the altitude. There is no semi-diameter to be applied and parallax is negligible so a separate Total Star Correction table (Burton's page 48) has been compiled and must be used instead of the sun's. (See worked example on Fig. 17.3 Sight Form.)

(6) *Plotting the Intercept and Position Line.* This may be done on a chart of suitably large scale, or graph or lined paper may be used, the paper representing part of a chart. Special Plotting Sheets can also be purchased.

To Construct a Plotting Chart. Turn the sight book sideways; the lines now represent meridians 1′ (or 2′ or any other number) apart. Draw a horizontal line to represent the parallel of Chosen Latitude. On a mercator chart the longitude and latitude scale are different (except on the equator), so a scale of latitude must be drawn. From one end of the parallel of latitude, or from any other parallel line, draw an angle equal to the latitude. The intersections of this line with the meridians mark the lengths of minutes of latitude, and miles of distance for this latitude. (Figure 17.4).

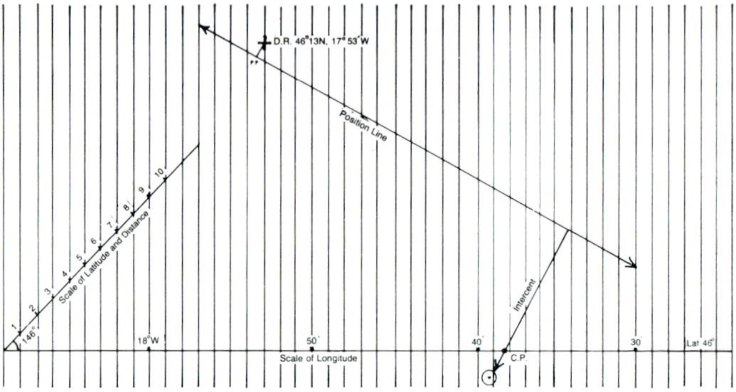

Fig. 17.4

Plotting the Position Line. On the Chosen Latitude mark a convenient meridian as the Chosen Longitude. From this Chosen Position draw the intercept—towards the body if the intercept is towards, or on the reciprocal of the bearing if the intercept is away. Measure the distance of the Intercept on the Latitude/Distance scale. At the end of the Intercept (the Intercept Terminal Position) draw a line at right angles, which is the Position Line.

The Probable Position. The position of the ship is somewhere on the Position Line. It is unlikely to be at the Intercept Terminal Position because the Chosen Position is often some distance from the Dead Reckoning Position—which is where the ship is assumed to be. So the D.R. position is now plotted from the C.P. The nearest point on the Position Line to the D.R. position is the Probable Position of the ship. Draw a perpendicular from the D.R. position to find the Probable Position of the ship.

Plotting Calculated Intercepts. When the intercepts have been found by calculation (not by Sight Reduction Tables), the D.R. position (and not the C.P.) is used so the intercept is of course plotted from the D.R. position and the C.P. does not come into the picture at all. Plotting is therefore slightly simpler. The intercepts are usually shorter and a different method of constructing a plotting chart may be prefered. Keep the lines of the sight book horizontal;

they then form the scale of latitude and distance. Draw a perpendicular to mark the meridian passing through the D.R. position. The difference of latitude and departure of the Probable Position from the D.R. position is measured on the latitude scale; the departure is then changed into difference of longitude by using the Traverse Table (page 122.) (Figure 17.5)

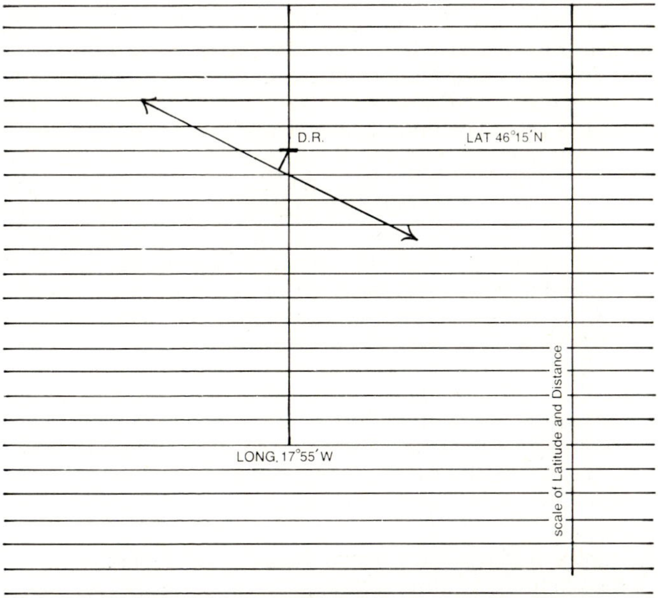

FIG. 17.5

Use of a Single Position Line. One sight gives one position line, but even a single position line can be of use. If the body observed is abeam the position line will lie parallel to the track and it can be seen whether the course has been made good. If the body is ahead or astern the intercept will indicate whether you are ahead or astern of the D.R. position and thus be a check on your speed made good. (*See* also page 32).

The Observed Position. To obtain a fix, two, or preferably more, position lines are required. If these can be obtained at the same time by, e.g. nearly simultaneous observations of two or more stars, or a cross between the sun and moon, the ship's position is where the position lines cut.

In a slowly moving vessel the distance covered between taking each sight is small enough to be disregarded, but if the speed is high and the interval between sights long, the ship's movement should be allowed for. This is done on the plot by advancing the position lines found from the earlier sights to the time of the last sight: the method followed in plotting a running-fix.

O

EXAMPLE: 1800 First star observation. Course 260°, speed 12 K. 1815 Second star observation. The Probable Position obtained by the first sight is moved 3 M., 260°, and the position line then drawn through it.

Star Sights by *Sight Reduction Tables for Air Navigation*, Volume I (A.P. 3270).

The navigator's tasks of calculation have been immeasurably lightened during the last fifty years by the skilful arrangement of Nautical Almanacs and Nautical Tables. Perhaps the most notable innovation is the *Sight Reduction Tables for Air Navigation*, Volume I for Selected Stars. The book provides the Tabulated Altitude (Hc.) to the nearest minute, and the True Bearing (Zn.) to the nearest degree, for each degree of Latitude and Local Hour Angle of Aries. The seven stars are selected from a total of 41, the brightest (stars of the first magnitude) being given in capital letters. They have been chosen with regard to their altitude and bearing as well as their brightness, and of each selection of seven stars three are marked with an asterisk as being suitable for a three-star fix. An extract from these Tables (Fig. 17.6) is given on page 189.

A full description of the tables and instructions as to how they are to be used is given in the book. The tables were designed primarily for air-navigation and with the use of the *Air Almanac*, but they can equally well be used at sea with the *Nautical Almanac*, and the special correction tables for, e.g. the motion of the observer between sights which must be taken into account in an aircraft, can be disregarded.

The tables are intended for two distinct operations:

(1) the planning of observations;

(2) the reduction of observations when taken.

(1) **The Planning of Observations.**

The problem is: what stars may be best observed? We need to know *at what time* we shall be making the observations and *where we shall be* (approximately) at that time. The altitude of stars must be observed while the horizon is still visible, and the stars at this time are not easily discerned as there is still a good deal of light in the sky.

Finding 'Star Time'. The time of 'civil twilight,' the best time for observing stars, for the day in question is found in the *Nautical Almanac*. This is strictly the mean time for an observer on the Greenwich meridian, but practically also the mean time for observers on other meridians. Find the G.M.T. corresponding to civil twilight by applying the longitude, in time, Conversion of Arc to Time, page 134 Burton, or in the *Nautical Almanac*, according to the rule 'Longitude East, Greenwich is least; Longitude West, Greenwich is best.' To the G.M.T. apply the zone (with the time reversed, since it is local time we need, not Greenwich time.) This is Zone Time of twilight and about the best time for star observations. (See page 122)

LAT. 36° N

LHA ♈	Hc.	Zn.	Hc.	Zu.	Hc.	Zn.	Hc.	Zn.	Hc.	Zn.	Hc.	Zn.	Hc.	Zn.	LHA ♈	Hc.	Zn.	Hc.	Zn.
	*Capella		Aldebaran		*Diphda		Fomalhaut		Altair		*Vega		Kochab			*Dubhe		Regulus	
0	32 11	054	26 42	088	34 57	168	22 34	195	27 15	261	27 50	299	23 51	348	90	37 51	035	29 50	097
1	32 50	054	27 31	089	35 07	169	22 21	196	26 27	261	27 08	300	23 42	349	91	38 19	035	30 38	097
2	33 30	054	28 19	090	35 16	170	22 07	197	25 39	262	26 26	300	23 32	349	92	38 47	035	31 26	098
3	34 09	054	29 08	090	35 24	171	21 53	198	24 50	263	25 44	301	23 23	349	93	39 15	035	32 14	099
4	34 49	055	29 56	091	35 31	172	21 38	199	24 02	263	25 02	301	23 14	349	94	39 43	036	33 02	099
5	35 28	055	30 45	091	35 37	173	21 22	199	23 14	264	24 21	301	23 05	350	95	40 12	036	33 50	100
6	36 08	055	31 33	092	35 42	175	21 05	200	22 26	265	23 39	302	22 56	350	96	40 40	036	34 38	101
7	36 48	056	32 22	093	35 46	176	20 48	201	21 37	265	22 58	302	22 48	350	97	41 08	036	35 25	101
8	37 28	056	33 10	093	35 49	177	20 36	202	20 49	266	22 17	303	22 40	350	98	41 36	036	36 13	102

Fig. 17.6—Sight Reduction Tables for Air Navigation (A.P. 3270) Volume I. Part of page.

Finding what Stars can be Observed. From the *Nautical Almanac* look out the G.H.A. Aries for the G.M.T. of star time. Apply the longitude to obtain the Local Hour Angle Aries.

What our D.R. position will be at star time must be calculated by Traverse table or plotted on the chart.

We can now turn to the Sight Reduction Tables where we can see at a glance the seven best stars to observe for that time and latitude, with their altitudes and azimuths.

EXAMPLE: What stars can be observed on the evening of 1st January 1973 in latitude 36° N., longitude 7° W.?
Civil twilight, 1st Jan.

Lat. 36° N.		1726
Longitude in time	W.	28
G.M.T.		1754
Zone		0
Zone Time		1754

G.H.A. Aries 17h.	356° 12′·1
inc. 54m.	13° 32′·2
G.H.A. Aries	369° 44′·3
Long. W.	7°
	362° 44′·3
	−360°
L.H.A. Aries	2° 44′·3

Referring to the extract on page 189, we see that, at this time,

*CAPELLA	ALDEBARAN	*diphda
Hc. 34° Zn. 054°	Hc. 29° Zn. 090°	Hc. 23° 20′ Zn. 171°
FOMALHAUT	ALTAIR	*VEGA
Hc. 22° Zn. 198°	Hc. 25° Zn. 263°	Hc. 26° Zn. 301°
	Kochab	
	Hc. 23° 30′ Zn. 349°	

To obtain the approximate altitude to set on the sextant we can add 5′ to the Tabulated Altitudes.

We might decide to observe Capella, Diphda and Vega; observing Capella first, as it is a bright star and the eastern horizon is usually the first to darken, then Vega to the west and Diphda last, as the least bright star and least quickly seen.

(2) **The Reduction of Observations.**

The observations have been made and recorded as follows (probably more than one observation of each star would be taken and the mean used).

D.R. 36° N., 7° W.			
Star Obs. Alt.	Capella 33° 41½′	Vega 26° 6′	Diphda 35° 28′
Cor. (H.E. 10 ft.)	— 4·5′	— 5′	— 4½′
True Alt. (Ho.)	33° 37′	26° 1′	35° 23½′
Chronometer	17h 50m 35s	17h 53m 6s	17h 59m 41s
error	+ 1m 5s	+ 1m 5s	+ 1m 5s
G.M.T.	17h 51m 40s	17h 54m 11s	18h 0m 46s
G.H.A. Aries 17h.	356° 12·1′	356° 12·1′	11° 4·5′
increment	12° 57·1′	13° 35′	11·5′
G.H.A. Aries	369° 9·2′	369° 47·1′	11° 16′
	−360°	−360°	
	9° 9·2′	9° 47·1′	11° 16′
Chosen longitude	7° 9·2′	6° 47·1′	7° 16′
L.H.A. Aries	2°	3°	4°
Tabulated Alt. (Hc.)	33° 30′	25° 44′	35° 31′
True Alt. (Ho.)	33° 37′	26° 1′	35° 23½′
Intercept. Bearing	7′ to 054°	17′ to 301°	7½′ away 172°

Plotting the Observations.

A horizontal line representing the Chosen Latitude (36°) is drawn and, at one end an angle of 36° is constructed to provide a

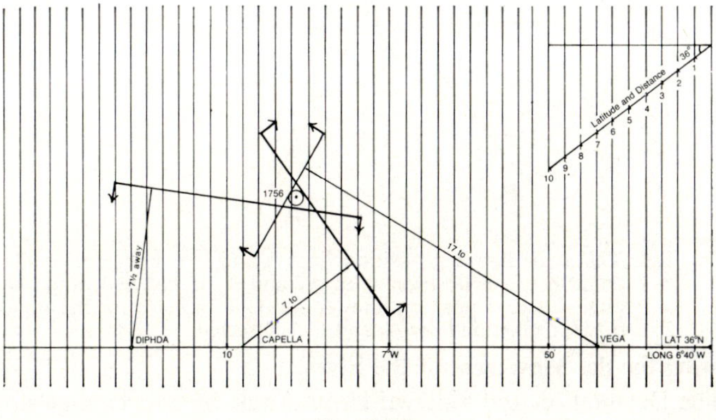

FIG. 17.7

scale of latitude and distance. The three star's intercepts are plotted from their Chosen Longitudes and the position lines drawn from the ends of the intercepts. The ship's position is considered to be in the centre of 'the cocked hat' made by the three position lines: it is 36° 7′ N., 7° 5½′ W. at 1756 G.M.T. (Fig. 17.7).

The 'Cocked Hat'

If all the observations were precisely correct in every detail, the position lines would meet at a point. This rarely happens, and even in good conditions, a triangle or 'cocked hat' of two or three miles may result. Although the ship is often considered to be at the centre of the cocked hat, a more accurate position is obtained by moving each position line the same way by the same amount until the cocked hat has shrunk to nothing.

The practical way to do this is to bisect the external angles between the position lines, the intersection of the bisectors then gives the position of the fix.

Proceed as follows:

1. Turn the ends of each position line 90° into the direction of the azimuth.

2. Bisect those angles made by position lines whose arrows point (a) towards each other, or (b) away from each other.

It will be noticed that the fix does not always lie inside the cocked hat. Fig. 17.8

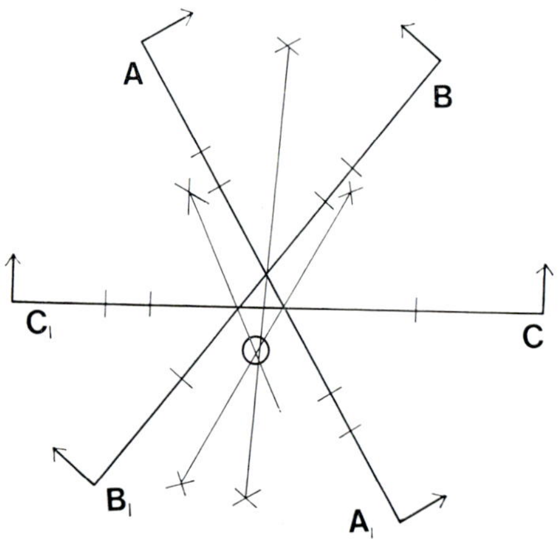

FIG. 17·8 The angle between position lines A and B is bisected, or the angle between position lines A_1 and B_1 is bisected, and the angle between position lines A and C is bisceted, or the angle between position lines A_1 and C_1 is bisected.

Allowing for the Movement of the Stars.

The Declination and Sidereal Hour Angle of stars change slowly with the years and the Tables in A.P. 3270, Volume I are re-calculated every five years to keep them up to date. For 1974 there is no correction and in 1975 it can be ignored. Table 5 *Correction for Precession and Nutation*, at the back of the book can be used to correct either a position line or a fix. Each entry in Table 5 consists of a single figure in bold type—the *distance* in nautical miles which a position line or fix is to be moved, and a group of three figures which is the *direction* in which to move it.

If the example was to be taken in 1979, the fix would have to be be moved 3 M., in the direction 060°—which would be best done on the plot.

Exercises in Plotting Position Lines.

1. Draw the position line and find the most probable position. In Chosen Position 50° 0′ N., 18° 8′ W., the sun bore 265°, intercept 13 M. towards. D.R. 49° 51′ N., 18° 20′ W.

2. In Chosen Position 45° 15′ S., 152° 13′ E., Sun's bearing was 132°, intercept 9 M. away. D.R. 45° 10′ S., 152° 15′ E. Find the most probable position.

3. Find the observed position of the ship. Altair bore 052°, intercept 10′ towards; Arcturus bore 330°, intercept 6 M. away. Chosen Position for Altair 23° N., 65° 55′ E. Chosen Position for Arcturus 23° N., 65° 30′ E.

4. In Chosen Position 49° N., 20° 15′ W., Rigel bore 105°, 5 M. towards. In Chosen Position 49° N., 19° 53′ W., Capella bore 063° 7 M. away. Find the Observed Position.

5. In a slow-moving yacht the following observations were made: Rigel: bearing 283°, intercept 23·5 M. away, Jupiter bearing 345°, intercept 7 M. away. Regulus bearing 037° intercept 3·6 M. away. Chosen Positions: Rigel: 16° S., 45° 13·5′ W. Jupiter: 16° S., 45° 3′ W. Regulus: 16° S., 44° 48·5′ W. D.R. 16° S., 45° W. Required the Observed Position.

6. Observed sun bearing 145°, intercept 7 M. away, in Chosen Position 50° N., 8° 14′ W. D.R. 50° 3′ N., 8° 22′ W. The ship then ran course: 262°, distance: 35 M., when a second observation of the sun gave a bearing of 245°, intercept 15 M. away in Chosen Position 50° N., 09° 32·5′ W. (*See* Fig. 00.0).

Required the Observed Position at the second observation.

Planning a Programme by Star Chart.

If we have no Volume I A.P. 3270 a Star Chart may be used to find the approximate altitudes and azimuths of suitable stars and planets. The planets—especially Venus and Jupiter, the brightest—are often useful. Whether they are 'morning or evening stars' can be found in *Planet Notes* in the *Nautical Almanac*, but their position cannot be tabulated for they are 'wanderers' as their name implies. Use of *Brown's Star Chart*. This is a star chart drawn on mercator projection. The position of the observer (strictly, the position of his zenith) is plotted on the chart, using the latitude and (for longitude) the L.H.A. Aries. The position of the planets, using their declination (as 'latitude') and the difference of G.H.A. Aries and G.H.A. planet (as 'longitude') can be plotted. What stars are suitably placed can be seen at a glance and their approximate altitudes measured directly on the chart.

Successive Observations of the Sun.

The sun is the heavenly body most often available for observation and a 'running fix' (page 39) can be obtained from two sights taken about 3 hours apart when the azimuth has changed not less than 30° (to get a good angle between the position lines). The larger the

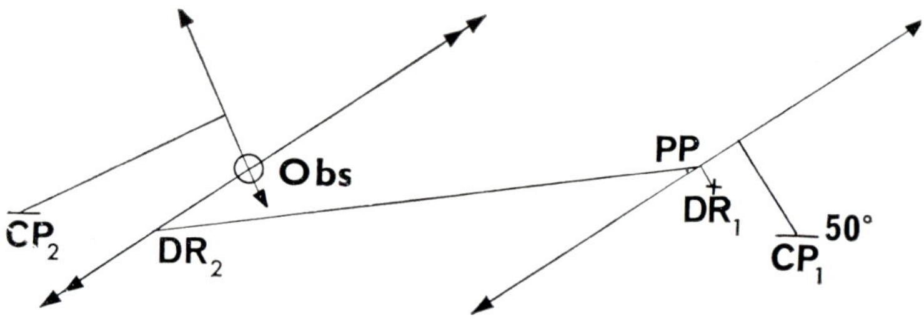

FIG. 17.9

angle, but the shorter the run between sights the better. Plotting can be done as in Fig. 17.9, the graphical problem then becoming that of a running fix. In this method, draw the run from the Probable Position (allowing for the current if known) and transfer the first position line to the end of the run. Where the transferred position line cuts the position line obtained from the second observation is the ship's position at the time of the second observation.

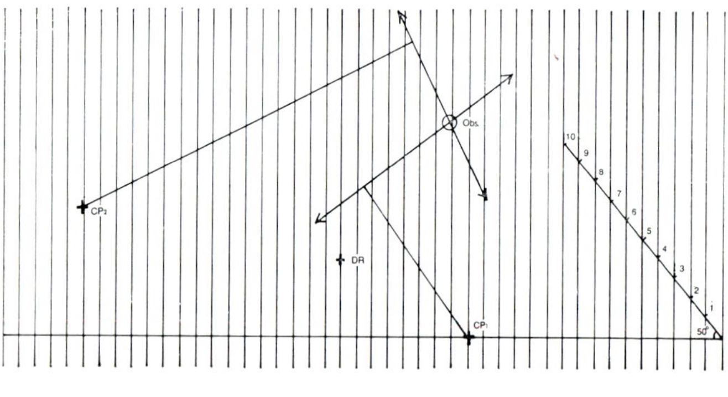

FIG. 17.10—Two sun sights with the run between them plotted

If the length of the run would bring the second D.R. position off the plot it is more convenient *not* to plot the run, but to use the Traverse Table to find the second D.R. position. Providing that the run allowed between sights is correct, the ship's observed position will be the same distance from a D.R. position at whatever time the D.R. position is plotted. One D.R. position will thus serve for both. No plotting is carried out until the second sight has been worked. A more or less central spot on the paper is taken for the D.R. position. The first Chosen Position is then plotted from the D.R. and the intercept and position line of the first observation is drawn. The second Chosen Position is then plotted from the D.R. (now the second D.R.) and the intercept and position line of the second sight drawn. The latitude and longitude of the resulting observed position is obtained by measurement from the (second) D.R. position. Figure 17.10 will make this clear.

D.R.	50° 3′ N.	8° 22′ W.	D.R.1	50° 3′ N.	8° 22′ W.
run 262° 35 M.	5 S.	54′ W. (34·7)	C.P.1	50°	8° 14′ W.
D.R.2	49° 58′ N.	9° 16′ W.	C.P.1 from D.R.	3′ S.	8′ E.
D.R.2	49° 58′ N.	9° 16′ W.	D.R.2	49° 58′ N.	9° 16′ W.
C.P.2	50° 0′ N.	9° 32·5′ W.	from D.R.2	5·5′ N.	7′ E.
C.P.2 from D.R.	2′ N.	16·5′ W.	Obs.	50° 3·5′ N.	9° 9′ W.

If the intercepts are calculated (and not taken out of A.P. 3270) they are drawn from the D.R. position and there are no Chosen Positions.

This method allows a larger scale to be used than when the run is plotted and will generally be preferred during an ocean passage.

When a landfall is to be made it is better to plot upon the approach chart using the method shown in Figure 17.9.

Exercises in Astronomical Navigation.

Sun Sights

Find the most probable position and draw the position line through it.

1. On Jan. 2nd, 1973, at 0912 Z.T. (in Zone +1), in D.R. Lat. 45° 35′ N., Long. 11° 17′ W., the chronometer read 10 hr. 10 min. 57 sec. Its error was 1 min. 13 sec. slow. The sextant altitude of the sun's lower limb was 12° 27′. The index error of the sextant was 2·5′ off the arc and the height of the observer's eye was 15 ft.

2. At 1513 Z.T. (Zone −10) on 23rd Sept., in D.R. Lat. 41° 17′ S., Long. 155° 7′ E., the chronometer showed 5 hr. 9 min. 57 sec. It was 1 min. 13 sec. fast of G.M.T. The sextant altitude of the sun's lower limb was 26° 8′. Index error was 1·5′ on the arc, height of eye 20 ft.

3. Sept. 24th, at 1418 Z.T. (Zone not known), D.R. position 51° 39′ N., 55° 17′ W. Chronometer 6 hr. 16 min. 6 sec: error, 4 min. 1 sec. fast. Sextant altitude of sun's lower limb 27° 32′. Index error 1′ on the arc, height of eye 20 ft.

4. Jan. 6th, 1002 Z.T. (Zone 0), D.R. position 41° 7′ S., 7° 15′ E. Chronometer 10 hr. 1 min. 16 sec. which was 3 min. 2 slow sec. of G.M.T. Altitude by sextant of the sun's lower limb 63° 44′. Index error 1′ off the arc. Height of eye 12 ft.

5. Jan. 3rd, 1002 Z.T. (Zone +12). D.R. position 55° 15′ N., 173° 30′ W. Chronometer 10 hr. 5 min. 15 sec., error 2 min. 13 sec. slow. Sextant altitude of the sun's upper limb 10° 29½′. Index error 2·5′ on the arc. Height of eye 16 ft.

6. Sept. 21st, 1402 Z.T. (Zone −5). D.R. position 42° 19′ S., 81° 19′ E. Sextant altitude of sun's lower limb 34° 14½′. Index error 2′ off the arc. Height of eye 15 ft. Chronometer 9 hr. 0 min. 5 sec., error 2 min. 13 sec. slow.

Planet Sights

1. On Jan. 3rd, 1973, at 0520 Z.T. (in Zone +4) in D.R. 54° 54′ S., 54° 54′ W., the chronometer showed 19 hr. 17 min. 20 sec. It was 4 min. 50 sec. fast of G.M.T. The sextant altitude of Venus at this time was 28° 5′. Height of eye was 10 ft and there was no index error.

2. On Jan. 1st at 0615 Z.T., D.R. 41° 16′ S., 30° 12′ W. Chronometer 8 hr. 14 min. 17 sec. Chronometer error 2 min. 13 sec. fast. Sextant altitude of Jupiter 10° 32′. Index error 2′ off the arc. Height of eye 15 ft.

3. On Jan. 3rd at 1830 Z.T. (Zone −5). D.R. 44° 5′ N., 77° 15′ E. chronometer 1 hr. 10 min. 11 sec. which was 1 min. 12 sec. slow of G.M.T. Sextant altitude of Saturn 38° 50′. No index error. Height of eye 15 ft.

Moon Sights

1. On Sept. 24th, 1306 Z.T. (Zone +7) in D.R. 48° 5′ N., 114° 25′ W., chronometer 8 hr. 7 min. 20 sec., error 1 min. 15 sec. fast. Sextant altitude of the moon's upper limb 37° 58′. Index error 1½′ on the arc. Height of eye 14 ft.

2. On Sept. 19th at 1213 Z.T. (Zone +3). D.R. 52° 3′ N., 38° 47′ W. Chronometer 3 hr. 12 min. 13 sec. Error 3 min. 3 sec. fast. Sextant altitude of the moon's lower limb 9° 44½′. Index error 1′ off the arc. Height of eye 15 ft.

3. On Jan. 5th, 1608 Z.T. (Zone −12). D.R. 45° 49′ N., 173° 2′ E., chronometer 4 hr. 15 min. 1 sec.; 7 min. 8 sec. fast. Sextant altitude of the moon's lower limb 8° 58′. Index error 3′ on the arc. Height of eye 16 ft.

Star Sights (Single)

Find the most probable position and draw the position line through it.

1. On Jan. 2nd, 1973, at 1715 Z.T. (Zone +1), in D.R. Lat. 40° 7′ N., Long. 10° 40′ W., the chronometer, which was 8 min. 5 sec. slow, showed 6 hr. 1 min. 57 sec. when the sextant altitude of Fomalhaut was 17° 47½′. Index error was 1′ on the arc, height of eye was 12 ft.

2. On Jan. 3rd at 1702 Z.T. chronometer showed 10 hr. 2 min. 15 sec. It was 5 min. 1 sec. slow of G.M.T. D.R. 42° 9′ S., 76° 15′ W. The sextant altitude of Hamal was 18° 4′. Index error 3′ off the arc. Height of eye 10 ft.

3. On Jan. 1st at 0314 Z.T. (Zone −3), in D.R. 51° 9′ S., 40° 51′ E., the sextant altitude of Sirius was 43° 7′ at 11 hr. 56 min. 3 sec. by chronometer which was 13 min. 58 sec. slow of G.M.T. Index error was 0 and the height of eye 18 ft.

4. On Jan. 2nd, 1613 Z.T. (Zone −11), in D.R. 49° 25′ N., 171° 3′ E. Chronometer showed 5 hr. 12 min. 1 sec. It was 5 sec. fast. Sextant altitude of Altair was 30° 5′. Index error 2′ off the arc. Height of eye 12 ft.

5. On Jan. 2nd, 0659 Z.T. (Zone +12) in D.R. 51° 0′ N., 174° 15′ W. Pollux's altitude by sextant was 17° 1½′. Index error ½′ on the arc, height of eye 10 ft. Chronometer showed 6 hr. 59 min. 5 sec. It was 3 min. 1 sec. slow.

6. On Jan. 3rd, 0802 Z.T. (Zone 0). D.R. 45° 58′ S., 0° 5′ W. Sextant altitude of Arturus was 24° 15′. Index error—3½′, height of eye 18 ft. Chronometer showed 8 hr. 1 min. 53 sec., error 2 min. 5 sec. slow.

Fix by (nearly) Simultaneous Sights of Two Bodies.

1. *Nearly simultaneous observations of two stars.*

1973 Jan. 1st at 1710 Z.T. in D.R. 48° 17′ N., 55° 10′ W. Sextant altitude Aldebaran 28° 49½′. Index error 1·5′ on the arc, height of eye 10 ft., chronometer 9 hr. 8 min. 5 sec., error 1 min. 15 sec. slow of G.M.T. At 1712 Sextant altitude Hamal 57° 0½′, chronometer 9 hr. 10 min. 2 sec. Zone +4.

2. *Nearly simultaneous observations of a planet and star.*

Jan. 2nd at 0705 Z.T. (Zone +3). D.R. 46° 46′ N., 45° 17′ W. Sextant altitude Mars 16° 45½′, chronometer 10 hr. 4 min. 58 sec. Index error 2′ off the arc. Height of eye 12 ft. Error of chronometer 2 min. 16 sec. fast. Two minutes later, sextant altitude Regulus 31° 54′ at chronometer 10 hr. 6 min. 13 sec.

3. *Nearly simultaneous observations of planet and sun.*

Jan. 3rd at 1006 Z.T. (Zone −12). D.R. 46° 6′ S., 179° 55′ E. Sextant altitude Mars 64° 42′, chronometer 10 hr. 4 min. 2 sec. Sextant altitude Sun's L.L. 57° 9′, chronometer 10 hr. 4 min. 58 sec. Chronometer error 5 min. 13 sec. slow. Index error 2·5′ on the arc. Height of eye 10 ft.

4. *Nearly simultaneous observations of moon and sun.*

Sept. 23rd at 0900 Z.T. (Zone −10). D.R. 45° 5′ S., 155° 10′ E. Sextant altitude Sun's L.L. 34° 16½′ at chronometer time 10 hr. 58 min. 5 sec. Sextant altitude Moon's upper limb 30° 5′ at chronometer time 11 hr. 0 min. 20 sec. Chronometer error 7 min. 30 sec. slow. Index error 2·5′ off the arc. Height of eye 25 ft.

Fix by Successive Observations.

5. *Two observations of the sun with a run between them.*

Sept. 24th at 0758 Z.T. (Zone 0). D.R. 49° 22′ N., 6° 0′ W. Sextant altitude Sun's L.L. 16° 7½′, chronometer 7 hr. 46 min. 15 sec. 1110 Sextant altitude Sun's L.L. 37° 23½′, chronometer 10 hr. 57 min. 45 sec. Run between sights: Course 065°, distance 31 m. Chronometer 14 min. 5 sec. slow on G.M.T. Index error 1′ off the arc. Height of eye 11 ft. Find the observed position at 1110.

6. *Two observations of the sun with a run between them.*

Jan. 6th, 1005 Z.T. (Zone 0). D.R. 48° 50′ N., 5° 10′ W. Sextant altitude Sun's L.L. 12° 27′, chronometer 10 hr. 5 min. 42 sec. 1405 Sextant altitude Sun's L.L. 14° 13½′, chronometer 2 hr. 5 min. 7 sec. Run between sights: Course 040°, distance 40 M. Chronometer 6 min. 13 sec. slow on G.M.T. Index error 2′ on the arc. Height of eye 10 ft. Find the observed position at 1405.

Star Sights (by A.P. 3270 Vol. I)

Find (*a*) what stars can be observed, their altitudes to set on the sextant and their bearings.

Find (*b*) after observations, the ship's position. *N.B.*—Omit annual Correction for Precession and Nutation.

1. (*a*) 1973 Jan. 3, p.m., in the North Atlantic in 1200 D.R. 42° 12′ N., 11° 36′ W. Course 260° T. Speed 6 kn. Zone +1. Height of eye 10 ft. No Index error.

(*b*) The following observations were taken:

Capella	36° 38½′	at	5 hr.	58 min.	2 sec.	by chronometer
Diphda	29° 21′	,,	6 ,,	2 ,,	5 ,,	,, ,,
Vega	29° 34′	,,	6 ,,	5 ,,	56 ,,	,, ,,

Chronometer was 4 min. 5 sec. slow of G.M.T.

2. (*a*) In the North Atlantic, Jan. 2nd, morning stars. D.R. Jan. 1st 1200, 51° 28′ N., 41° 57′ W. Course 065°, 5 kn. Zone +2.

(*b*) The following sights were taken:

Deneb	29° 55½′	at	9 hr.	59 min.	47 sec.	by chronometer
Arcturus	58° 16′	,,	10 ,,	2 ,,	51 ,,	,, ,,
Pollux	16° 13′	,,	10 ,,	4 ,,	45 ,,	,, ,,

No Index error. Height of eye 12 ft. Chronometer 2 min. 18 sec. slow.

3. (*a*) In the Mediterranean on the evening of Jan. 1st, Zone 0, 1200 D.R. 41° 14½′ N., 3° 46½′ E. Course 220°. Speed 6½ kn.

(*b*) The following sights were taken:

Capella	34° 28½′	at	5 hr.	0 min.	17 sec.	by chronometer
Aldebaran	26° 49′	,,	5 ,,	2 ,,	1 ,,	,, ,,
Fomalhaut	18° 14′	,,	5 ,,	3 ,,	9 ,,	,, ,,
Altair	25° 47½′	,,	5 ,,	5 ,,	2 ,,	,, ,,
Vega	29° 8′	,,	5 ,,	6 ,,	8 ,,	,, ,,

Index error +1½′. Height of eye 10 ft. Chronometer error 7 min. 15 sec. fast.

4. (*a*) In the Mediterranean, Jan. 3rd, a.m. Zone −2, 1200 (Jan. 2nd). D.R. 35° 50′ N., 22° 3′ E. Course 105°. Speed 5·5 kn.

(*b*) The following sights were taken:

Regulus	39° 0′	at	5 hr.	1 min.	20 sec.
,,	38° 51½′	,,	5 ,,	2 ,,	10 ,,
,,	38° 40′	,,	5 ,,	2 ,,	30 ,,
Antares	15° 3′	at	5 hr.	5 min.	16 sec.
,,	15° 16′	,,	5 ,,	5 ,,	42 ,,
,,	15° 26′	,,	5 ,,	5 ,,	59 ,,

Kochab	49° 49′	at	5 hr.	6 min.	50 sec.
,,	49° 54′	,,	5 ,,	6 ,,	53 ,,
,,	50° 2′	,,	5 ,,	6 ,,	59 ,,

No Index error. Height of eye 12 ft. Chronometer 1 min. 3 sec. slow.

5. (*a*) In the Caribbean Sea, Jan. 1st, p.m., Zone +5, 1200. D.R. 14° 10′ N., 76° 13′ W. Course 195°. Speed 6 kn.

(*b*) The following stars were observed:

Capella	25° 44′	at	11 hr.	1 min.	6 sec.
Achernar	18° 23½′	,,	11 ,,	2 ,,	15 ,,
Fomalhaut	38° 52½′	,,	11 ,,	3 ,,	9 ,,
Deneb	28° 56½′	,,	11 ,,	4 ,,	42 ,,

Index error 2′ off the arc. Height of eye 15 ft. Chronometer error 6 min. 7 sec. slow.

6. (*a*) In the North Pacific, Jan. 2nd, a.m., Zone +8, 1200 (Jan. 1st). D.R. 48° 12′ N., 135° 0′ W. Course 085°. Speed 6·2 kn.

(*b*) Sights were taken of:

Deneb	27° 59½′	at	4 hr.	13 min.	33 sec.
Vega	42° 36½′	,,	4 ,,	14 ,,	7 ,,
Arcturus	61° 2½′	,,	4 ,,	15 ,,	52 ,,
Regulus	27° 35′	,,	4 ,,	17 ,,	59 ,,

Index error 1½′ off the arc. Height of eye 15 ft. Chronometer 4 min. 2 sec. fast.

7. (*a*) In the Tasman Sea, Jan. 2nd, p.m., Zone −10, 1200. D.R. 40° 5′ S., 165° 3′ E. Course 105°. Speed 7 kn.

(*b*) Observed:

Aldebaran	28° 58½′	at	9 hr.	39 min.	18 sec.
Rigel	46° 22½′	,,	9 ,,	41 ,,	33 ,,
Sirius	36° 50½′	,,	9 ,,	43 ,,	5 ,,
Canopus	54° 15′	,,	9 ,,	44 ,,	3 ,,
Fomalhaut	38° 31′	,,	9 ,,	49 ,,	8 ,,
Hamal	24° 38½′	,,	9 ,,	50 ,,	23 ,,

Index error 2′ on the arc. Height of eye 12 ft. Chronometer error 37 min. 3 sec. fast.

8. (*a*) In the Gilbert Islands on Jan. 3rd, a.m., Zone −12, Jan. 2nd, 1200. D.R. 2° 5′ S., 171° E. Course 000°. Speed 6 kn.

(*b*) Sights were taken of:

Arcturus	55° 21′	at	6 hr.	7 min.	50 sec.
Acrux	28° 32′	,,	6 ,,	9 ,,	25 ,,
Regulus	53° 23½′	,,	6 ,,	10 ,,	55 ,,

Index error 1′ off the arc. Height of eye 10 ft. Chronometer error 10 sec. slow.

9. (*a*) Off Cape Horn on Jan. 1st, a.m., Zone +4, Dec. 31st, 1200. D.R. 55° 10′ S., 71° 35′ W. Course 115°. Speed 8 kn.

(*b*) Observations were obtained of:

Regulus	20° 59½′ at	6 hr.	45 min.	52 sec.
Spica	24° 26′ ,,	6 ,,	47 ,,	59 ,,
Peacock	23° 36′ ,,	6 ,,	52 ,,	58 ,,
Sirius	40° 42′ ,,	6 ,,	56 ,,	5 ,,

Index error 2′ on the arc. Height of eye 15 ft. Chronometer 16 min. 3 sec. slow.

10. (*a*) South Indian Ocean on Jan. 2nd, p.m., 1200. D.R. 35° 0′ S., 71° 0′ E. Course 096°. Speed 6 kn.

(*b*) The following sights were taken:

Aldebaran	31° 20½′ at	2 hr.	54 min.	4 sec.
Sirius	32° 59′ ,,	2 ,,	55 ,,	59 ,,
Achernar	65° 28½′ ,,	2 ,,	58 ,,	35 ,,
Alpheratz	15° 51½′ ,,	3 ,,	1 ,,	53 ,,

No Index error. Height of eye 20 ft. Chronometer error 7 min. 16 sec. slow.

For further examples see Chapter 19 'Days at Sea.'

CHAPTER 18

SPECIAL CASES

'Lighthouses in the sky.'

George W. Mixter.

Special Cases.

1. Latitude by meridian Altitude and Ex-meridian Altitude.
2. Latitude by Pole Star.
3. Position by Double Altitude and Noon Curve.

Sometimes a position line can be obtained more simply than by the general method described in the previous chapter.

1. Latitude by Meridian Altitude. Once a day, at noon, the sun offers us a ready means of finding our latitude.

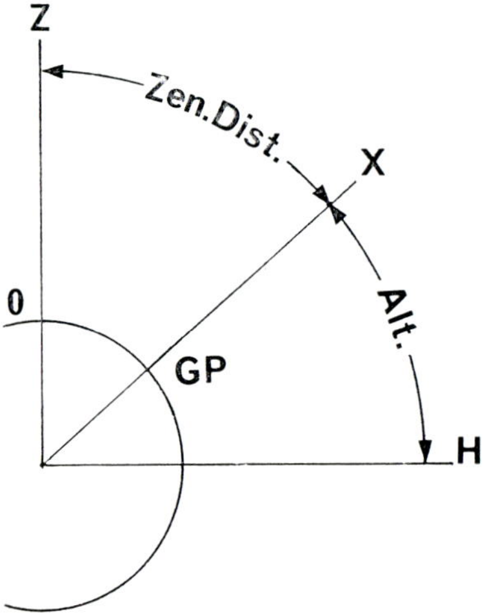

FIG. 18.1

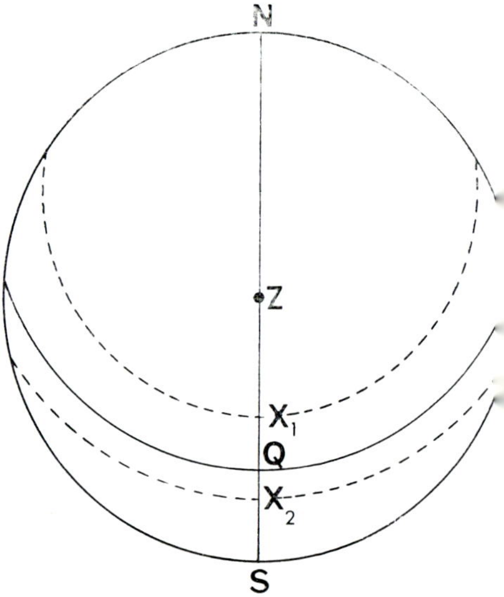

FIG. 18.2

The circle represents the horizon and the straight line *NZS* the observer's meridian rising from *N* on the horizon, through *Z* overhead down to *S* on the horizon. $ZS = 90°$, XS = altitude, XZ = zenith distance. *Q* is the point at which the celestial equator cuts the meridian, so ZQ = latitude of the observer.

X_1 is a body on the meridian north of the equator.

X_2 is a body on the equator south of the equator, in each case QX is the declination.

In working meridian-altitude sights always draw a figure (the straight line is sufficient) with Z at the centre to see whether D.R. Lat. should be added or subtracted to Dec. to obtain the Z.D.

As the sun rises in the sky and approaches the meridian the triangle *PZX* (Fig. 16.2) shrinks until, at noon, when the Hour Angle is 0° it becomes a straight line. The sun is then at its greatest altitude and bears exactly either north or south.

If we then find how far we are from the sun (the zenith distance) and how far the sun is from the equator (the declination), we can find our distance from the equator (our latitude) simply by addition or subtraction of the Zenith Distance and Declination. The method is shown in the worked example below. (Fig. 18.2).

EXAMPLE: On Sept. 19th, D.R. 40° N., 8° E., the sun's altitude (lower limb) when on the meridian was 51° 30·5′. Index Error $1\frac{1}{2}'$ on the arc. Height of eye 20 ft. Appoximate G.M.T. (necessary for finding the sun's declination) 11 hr. 5 min.

Obs. Alt.		51° 29′
cor.	+	10·8′
True Alt.		51° 39·8′
		90°
Zen. Dist.		38° 20·2′ (*ZX*)
Dec. N		1° 26·9′ (*QX*)
Lat.		39° 47′ N. (*ZQ*)

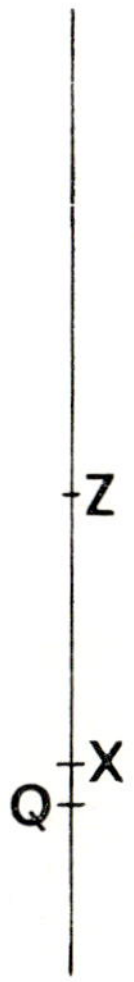

The figure shows us that, in this case, zenith distance and declination should be added to give the latitude.

Ex-meridian Sights ('ex' implies 'out of, but near'). In cloudy weather it sometimes happens that the sun is obscured when it is *on* the meridian but can be observed shortly before or after. A correction can be added to the altitude then taken to bring it up to what the meridian altitude should be.

METHOD: Note the G.M.T. (precisely this time) at the moment of observation and find the L.H.A. of the sun (*see* page 86). Enter Ex-meridian Table 1 (Burton, page 53) with the D.R. latitude and Declination, and extract the Factor F.

P

Enter Table 2 with the L.H.A. and factor and add the correction thus found to the observed altitude. The latitude is then found as in the meridian altitude problem, the latitude being at the time of observation (not at noon).

EXAMPLE: In the previous example, the sun was observed before the sun had passed the meridian, at 11 hr. 10 min. 10 sec. G.M.T. What was the latitude at that time?

G.M.T. 11h 10m 10s		Dec. N.	1° 26·9′	Table 1
G.H.A. 11h	346° 33·1′	Lat.	40° N.	Factor=2·41
inc.	2° 32·5′			
		Obs. alt.	51° 26·5′	
G.H.A.	349° 5·6′	cor. +	10·8′	
Long. E.	8°			
			51° 37·3′	
L.H.A.	357° 5·6′	Table 2·0	4·7′	
		·4	·9′	
		·01	·0′	
		True alt.	51° 42·9′	
			90°	
		Zen. dist.	38° 17′	
		Dec. N.	1° 27′	
		1110 G.M.T. Lat.	39° 44′ N.	

Other examples, using the intercept method, can be found in *Burton* page vii.

There is a limit to the time from noon within which this method may be used: the number of minutes from noon should not exceed the number of degrees in the zenith distance. In the example above the zenith distance is 38° and the H.A. should therefore not exceed 38 minutes (i.e. 9° 30′) and must lie between 350° 30′ and 9° 30′.

Meridian Altitudes of Venus and Jupiter. The meridian altitudes (and ex-meridian altitudes) of these planets can sometimes be observed in daylight, even when they cannot be seen by the unaided eye. If the sun then bears about east or west an excellent cross of position lines can be obtained.

In order to be able to see the planet it is necessary first to set the approximate altitude on the sextant. Then, if the observer looks towards the southern (or northern) horizon the reflected planet can be seen through the telescope. Contact is made; the time noted; and the problem worked in the same manner as when observing the sun.

EXAMPLE: Sept. 20th. D.R. 45° 15′ N., 12° 10′ W. Meridian passage of Venus is about 1525 G.M.T. What altitude should be set on the sextant? Index Error 2′ off the arc. Height of eye 10 ft.

D.R. Lat.	45° 15′ N.	(draw the figure now)
Dec. S.	15° 19½′	
Zen. dist.	60° 34½′	
	90°	
True. alt.	29° 25½′	
correction	+ 5′	(sign reversed)
	29° 30½′	
I.E.	− 2′	(sign reversed)
Sext. alt.	29° 28½′	

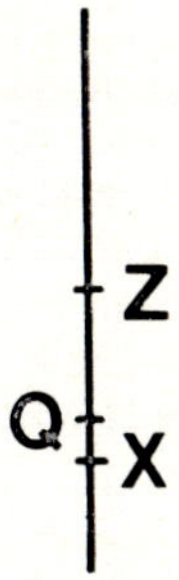

Observed altitude of Venus at 1523 G.M.T. was 29° 25′. What is the latitude?

G.H.A. 15 hr.	8° 37·2′	Obs. alt.		29° 25′
inc. 23m	5° 45′	cor.		− 5′
G.H.A. Venus	14° 22·2′			29° 20′
Long. W.	12° 10′	Table 2	1·0	1·1′
			0·5	·5′
L.H.A.	2° 22′		0·04	·0′
		True alt.		29° 21·6′
Lat. 45° N. / Dec. S. 15° 19′	Factor 1·54			90°
		Zen. dist.		60° 38½′
		Dec. S.		15° 19½′
		Lat.		45° 19′ N.

Time of the Sun's Meridianal Passage. It is clear that to take a meridian altitude sight we must know when the body will be on the meridian. In the case of the sun, the simplest way to do this is to add the easterly H.A. (in time) at the morning sight to the clock time at which it was taken. (To change Arc in Time the table in the *Nautical Almanac* can be used—it is the first of the yellow pages—or that on page 134, Burton.)

EXAMPLE: E.H.A. 50° 30′ at 0900 *Burton*

50° =	3h 20m
30′ +	2m
50° 30′	3h 22m
Clock	09 00
Noon	12 22

This is correct if the vessel is in the same longitude, but she will probably not be, so the time of noon is approximate only. Observations of the sun should thus be made a little before the predicted time of noon and continued at short intervals until the altitude ceases to rise. The sun is then on the meridian and, afterwards, the altitude will decrease.

If no morning sight of the sun has been taken, the time of the sun's meridianal passage to the nearest minute is given at the foot of the page of the *Nautical Almanac* for each day. This is strictly the mean time at which the sun crosses the meridian at Greenwich, but it is near enough for other places in other longitudes.

To find the Zone Time of transit at ship proceed as follows:

Mer. Pass. Sept. 19th	11h 54m	
D.R. Long. 8° E.	— 32m	
G.M.T.	11h 22m	
Zone −1	+1h	(Note that the zone sign is reversed.)
Approx. Mer. Pass. at ship Z.T.	1222.	

This will be correct if we were in Long. 8° E. at 1222. If the D.R. was for 0900 on 19th we must make a second approximation, allowing for the run between 0900 and 1222 and using the longitude at 1222:

Run: Course 105° Speed 15 knots = dist. 50·5 gives dep = 48·5 = d. long. (Lat. 40°) 1° 3′ E. Long. at 1222 9° 3′ E.

Mer. Pass. Sept. 19th	11h 54m
D.R. Long.	— 36m
G.M.T.	11h 18m
Zone −1	+1
2nd Approx. Mer. Pass.	12h 18m

This allowance for difference of longitude can conveniently be done on the chart, no second calculation then being necessary. The

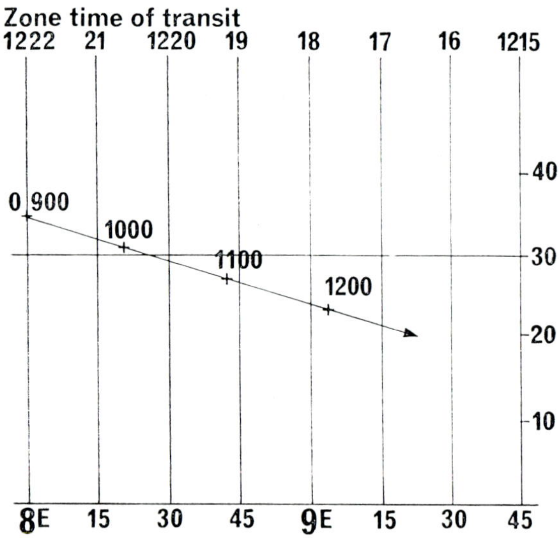

FIG. 18.3

sun moves westward through 1° of longitude in every 4 minutes (or 15′ per minute). Find the Z.T. of meridianal passage at some meridian near the ship's expected position at transit and write the Zone Times against this meridian and the meridians near it. (*See* Fig. 18.3). It can then be seen when the ship and the sun will be on the same meridian.

Time of Planets' Meridianal Passage. The time of meridianal passage is listed in the *Nautical Almanac* for the middle day on each page. Proceed as with the sun:

EXAMPLE: Find the Zone Time (Zone −3) at which Venus will be on the meridian on Jan. 2nd in Longitude 48° 30′ E.

Mer. Pass.	14h 24m
Long. E.	3h 14m
G.M.T.	11h 10m
Zone −3	+3h
Approx. Z.T.	14h 10m

This approximate time may be up to 3 min. in error according to the daily differences in times of meridional passage. This approximation should be sufficient for normal navigational practice, for the altitude changes slowly at the transit and it is more convenient to observe an ex-meridian sight of a planet than to wait for it to be exactly on the meridian.

The symbols for the navigational plants are:

♀ Venus ♂ Mars ♃ Jupiter ♄ Saturn

Exercises in Finding the Time of Meridianal Passage and Latitude from Meridian Altitude Observations and Ex-meridian Observations.

1. Find the Zone Time at which the sun crosses the meridian on 1958 Jan. 3rd becalmed in D.R. 17° 17′ N., 105° 15′ W. If, at that time, the sextant altitude of the sun's lower limb was 49° 47′, index error 1·5′ off the arc, height of eye 12 ft, what was the observer's latitude?

2. Sept. 22nd 0840 D.R. 35° 10′ S., 26° 16′ E. Course 340°, speed 7 knots. When will the sun cross the meridian (Zone Time)? Zone −2. What altitude should be set on the sextant for a meridian altitude observation? No index error, height of eye 15 ft. The sextant altitude at transit was 54° 41′. What was the latitude then?

3. At what Zone Time on Jan. 2nd was Jupiter on the meridian? D.R. 40° 0′ N., 14° 0′ E. What altitude should be set on the sextant at this time? Index error 2′ off the arc, height of eye 10 ft. If the meridian altitude observed by sextant was 27° 44′, what was the observed latitude?

4. At what Zone Time (+12) will Venus be on the meridian on

2nd Jan. in D.R. 36° 30′ S., 177° 15′ W? What will the altitude then be to set on the sextant for a meridian altitude observation? Index error 2′ off the arc. Height of eye 20 ft. The altitude observed on the sextant was 75° 57′ on the meridian. What was the latitude?

5. Sept. 20th 0820 D.R. 25° 9½′ N., 63° 36½′ W. Course 250°, speed 8 knots. E.H.A. at 0820 41° 45′. At what time will the sun be on the meridian?

At 16 hr. 13 min. 50 sec. G.M.T. the altitude of the sun was 65° 46′. Index error was −1′. Height of eye 15 ft. What was the latitude at 1114 Z.T.?

6. On Jan. 1st in D.R. 40° N., 5° 15′ E., Venus was observed to get a cross with the sun's position line. At what time was Venus on the meridian?

At 10 hr. 13 min. 30 sec. G.M.T. the altitude of Venus was 28° 10′. There was no index error and height of eye was 10 ft. What was the latitude at that time?

2. Latitude by Pole Star.

The star Polaris, the last in the tail of the Little Bear, is so close to the north celestial pole that whenever it can be seen it gives the traveller an immediate indication of how he is heading, while, if the altitude is observed, his latitude may simply be found.

Polaris is not a bright star but can be easily recognised. The two bright stars, Merak and Dubhe in the constellation Ursa Major, the Great Bear, point to it and, when the Great Bear cannot be seen, Cassopeia on the other side of the pole provides a guide. (*See* Fig. 18.4). The approximate correction to the altitude, to obtain the latitude, is also shown in the figure, but as stars change their position slowly the special Pole Star Tables for the current year should be normally always used.

An observer at the earth's north pole would have the celestial pole directly above him, in his zenith. Its altitude would be 90°. If he then travelled southwards the pole would decrease in altitude until he reached the equator, when it would be 0°. The altitude of the pole is thus equal to the observer's latitude. (*See* Fig. 18.2). In Fig. 18.2, take 90° (the distance *ZN* or *ZS*) on compasses or dividers and, placing one point on *Q* mark the position of the elevated pole *P*. Now measure the distance *PN* (the altitude of the pole): it will be equal to *QZ* (the latitude of the observer).

Polaris, the Pole Star, is situated within 1° of the north celestial pole and, like all stars, appears to revolve round it.

METHOD: The altitude is observed with the sextant and the usual corrections applied to obtain the true altitude. A further correction to allow for the star's height above or below the pole at the time of observation is then applied, and the result is the observer's latitude.

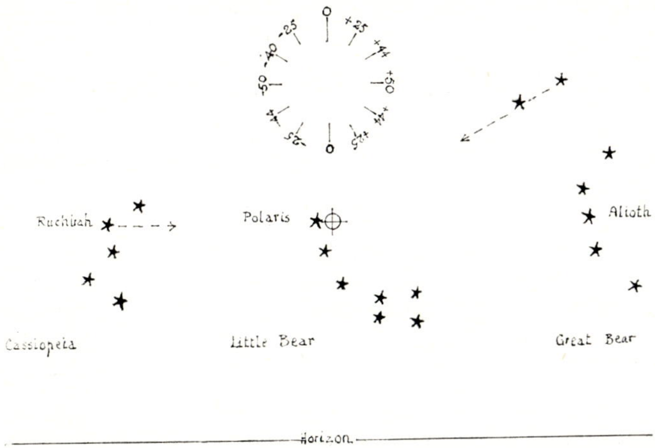

FIG. 18.4

1. *Identifying the Pole Star, Polaris.* Lines from the Pointers of the Great Bear, and from Ruchbah between its neighbouring stars, lead to Polaris.

2. *Estimating the approximate correction to the altitude to give the latitude.* Turn the page to suit the situation and read the correction at the top of the dial. Examples: The figure shows Polaris at the same height as the pole: correction zero. Turn the page until the Pointers are vertically above Polaris: correction about +45′. Turn further until Alioth is vertically above Polaris: correction about +50′.

The special corrections necessary are given in the *Nautical Almanac* (in a table near the end), part of which is reproduced in this book. Another, and even simpler table, is given in *Reed's Almanac*.

When the Pole Star bears exactly north the position line found by the observation is a parallel of latitude. When the position line has to be plotted with those obtained from other stars to get a fix, it should be plotted at right angles to the azimuth and the azimuth is given in the same table for this purpose. The azimuth is never more than $1\frac{1}{2}°$ from 360° in navigable latitudes.

Polaris is not a bright star and it is best to calculate the approximate altitude and set this on the sextant before trying to observe the star. Polaris should then be seen in the telescope close to the horizon when the sextant is pointed at the northern horizon, and the exact altitude can then be measured.

The problem thus falls into two parts:

(*a*) to find the altitude to set on the sextant,

(*b*) to find the latitude when the correct altitude has been measured.

In each case we need to know the approximate G.M.T. at the time of the observation and our D.R. position.

To Find the Latitude from a Sextant Altitude of Polaris.

Find the L.H.A. Aries corresponding to the G.M.T. of the observation. This is done in the same manner as finding the L.H.A. Sun. G.H.A. Aries is listed for every hour in the *Nautical Almanac*. (*See* page 171). To the G.H.A. for the hours, G.M.T., add the increment for the minutes G.M.T. Then apply the Longitude ('Longitude East, Greenwich is East; Longitude West, Greenwich is best') to obtain the L.H.A. Aries. In short:

L.H.A. Aries = G.H.A. Aries ± Longitude.

To the altitude obtained by sextant apply the Index Error and the correction for dip and refraction in the usual way, thus obtaining the true altitude.

Subtract 1° *from the True Altitude.* This is done to make the 'a' corrections next to be applied always additive.

Now add the three 'a' corrections. These give the distance that the pole star is above, or below, the pole. The table is entered with L.H.A. Aries. Choose the column showing the correct tens of degrees of L.H.A. Aries. Run the finger down to the unit degrees, interpolating by eye for the minutes. This is correction a_0.

Two further small corrections are made. Using the same column, find the correction for the latitude (a_1) and the month (a_2). These are all added to the True Altitude and the sum is the observer's latitude.

The *Azimuth* is given at the bottom of the same column.

WORKED EXAMPLES: The sextant angle of Polaris at 0h 2m 33s G.M.T. Jan. 2nd, 1973, was 55° 30′. Index error 3′ off the arc. Height of eye 12 ft. D.R. 54° 45′ N., 30° 0′ W.

G.H.A. Aries 00h	101° 29·3′	Sext. Alt.	55° 30′
inc. 4m	0° 38·4′	I.E.	+ 3′
G.H.A. 0004	102° 7·7′		55° 33′
Long. W.	30° 0′	cor.	− 4·2′
L.H.A. Aries	72° 7·7′	T. Alt.	55° 28·8′
			−1°
			54° 28·8′
		a_0	18·5′
		a_1	·7′
		a_2	·7′
Az.	358·8°	Obs. Lat.	54° 48·7′ N.

To Find the Altitude to Set Upon the Sextant.

This is the same procedure, but worked in reverse.

To the D.R. Latitude *subtract* $(a_0+a_1+a_2)$. *Add* 1°. *Add* star's correction and apply the index error in reverse. The result will be the expected altitude to be set on the sextant.

WORKED EXAMPLE: What altitude should be set on the sextant to observe Polaris at 2310½ Z.T. (Zone +2) on Jan. 2nd, in D.R. 45° 45′ N., Long. 40° 40′ W? Index error 2′ on the arc, height of eye 15 ft.

G.H.A. Aries 01h	116° 31·8′	D.R. Lat.	45° 45′
inc. 8m	2° 37·5′	$(a_0+a_1+a_2)$	− 24′
G.H.A. 0108	119° 9·4′		45° 21′
Long. W.	40° 40′		+1°
L.H.A. Aries	78° 29·4′		46° 21′
		cor 15 ft	4·7′
a_0	23·1		
a_1	·6		46° 25·7′
a_2	·7	I E	+ 2′
	24·4	Sext alt	46° 27·7′

Latitude by Pole Star.

1. (*a*) Find what altitude to set on a sextant (index error 3′ off the arc, height of eye 12 ft.), to find the latitude on Jan. 1st, 1973, G.M.T. 2105. D.R. Lat. 55° 15′ N., Long. 6° 36′ W.

(*b*) The sextant altitude of Polaris at 2107 G.M.T. was 56° 12′. What was the latitude and the azimuth of the pole star?

2. (*a*) Find what altitude to set on a sextant (index error 2′ on the arc, height of eye 10 ft.), on Jan. 2nd, 0212 G.M.T.

(*b*) Sextant altitude of Polaris at 0208 was 50° 1′. D.R. Lat. 49° 15′ N., 47° 58′ W. What was the observed latitude?

3. (*a*) Find the sextant setting at 1404 G.M.T. Jan. 2nd. D.R. Lat. 35° 17′ N., Long. 150° E. Index error 5′ off the arc height of eye 18 ft.

(*b*) Sextant read 35° 12′ at 1411 G.M.T. What was the latitude?

4. (*a*) Find the sextant altitude at 2210 Z.T. on Jan. 2nd (Zone +5). D.R. 32° 9′ N. 70° 37′ W.

(*b*) Find the latitude and azimuth of the pole star at 2206 Z.T. Index error nil, height of eye 15 ft. Sextant read 33° 0′.

5. (*a*) Find the sextant setting at 2112 Z.T. (Zone −12) on Jan. 3rd. D.R. 45° 0′ N., 179° 30′ E.

(*b*) Find the latitude at 2112 Z.T. Sextant read 46° 39′. Index error 3·5′ on the arc, height of eye 25 ft.

6. (*a*) Find the sextant setting at 2230 L.M.T. Jan. 2nd. D.R. 38° 15′ N., 38° 15′ W.

(*b*) Find the latitude at 2235 L.M T Sextant read 39° 2′. Height of eye 20 ft. Index error 0.

Position by Equal Altitudes.

When the conditions are suitable; that is, when the sun is high and the sky clear and good sights are obtainable, an observed position can be found within about half an hour from noon with very little calculation, no tables other than the *Nautical Almanac* being required. The G.M.T. must be accurately known.

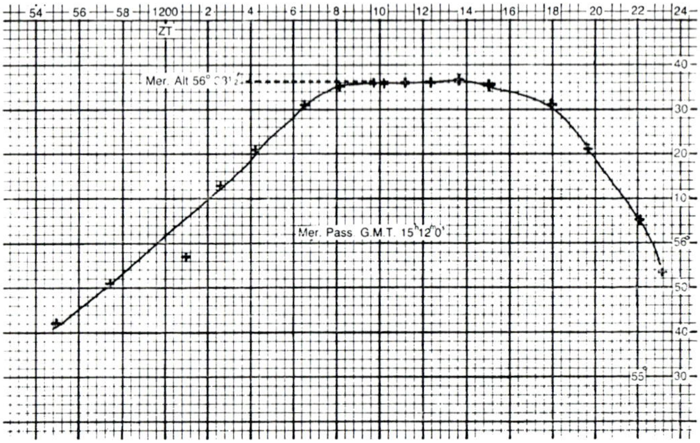

Fig. 18.4

METHOD: At about 15 minutes before noon an altitude of the sun is taken and the time by chronometer noted. (The meridian altitude of the sun is next observed for latitude as previously described.) After noon the sextant is set to the altitude of the first observation and the sun watched as its altitude slowly decreases. The exact time that it reaches the horizon is recorded.

The times of the first and second altitudes are now added and divided by 2 to get the time half way between them. Since the altitudes are equal it must have been at that time that the sun reached its greatest height as it crossed the meridian. At this moment the Local Hour Angle is 0°: we have the G.M.T. and can look up the Greenwich Hour Angle for this time. The G.H.A. at this moment equals the longitude (when the longitude is west) or 360°—longitude (when the longitude is east). (*See* Fig. 16.1.)

The position is strictly correct if the ship remained in the same latitude during the interval between sights and a yacht's movement can usually be disregarded. Should the ship have moved north or south between sights a correction to allow for this can be easily made. Find from the Traverse Table the difference of latitude made good between sights. Add this to the first altitude if you are steering *towards* the sun to find the altitude to set on the sextant for the second observation If you are steering *away* from the sun, subtract the difference of latitude

The reason why the sun must be high to get a good result is that the altitude changes only slowly when the sun is low, and this makes the exact time of contact difficult to determine. Since the longitude depends upon the G.H.A., the Greenwich Mean Time must be known precisely.

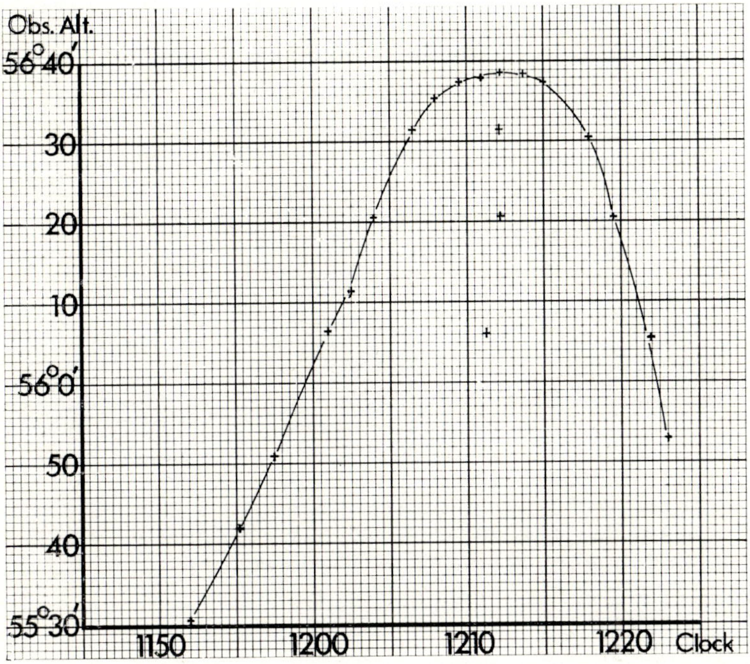

FIG. 18.5

3. The Noon Curve

A more accurate result will be obtained if a whole series of altitudes are taken and timed, both before and after the meridian passage, and the results are then plotted, time against altitude, on graph paper. A fair curve drawn through the different sights will give both the time and altitude of the sun's meridian passage. Start not less than 15 minutes before meridian passage. The intervals between sights need not be equal and should be shorter as the sun approaches and passes the meridian. The highest altitude—the meridian altitude—can be easily read off the graph and used to find the latitude. Bisect several horizontal lines to find the exact time of meridian passage and, from this, the longitude.

An added advantage of this method is that the precision of the result can be estimated.

EXAMPLES

(1) *Position by Equal Altitudes.*

On Jan. 2nd in D.R. 10° 12′ N., 47° 5′ W. Height of eye 8 ft, the following observations were made:

(a.m.) 56° 20′ at 15 hr. 3 min. 58 sec. G.M.T. Noon, observed altitude of the sun's lower limb 56° 38½′. (p.m.) 56° 20′ at 15 hr. 20 min. 4 sec. G.M.T. Required the position at noon.

Latitude		*Longitude*	
Obs. Mer. Alt.	56° 38½′	1st Obs.	15h 3m 58s
cor. for 8 ft.	+ 13′	2nd Obs.	15h 20m 4s
True Alt.	56° 51½′	2	30h 24m 2s
	90°		
		Noon G.M.T.	15h 12m 1s
Zen. Dist.	33° 8½′		
Dec. S.	22° 53½′	G.H.A.h 15 =	43° 57·5′
		inc. 12m 1s	3° 0·3′
G.M.T. 1512			
Latitude	10° 15′ N.	G.H.A. noon	46° 57·8′
		= Long.	46° 58′ W.

(2) *Position by Noon Curve.* (Fig. 18·5).

On Jan. 2nd, D.R. 10° 12′ N., 47° 5′ W., height of eye 8 ft., the following observations were made. Clocks were correct to Z.T. +3.

	h	m	s		Obs. Alt. °	′		
Clock	11	52	10	Obs. Alt.	55	30½		
	11	55	2		55	42		
	11	57	38		55	51		
	12	1	5		55	56½		
	12	2	41		56	11½		
	12	4	15		56	20½	Mer. Pass	1204
	12	6	32		56	31½	Long. W.	308
	12	8	10		56	35½		
	12	9	40		56	37½	G.M.T. Mer.	
	12	10	10		56	37½	Pass.	1512
	12	11	15		56	38	Zone	3
	12	12	26		56	38½		
	12	13	41		56	38½	Z.T. Mer. Pass.	1212
	12	15	2		56	37½		
	12	17	50		56	30½		
	12	19	35		56	20½		
	12	22	0		56	5½		
	12	23	0		55	52½		

The time of Meridian Passage by clock was first found in order to see at what time observations should be begun. The results were graphed and the G.M.T. Mer. Pass. estimated to be 15 hr. 12 min 0 sec. The problem was then completed as in Example 1.

Exercises:

Position by Equal Altitudes.

In each question find the ship's position at noon.

1. Jan. 2nd, D.R. 42° 25′ S., 178° 0′ E. Height of eye 10 ft. Index error 1′ off the arc. The following observations were made:

a.m. obs. Sextant alt.	70° 19′	at G.M.T.	00h 4m 35s
noon ,,	70° 27′		
p.m. ,,	70° 19′		00h 19m 25s

2. Sept. 23rd, D.R. 11° 15′ N., 65° 5′ W. Height of eye 12 ft. Zone +4, clock correct.

a.m. obs.	78° 38′	at Z.T.	12h 4m 50s
noon obs.	78° 46′		12h 12m
p.m. obs.	78° 38′		12h 19m 5s

3. Jan. 3rd, D.R. 30° 0′ S., 15° 5′ W. Height of eye 8 ft.

a.m. obs.	82° 12′	at G.M.T.	13h 0m 50s
noon obs.	82° 17½′		
p.m. obs.	82° 12′		13h 9m 12s

Position by Noon Curve.

1. Sept. 19th, D.R. 43° 15′ N., 18° 15′ W. Height of eye 8 ft. Index error 2′ off the arc. Zone +1, clock correct Z.T. The following observations were taken:

Z.T.	11h	41m	45s	Sextant	Altitude	47°	28½′
	11	43	51	,,	,,	47	35
	11	52	55	,,	,,	47	48
	11	56	53	,,	,,	47	51
	11	57	55	,,	,,	47	51½
	11	59	24	,,	,,	47	53½
	12	2	53	,,	,,	47	57½
	12	6	51	,,	,,	47	58
	12	10	11	,,	,,	47	57
	12	12	41	,,	,,	47	55
	12	16	33	,,	,,	47	54
	12	23	22	,,	,,	47	46
	12	26	27	,,	,,	47	40

2. Sept. 20th, D.R. 39° N., 65° W. Zone +4. Index error 0 Height of eye 12 ft. Clock correct Zone Time.

Z.T.	11h	58m	30s	Sextant	Altitude	51°	27½′
	12	0	25	,,	,,	51	32
	12	2	10	,,	,,	51	37½

Z.T.	12h	5m	0s	Sextant Altitude		51	42½
	12	6	55	,,	,,	51	45
	12	9	0	,,	,,	51	46
	12	10	30	,,	,,	51	46½
	12	11	30	,,	,,	51	47
	12	13	0	,,	,,	51	47
	12	14	20	,,	,,	51	47
	12	16	25	,,	,,	51	46½
	12	19	0	,,	,,	51	45
	12	21	20	,,	,,	51	42
	12	23	15	,,	,,	51	38
	12	25	45	,,	,,	51	30

3. Jan. 3rd, D.R. 31° 12′ S., 166° 5′ E. Zone −11. Index error 2′ on the arc. Height of eye 15 ft.

G.M.T.	00h	49m	10s	Sextant Altitude		81°	7′
	00	50	30	,,	,,	81	11½
	00	51	50	,,	,,	81	17
	00	53	0	,,	,,	81	20
	00	54	2	,,	,,	81	23
	00	54	58	,,	,,	81	24½
	00	56	30	,,	,,	81	27½
	00	58	0	,,	,,	81	29
	00	58	40	,,	,,	81	29½
	00	59	20	,,	,,	81	30
	01	0	2	,,	,,	81	31
	01	1	1	,,	,,	81	30½
	01	3	0	,,	,,	81	28½
	01	5	0	,,	,,	81	25½
	01	7	10	,,	,,	81	21½
	01	9	30	,,	,,	81	13½
	01	10	32	,,	,,	81	7½

CHAPTER 19

DAYS AT SEA

'Sailing along in great felicity.'
Captain Joshua Slocum.

Not all the sights which follow would probably be taken on the day in question, though any of them might be. It is suggested that all should be worked for practice and the results compared with the working shown in 'Answers.'

The lower limb of the sun was observed and in all cases the Index Error of the sextant was 2′ off the arc and the height of eye above sea level 8 ft.

I

Clear sky. 1973 Jan. 5th. 1200 Z.T. (Zone +1). Obs. 45° 15′ N., 12° 17′ W. Log set 0. Course 215° T. Speed about 6 knots.

Compass Error and Deviation by Amplitude.

Sunset at 1633 clock time, bearing 247° C. Log 27. Variation 11¾′ W. What is the compass error and deviation?

P.M. Star Observations.

What stars can be observed this evening? What will be their approximate altitudes and azimuths at star time? (Include the Pole Star.)

The following 4 bright stars and Polaris were observed. Chronometer error 2 min. 5 sec. slow of G.M.T. What is the position at 1700 Z.T.?

Vega	Sextant	Altitude	30°	12½′	Chronometer	5h	59m	17s.
Altair	,,	,,	23	32	,,	6	0	31
Aldebaran	,,	,,	28	47½	,,	6	0	56
Capella	,,	,,	39	10	,,	6	2	10
Polaris	,,	,,	45	33	,,	6	5	0

1700 Z.T. log 31. Course from 1200 215° T.

A.M. Star Observations

Jan. 6th. The following stars were observed. Required the observed position at 0700 and the Course, Distance and Speed made good since Jan. 5th 1700.

Deneb	Sextant	Altitude	24°	45′	Chronometer	7h	58m	58s.
Vega	,,	,,	40	54½	,,	7	59	45
Antares	,,	,,	13	17	,,	8	2	1
Arcturus	,,	,,	65	36½	,,	8	3	15
Regulus	,,	,,	28	55	,,	8	3	50

0700 Z.T. log 115. Chronometer 2 min. 6 sec. slow of G.M.T.

A.M. Sun Observation.

At 0900 Z.T. 6th Jan., log 127, the altitude of the sun by sextant was 11° 55′ at 10 hr. 1 min. 12 sec. by chronometer which was 2 min. 6 sec. slow of G.M.T. Required the resulting position line and bearing of the sun.

Latitude by Meridian Altitude of the Sun and Noon Position and Day's Run.

The meridian altitude of the sun was observed and found to be 23° 58′ by sextant. Log 145. Using the position line found at 0900 and allowing course 218° and log distance since 0900, find the ship's position at 1200 Z.T. 6th Jan. and the course, distance and average speed since 1200 Z.T. 5th Jan.

II

Sept. 23rd 1973. Weather partly cloudy. 1200 Z.T. (Z −2). D.R. 40° 53′ N., 18° 31′ E. Log set 0. Course 282° T., speed about 7 knots.

Reference to the *Nautical Almanac* showed that Venus was on the meridian at 1427, so it was hoped to obtain a cross with the sun about that time. Required the altitude of Venus to set on the sextant for a meridian altitude observation.

Observation of Planet and Sun.

At 1504 Z.T. log 22, sextant altitude Venus 32° 33′ at chronometer time 1 hr. 3 min. 2 sec. (error of chronometer 4 min. 36 sec. slow of G.M.T.) and sextant altitude of the sun 36° 38′ at chronometer time 1 hr. 5 min. 17 sec. Required the ship's position.

Observation of Moon and Sun.

In a break in the clouds at 0805 Z.T. (−1), 24th Sept., the sun's sextant altitude was 23° 45½′ at chronometer time 6 hr. 59 min. 36 sec. (error 4 min. 36 sec. slow). Log 148. Course since previous noon 282° T. Clocks had been put back 1 hr. to Zone 1 during the night. At 0858 the sextant altitude of the moon's upper limb was 52° 54½′, chronometer 7 hr. 58 min. 2 sec., log 154. Required the position at 0858

Noon Position by Equal and Meridian Altitudes.

It was hoped to get a position at noon and sights were obtained as follows:

Chronometer	10h	28m	10s	Sextant Altitude Sun		47°	24′
,,	10	43	55	,, ,, ,,		47	50½
,,	10	46	15	,, ,, ,,		47	54
,,	10	53	40	,, ,, ,,		47	55

Assessed as meridian altitude. Log 175.
Sun obscured.

Chronometer	11h	19m	12s	,, ,, ,,	47°	24′

Required the position at the sun's meridian passage.

ANSWERS

Paper I. Preliminary Navigation. Pages 14-15

2. The International Nautical Mile = 1852 m. or 6,076 ft. A cable is one tenth of a nautical mile and assumed to be 200 yds.
4. (*a*) 151° T. (*b*) 154½° T. (*c*) 200° T. (*d*) 314° T. (*e*) 097½° T.
5. (*a*) 224° C. (*b*) 335° C. (*c*) 110° C. (*d*) 349° C. (*e*) 351° C.
6. (*a*) 056° C. (*b*) 282° C. (*c*) 114° C. (*d*) 355½° C.

True Course	Varia-tion	Mag. Co.	Devia-tion	Comp. Co.	Comp. Error s	Bearing of Object	
						Comp.	True s
°	°	°	°	°	°	°	°
		030		036	5 E.		158
292	22 W.	314				169	
227		242	5 E.				065
		252		248	23 E.		329
	6 E.		3 W.		3 E.	177	
	10 W.			222	15 W.		007
044		032	2 E.		14 E.		
080	09 W.			194	13 W.		
	12 E.	345		336			189

For further practice this table can be used as a question and the answer read on page 16.

8. (*a*) The latitude scale. (*b*) As curves 'turning their backs' to the pole. (*c*) For short distances, no. A rhumbline. (*d*) Because it must be expanded in the same proportion as the meridians converge.
9. (*a*) 220° T. (*b*) 301° T.

Exercises in Chartwork. Page 43

1. (*a*) 091½° 9·8 K. (*b*) 090° T.
2. (*a*) 101½° 10·2 K. (*b*) 080° T.
3. 004° T., 015° C.
4. 085° C. 7·5 K.
5. Ship bears 158° T., 4·4 M. from B.
6. B bears 261° T., 2·1 M. from the ship. Compass Error 9° W., Deviation 5° W.
7. Lat. 50° 37½′ N. Long. 0° 33′ E.
8. Lat. 49° 37½′ N. Long. 0° 5¼′ W.
9. Lat. 50° 38½′ N. Long. 0° 8¼′ E.
10. Course 257° T. distance 49 M.
11. Position at 1300. Lat. 50° 18′ N., Long. 2° 35′ W. Co. 080° T., Distance 113 M.
12. Speed 8¼ Kn. Course 084° T.
13. Distance off 1·75 M.
14. 0° 36′. On large scale charts and in the Admiralty List of Lights, Vol. I.
15. Distance 4·1 M. Lat. 50° 39·7′ N., Long. 0° 14½′ E.
16. Fix 50° 32·2′ N., 5° 7½′ W. Distance when abeam 2 M.
17. 0835 fix 50° 20·2′ N., 5° 26½′ W. Distance off buoy 4½ M.
18. Distance off at 0345: 5¾ M.
19. Distance off when abeam: 4 M.
20. Distance off 0215: 10½ M. Distance off abeam: 8¼ M.

21. Distance off 2215: 10½ M. Distance off abeam: 7¼ M. at 2330.
22. E.P. 50° 28½′ N., 5° 55′ W. Co. made good 082° T., speed 11¾ Kn.
23. 1500 E.P. 50° 16·8′ N., 5° 38′ W.
24. 1500 E.P. 49° 50′ N., 4° 52′ W. Co. made good 084° T., distance 11·8 M., speed 3·9 Kn.
25. 1200 E.P. 50° 2¼′ N., 5° 48′ W. Longships bearing 050°, distance 2·8 M.

Chart Paper A. Page 46

1. A. 212° T., 6 M. from Beachy Hd. Lt. Ho.
   B. 294°, 14 M. from C. de la Heve Lt. Ho.
   C. 286°, 7 M. rom Casquets Lt. Ho.
   D. 229°, 5½ M. from Wolf Rk. Lt. Ho.
2. E. Lat. 51° 5′ N., Long. 4° 54′ W.
   F. Lat. 49° 48′ N., Long. 5° 21′ W.
   G. Lat. 49° 29′ N., Long. 3° 2′ W.
   H. Lat. 50° 29′ N., Long. 1° 9′ W.
3. Course 333° T., distance 81½ M.
4. 008° T., 12½ M. and 032° T., 70 M.
5. Course 243° T., 117 M.
6. Bearings 210° T., 327° relative or 33° on the port bow.

Chart Paper B. Page 46

1. Fix: 50° 29·6′ N., 5° 7·5′ W.
2. Fix: 49° 58·2′ N., 5° 3·8′ W.
3. Fix: 50° 3·8′ N., 5° 48·2′ W.
4. Distance off: 5 M.
5. Distance off 2215: 4·2 M. Distance off abeam: 2·4 M.

Chart Paper C. Page 47

1. Fix 49° 35½′ N., 3° 5′ W.
2. Fix: 48° 45′ N., 6° 23′ W.
3. Distance from Wolf Rock about 4½ M.
4. I. de Bas Lt. 185° T., 7 M.
5. Fix: 51° 5½′ N., 4° 45′ W. Set 076°, drift 4½ M.
6. 6 fathoms – – – – – – – – – – – – – – –
   10 fathoms · – · – · – · – · – · – ·
   100 fathoms · – – · – – · – – · – – · – – · – –

Chart Paper D. Page 47

1. 50° 9′ N., 4° 49¾′ W.
2. 49° 42′ W., 5° 50·8′ W.
3. 50°.
4. 083°, that is, parallel to the coast. A/c to North magnetic at 0700.
5. 49° 55′ N., 6° 29.2′ W. Distance 3½ M.
6. Flashing light: the duration of light is shorter than the duration of darkness. Occulting light: the duration of light is longer than the duration of darkness.

Chart Paper E. Page 48

1. 106½° T., distance 23½ M. Course 052° T., 055° C.
2. Fix: 48° 52′ N., 5° 12′ W.
3. Fix: 49° 30′ N., 3° 54′ W. Yes.
4. 1300 Fix: 50° 22′ N., 1° 5½′ W. Set 085°, drift 12 M.
5. Course to steer 025° T., 028° C. Distance to go 20 M. E.T.A. 1420.
6. (*a*) The note sounded falls sharply at the end in a 'grunt'.
   (*b*) Red.

Chart Paper F. Page 48

1. Distance off 3·2 M. 0730 Fix: 49° 54·5′ N., 5° 12′ W. Compass error 10° W., deviation 2° W.
2. Courses 285½° C. and 296° C. Sight Runnel Stone buoy at 1020.
3. Fix: 50° 1·8′ N., 5° 29·4′ W. Deviation 2° E.
4. Alter course at 1105 to 183° C.
5. (i) Distance off Seven Stones Lt. V. 6 M.
   (ii) 50° 6′ N., 6° 14′ W.
   (iii) Distance off when abeam 4 M., at 1130. (Speed made good 4·3 Kn.)
6. The alternating light flashes first white and then red.
   The light 'Fl.W.R.' gives white and red flashes *over different sectors.*

Chart Paper G. Page 49

1. Chart bearing of Land's End radio station 353½° T. This is the bearing on which you can safely approach Falmouth. Steer your present course until your true track cuts the bearing 353½° drawn 3 M. off Manacles buoy, then alter course to 353½°. E.T.A. a/c position 0008.
2. Lat. 50° 55′ N., Long. 5° 57½′ W. Possibly, Night Effect.
3. Time to a/c 2036. Course to steer 121° C. E.T.A. Lizard 2224.
4. Courses and distances: 095° C., 41 M.: 089° C., 112 M.; 123° C., 47 M. E.T.A. a/c position 0915, at destination 1222.
5. At about 2006, distance 7 M.
6. (*a*) Position approximate (*b*) Existence doubtful (*c*) Unwatched.

Cross-Channel Tides. Page 56

1. Track 155°. Total tide assessed as 250°, 7 M. Wake co. 150°, course to steer 162° C. Distance 78 M.
2. Track 332°. Total tide assessed as 080°, 8 M. Wake co. 325°, course to steer 332° C. Distance 65 M.
3. Track 350°. Total tide assessed as 080°, 10½ M. Wake co. 334°, course to steer 334° C. Distance 57 M.

Deviation Card. Page 82

Ship's Head by Compass	Deviation	Ship's Head Magnetic
000°	3° E.	003°
030°	5½° E.	035½°
060°	7° E.	067°
090°	5½° E.	095½°
120°	2½° E.	122½°
150°	0°	150°
180°	2½° W.	177½°
210°	5° W.	205°
240°	6½° W.	233½°
270°	5° W.	265°
300°	2½° W.	297½°
330°	0°	330°
360°	3° E.	003°

(Magnetic Bearing 123°)

Compass Error by Amplitude. Page 85

	Declination	*Amplitude*	*Bearing*	*Error*	*Deviation*
1.	S. 22·9°	W. 33·8° S.	236·2° T.	8·8° W.	0·2° E.
2.	S. 22·½°	E. 36·5° S.	126·5° T.	18° W.	0°
3.	S. 23°	W. 37·4° S.	232·6° T.	8·9° W.	0·9° W.

4. S. 0·3°. When the declination is 0° the sun rises and sets bearing 090° and 270°. Error 2·4° W., Deviation 4·4° W.

5.	N. 1·2°	W. 1·2° N.	271·2° T.	6·2° E.	3·8° W.

6. N. 0·7°. When the Latitude is 0° the amplitude is equal to the declination. Bearing 089·3° T. Error 3·3° E. Deviation 5·7° W.

Local Hour Angle and Declination. Page 88

1. L.H.A. 257° 46·6′. Dec. N. 1° 30·7′.
2. L.H.A. 314° 6·6′. Dec. S. 22° 52·9′.
3. L.H.A. 8° 9·9′. Dec. S. 22° 31·7′.
4. L.H.A. 152° 55′. Dec. N. 0° 23·6′.

Compass Error by Azimuth. Page 90

	L.H.A.	*True Bearing*	*Error*	*Deviation*
1.	318° 25·8′	142°	36° W.	1° W.
2.	27° 39′	212°	23° W.	4½° E.
3.	325°	147½°	13½° W.	2½° W.
4.	320° 31′	125¾°	13¼° W.	3¼° W.
5.	304° 38′	112½°	30° W.	3° W.
6.	226° 7′	113°	2° W.	2° E.

Tides at Secondary Ports. Page 103

1. Change in difference ·7. Ht. of H.W. St. Nazaire 14·6 ft.
2. Change in difference ·8. Ht. of L.W. Le Palais 3·7 ft.
3. Change in difference 4·1. Ht. of L.W. Le Pellerin 5·0 ft.
4. Change in difference 7 minutes. Time of H.W. La Trinité 0652.
5. Change in difference 27 minutes. Time of H.W. Les Sables d'Olounе 1522.
6. Change in difference 13 minutes. Time of L.W. Le Pouliguen 1434.

Heights of Tide at Times other than High Water. Page 103

Standard Ports

1. Factor ·8. Height at 1630 20·7 ft.
2. Factor ·3. Correction to Sounding −12·3 ft.
3. L.W. 5·5 ft. Water under keel at 2000, 33·9 ft., at L.W. 23·5 ft.
4. Factor ·94. 29 ft. on shoal.
5. Factor ·52. Pass at 0422.
6. Factor ·24. Pass under bridge 1603.

Secondary Ports Page 104

		Admiralty Time Tables	*Reed's Tables*
1.	Factor ·3. Height at 0943	7·0 ft.	6·1 ft.
2.	Factor ·08. Height at 1615.	6·9 ft.	7·1 ft.
3.	Factor ·65. 10 ft. at	1246	1246
4.	Depth beneath keel at 1045: 14·8 ft., at L.W. 4 ft.		
5.	Depth beneath keel at L.W.: 4 ft.		
6.	Factor ·25. Should float at:	0916	0900

Chronometer Correction. Page 109

1. Daily rate 5·4 sec. losing. Error 1 min. 46 sec. slow.
2. Daily rate 1·62 sec losing. G.M.T. 12 hr. 0 min. 2 sec.
3. Daily rate 4·11 sec. gaining. G.M.T. 17 hr. 59 min. 46 sec.
4. Daily rate 3·055 sec. gaining. 10 hr. 2 min. 46 sec.

Times of Sunrise, Moonrise, etc. Page 112

1. L.M.T. 0605. 0629.
2. Z.T. 0420. 2013.
3. L.M.T. 1803, G.M.T. 2133, S.T. 1803, Summer Time 1903.
4. S.T. 0958, 1733, 1000, 1724.
5. S.T. 1806, 1828, 0124.
6. Z.T. 0640, 1824, 0426, 1621.

Exercises in the Use of the Traverse Table. Page 124

Finding New Positions

1. Lat. 0° 13·9′ N., Long. 12° 28·3′ W.
2. Lat. 50° 27·4′ N., Long. 0° 59′ W.
3. Lat. 39° 29′ S., 146° 42′ E.

Finding Course and Distance

4. Course 084° T., distance 202 M.
5. Course 221° T., distance 45 M.
6. Course 306° T., distance 152 M.

Day's Work. Page 126

1. Course and distance made good: 029° 17·5 M.
   D.R. at 0000 45° 27·3′ N., 55° 51′ W.
2. (*a*) Course and distance made good: 219° 78·5 M.
   (*b*) E.P. 0000 39° 34′ N., 22° 19·5′ W.
3. E.P. 36° 23·8′ N., 15° 59.·3′ W.
4. (*a*) Course and distance made good: 085° 62 M.
   (*b*) E.P. 0000 41° 25·2′ N., 10° 39·3′ E.
5. (*a*) Course and distance made good: 099°, 125 M.
   (*b*) E.P. 1200 35° 24·3′ S., 22° 39′ E.
6. Course and distance made good: S. 68° W. (248° T.), 56 M
   1200 D.R. 36° 47′ N., 16° 57′ W.

Position Lines and Positions. Page 193

1. Most probable position 49° 50½′ N., 18° 27′ W.
2. Most probable position 45° 5·5′ S., 152° 8′ E.
3. Obs. position 23° 10′ N., 66° 0½′ E.
4. Obs. position 49° 1·8′ N., 20° 6½′ W.
5. Obs. position 16° 3·8′ S., 44° 50′ W.
6. Obs. position 50° 3½′ N., 9° 9′ W.

Sun Sights. Page 196

1. G.M.T. 2d 10h 12m 10s. True Altitude 12° 38′. Intercept (A.P. 3270) 6′ towards (versine formula) 8′ away. Bearing 143° T. Probable Position Lad. 45° 41½′ N., Long. 11° 25′ W.
2. G.M.T. 23d 5h 8m 44s. True Altitude 26° 17′. Intercept (A.P. 3270) 3′ away (versine formula) 11′ towards. Bearing 296° T. Probable Position Lat. 41° 12′ S., Long. 154° 52½′ E.
3. G.M.T. 24d 18h 12m 5s. True Altitude 27° 40½′. Intercept (A.P. 3270) 5½′ towards (versine formula) 16½′ away. Bearing 227° T. Probable Position Lat. 51° 51½′ N., Long. 54° 59′ W.
4. G.M.T. 6d 10h 4m 18s. True Altitude 63° 25′. Intercept (A.P. 3270) 4′ away (versine formula) 6′ away. Bearing 053° T. Probable Position Lat. 41° 10½′ S., Long. 7° 7½′ E.
5. G.M.T. 3d 22h 7m 28s. True Altitude 10° 2′. Intercept (A.P. 3270) 15′ towards (versine formula) 27′ towards. Bearing 158° T. Probable Position Lat. 54° 51′ N., Long. 173° 12½′ W.
6. G.M.T. 21d 09h 2m 18s. True Altitude 34° 27½′. Intercept (A.P. 3270) 14½′ away (versine formula) 15′ away. Bearing 310° T. Probable Position Lat. 42° 29′ S., Long. 81° 35′ E.

Planet Sights. Page 196

1. G.M.T. 3d 09h 12m 30s. True Altitude 28° 0′. Intercept (A.P. 3270) 17½′ towards (versince formula) 1′ towards. Bearing 089° T. Probable Position Lat. 54° 54′ S., Long. 54° 52½′ W.
2. G.M.T. 1d 08h 12m 4s. True Altitude 10° 25′. Intercept (A.P. 3270) 13′ towards (versine formula) 3′ away. Bearing 111° T. Probable Position Lat. 41° 15′ S., Long. 30° 15′ W.
3. G.M.T. 3d 13h 11m 23s. True Altitude 38° 45′. Intercept (A.P. 3270) 5′ towards (versine formula) 3′ towards. Bearing 098° T. Probable Position Lat. 44° 4½′ N., Long. 77° 19′ E.

Moon Sights Page 197

1. G.M.T. 24d 20h 06m 05s. True Altitude 38° 16′. Intercept (A.P. 3270) 10′ away (versine formula) 3½′ towards. Bearing 221° T. Probable Position Lat. 49° 2½′ N., Long. 114° 29′ W.
2. G.M.T. 19d 15h 09m 10s. True Altitude 10° 51′. Intercept (A.P. 3270) 3′ towards (versine formula) 4½′ towards. Bearing 294° T. Probable Position Lat. 52° 5′ N., Long. 38° 54½′ W.
3. G.M.T. 5d 04h 07m 53s. True Altitude 9° 55′. Intercept (A.P. 3270) 14′ towards (versine formula) 4′ away. Bearing 225° T. Probable Position Lat. 45° 52′ N., Long. 173° 7′ E.

Star Sights. Page 197

1. G.M.T. 2d 18h 10m 2s. True Altitude 17° 40′. Intercept (A.P. 3270) 8′ away (versine formula) 1′ away. Bearing 198° T. Probable Position Lat. 40° 8½′ N., Long. 10° 39½′ W.
2. G.M.T. 3d 22h 07m 16s. True Altitude 18° 1′. Intercept (A.P. 3270) 16′ away (versine formula) 3′ towards. Bearing 031° T. Probable Position Lat. 42° 7′ S., Long. 76° 12′ W.
3. G.M.T. 1d 00h 00m 1s. True Altitude 43° 3′. Intercept (A.P. 3270) 10′ away (versine formula) 6′ towards. Bearing 303° T. Probable Position Lat. 51° 5½′ S., Long. 40° 43′ E.
4. G.M.T. 2d 5h 11m 56s. True Altitude 30° 2′. Intercept (A.P. 3270) 20′ away (versine formula) 3′ towards. Bearing 246° T. Probable Position Lat. 49° 24′ N., Long. 170° 59½′ E.
5. G.M.T. 2d 19h 02m 6s. True Altitude 16° 55′. Intercept (A.P. 3270) 8′ towards (versine formula) 5′ away. Bearing 294° T. Probable Position Lat. 50° 58′ N., Long. 174° 7′ W.
6. G.M.T. 3d 08h 03m 58s. True Altitude 24° 5′. Intercept (A.P. 3270) 2′ towards (versine formula) 1′ towards. Bearing 350° T. Probable Position Lat. 45° 56½′ S., Long. 0° 5′ W.

Fix from Two Stars. Page 198

Intercepts, Azimuths and Positions

1.	G.M.T. Aldebaran	21h 9m 20s	Hamal	21h 11m 17s
	L.H.A.	295°		332°
	Cor. Tab. Alt.	28° 47·5′		56° 48½′
	Bearing	098°		128°
	Intercept	4½′ away (from D.R. 0′)		6½′ towards (from D.R. 2′ away)
	Observed position:	48° 20½′ N., 55° 9½′ W.		

Fix by Planet and Star. Page 198

2. G.M.T. Mars	10h 2m 42s	Regulus	10h 3m 57s
L.H.A.	327°		56°
Cor. Tab. Alt.	16° 26′		31° 37′
Bearing	148°		318°
Intercept	15′ towards (from D.R. 7′ away)		14′ towards (from D.R. 4′ away)
Observed position:	46° 50½′ N., 45° 26′ W.		

Fix by Planet and Sun. Page 198

3. G.M.T. Mars	22h 9m 15s	Sun	22h 10m 11s
L.H.A.	374° i.e. 14°		331°
Cor. Tab. Alt.	61° 54½′		57° 5′
Bearing	332°		054°
Intercept	18½′ away (from D.R. 4′ away)		15′ towards (from D.R. 6′ towards
Observed position:	46° 7′ S., 179° 55′ W.		

Fix by Sun and Moon. Page 198

4. G.M.T. Sun	22d 23h 5m 35s	Moon	23h 7m 50s
L.H.A.	323°		369°
Cor. Tab. Alt.	34° 19′		30° 36′
Bearing	047°		350°
Intercept	5′ towards (from D.R. 4′ away)		5′ away (from D.R. 1′ towards)
Observed position:	45° 4½′ S., 155° 1′ E.		

Two Observations of the Sun. Page 198

5. G.M.T. Sun 1st Obs.	08h 0m 20s	2nd Obs.	11h 11m 56s
L.H.A.	296°		342°
Cor. Tab. Alt.	16° 22′		38° 7′
Bearing	110°		157°
Intercept	4′ away (from D.R. 2′ towards)		31′ away (from D.R. 2′ away)

Observed position at 2nd observation: 49° 38′ N., 5° 13′ W.

6. G.M.T. Sun 1st Obs.	10h 11m 55s	2nd Obs.	14h 11m 20s
L.H.A.	326°		27°
Cor. Tab. Alt.	12° 20½′		14° 35′
Bearing	148°		206°
Intercept	13½′ towards (from D.R. 3′ away)		14′ away (from D.R. 2′ towards)

Observed position at 2nd Observation: 49° 20½′ N., 4° 38′ W.

Star Sights. A. 3270. Page 199

1. G.M.T. 3d 1800. L.H.A. Aries 0° 56·8′.

Predicted alt.	*Capella	36° 14′	Bearing 057°	Intercept	20′ towards
,, ,,	Aldebaran	27° 27′	,, 092°		
,, ,,	*Diphda	29° 14′	,, 170°	,,	5′ away
,, ,,	Fomalhaut	16° 34′	,, 195°		
,, ,,	Altair	25° 24′	,, 258°	,,	
,, ,,	*Vega	29° 59′	,, 297°	,,	11′ away
,, ,,	Kochab	29° 34′	,, 348°		

Fix: 1700 42° 8′ N., 12° 20′ W.

2. G.M.T. 2d 1004. L.H.A. Aries 213° 20′.

Predicted alt.		*Deneb	29° 45′	Bearing	054°	Intercept	5½′ towards
,,	,,	Vega	43° 19′	,,	078°		
,,	,,	Rasalhague	34° 5′	,,	115°		
,,	,,	*Arcturus	58° 18′	,,	179°	,,	7′ away
,,	,,	Regulus	27° 19′	,,	255°		
,,	,,	*Pollux	17° 19′	,,	293°	,,	7½′ away
,,	,,	Capella	14° 41′	,,	329°		

Fix: 0806 52° 6½′ N., 39° 22′ W.

3. G.M.T. 1d 1701. L.H.A. Aries 359° 44′.

Predicted alt.		*Capella	35° 04′	Bearing	056°	Intercept	34′ away
,,	,,	Aldebaran	26° 44′	,,	091°		6½′ towards
,,	,,	*Diphda	30° 04′	,,	168°	,,	
,,	,,	Fomalhaut	17° 43′	,,	194°	,,	32½′ towards
,,	,,	Altair	26° 20′	,,	258°	,,	14′ towards
,,	,,	*Vega	30° 12′	,,	297°	,,	22½′ away
,,	,,	Kochab	28° 45′	,,	348°		

Fix: 1701 40° 22′ N., 3° 23′ E.

4. G.M.T. 3d 0505. L.H.A. Aries 203° 4′.

Predicted alt.		*Kochab	49° 35′	Bearing	008°	Intercept	17½′ towards
,,	,,	Vega	30° 53′	,,	062°		
,,	,,	Rasalhague	31° 17′	,,	097°		
,,	,,	*Antares	15° 52′	,,	140°	,,	35′ away
,,	,,	Spica	43° 56′	,,	183°		
,,	,,	*Regulus	38° 26′	,,	257°	,,	20′ towards
,,	,,	Dubhe	54° 20′	,,	331°		

Fix: 0705 35° 20′ N., 24° 2′ E.

5. G.M.T. 1d 2312. L.H.A. Aries 13° 4′.

Predicted alt.		Capella	26° 50′	Bearing	045°	Intercept	27½′ away
,,	,,	*Aldebaran	36° 26′	,,	080°		
,,	,,	Acamar	28° 19′	,,	153°	,,	
,,	,,	Achernar	18° 02′	,,	174°	,,	24′ towards
,,	,,	*Fomalhaut	38° 07′	,,	212°	,,	12½′ towards
,,	,,	Altair	15° 59′	,,	275°	,,	
,,	,,	*Deneb	28° 55′	,,	314°	,,	2′ away

Fix: 1812 13° 35½′ N., 76° 30′ W.

6. G.M.T. 2d 1602. L.H.A. Aries 210° 36′.

Predicted alt.		Deneb	26° 22′	Bearing	051°	Intercept	30½′ towards
,,	,,	*Vega	40° 42′	,,	074°	,,	5′ away
,,	,,	Rasalhague	33° 25′	,,	110°	,,	
,,	,,	*Arcturus	61° 10′	,,	173°	,,	18½′ away
,,	,,	Denebola	46° 59′	,,	231°		
,,	,,	Regulus	30° 00′	,,	254°	,,	32′ away
,,	,,	*Dubhe	61° 43′	,,	316°		

Fix: 0802 48° 19′ N., 131° 56′ W.

7. G.M.T. 2d 0900. L.H.A. Aries 47° 7′.

Predicted alt.		Hamal	25° 45′	Bearing	348°	Intercept	24½′ away
,,	,,	*Aldebaran	28° 42′	,,	028°	,,	7½′ away
,,	,,	Rigel	45° 17′	,,	054°	,,	26½′ towards
,,	,,	Sirius	34° 59′	,,	083°	,,	18½′ towards
,,	,,	*Canopus	52° 20′	,,	128°	,,	40′ towards
,,	,,	Peacock	29° 06′	,,	218°		
,,	,,	*Fomalhaut	41° 28′	,,	263°	,,	43′ away

Fix: 1912 40° 21′ S., 167° 8′ E.

8. G.M.T. 2d 1810. L.H.A. Aries 185° 44′.

Predicted alt.	Alioth	32° 36′	Bearing	005°		
,, ,,	*Arcturus	56° 07′	,,	052°	Intercept	2′ away
,, ,,	Antares	26° 16′	,,	119°	,,	
,, ,,	*Acrux	28° 02′	,,	180°	,,	26′ towards
,, ,,	Suhail	29° 08′	,,	219°		
,, ,,	Alphard	45° 09′	,,	259°		
,, ,,	*Regulus	53° 34′	,,	292°	,,	13½′ away

Fix: 0611 0° 26′ S., 171° 4′ E.

9. G.M.T. 1d 0704. L.H.A. Aries 138° 20·6′.

Predicted alt.	Regulus	20° 57′	Bearing	014°	Intercept	6′ away
,, ,,	*Spica	24° 05′	,,	073°	,,	13′ towards
,, ,,	Rigil Kent	49° 48′	,,	131°		
,, ,,	*Peacock	23° 14′	,,	173°	,,	5′ towards
,, ,,	Achernar	35° 12′	,,	217°		
,, ,,	Canopus	65° 36′	,,	260°		
,, ,,	*Sirius	41° 45′	,,	309°	,,	17′ away

Fix: 0304 56° 6′ S., 68° 20′ W.

10. G.M.T. 2d 1503. L.H.A. Aries 40° 56′.

Predicted alt.	Hamal	30° 57′	Bearing	350°		
,, ,,	Aldebaran	32° 15′	,,	032°	Intercept	33½′ away
,, ,,	*Sirius	33° 50′	,,	087°	,,	8′ away
,, ,,	Canopus	47° 54′	,,	132°		
,, ,,	*Achernar	64° 57′	,,	202°	,,	26½′ towards
,, ,,	Fomalhaut	42° 17′	,,	260°		
,, ,,	*Alpheratz	16° 07′	,,	325°	,,	5½′ towards

Fix: 2007 35° 18′ S., 71° 54′ E.

Times of Meridian Passage and Latitudes by Meridian Altitudes and Ex-meridian Altitudes. Page 207

1. Zone Time of transit 1206, Lat. 17° 14′ N.
2. Z.T. of transit 1208½ Alt. to set on sextant 54° 43½′. Obs. Lat. 34° 49·6′ S.
3. Z.T. of transit 1235 Alt. to set on sextant 27° 39′. Obs. Lat. 39° 55·8′ N.
4. Z.T. of transit 1011 Alt. to set on sextant 75° 42′. Obs. Lat. 36° 15′ S.
5. Z.T. of transit 1109 Factor 4·37 1114 Obs. Lat. 25° 0½′ N.
6. Z.T. of transit 1101 Factor 1·58 1113½. Obs. Lat. 39° 53·8′ N.

Latitude by Pole Star. Page 211

1. (*a*) L.H.A. ♈ 51° 1′. Set sextant 56° 10′.
   (*b*) L.H.A. ♈ 51° 31′. Lat. 55° 23′ N., Azimuth 359·3.
2. (*a*) L.H.A. ♈ 86° 38′ Set sextant 49° 50′.
   (*b*) L.H.A. ♈ 85° 37′. Lat. 50° 25′ N., Azimuth 359.
3. (*a*) L.H.A. ♈ 103° 4′. Set sextant 35° 34′.
   (*b*) L.H.A. ♈ 102° 49′ Lat. 34° 55′ N. Azimuth 359.
4. (*a*) L.H.A. ♈ 79° 29′. Set sextant 32° 48′.
   (*b*) L.H.A. ♈ 78° 29′ Lat. 32° 20′ N. Azimuth 359.
5. (*a*) L.H.A. ♈ 60° 26½′. Set sextant 45° 47′.
   (*b*) L.H.A. ♈ 60° 26½′ Lat. 49° 45′ N., Azimuth 359½.
6. (*a*) L.H.A. ♈ 80° 1′. Set sextant 38° 54′.
   (*b*) L.H.A. ♈ 81° 16′ Lat. 39° 23½′ N. Azimuth 359.

Position by Equal Altitudes. Page 215

1. Meridian Passage (Z.T.) 1212 Obs. Lat. 42° 16′ S., Long. 177° 58′ E.
2. Meridian Passage (Z.T.) 1212 Obs. Lat. 11° 13′ N., Long. 64° 54·7′ W.
3. Meridian Passage (Z.T.) 1205 Obs. Lat. 30° 5′ S., Long. 14° 53′ W.

Position by Noon Curve. Page 215

1. Meridian Passage (Z.T.) 1207 Obs. Lat. 42° 13′ N., Long. 18° 22½′ W.
2. Meridian Passage (Z.T.) 1213 Obs. Lat. 38° 59½′ N., Long. 64° 54½′ W.
3. Meridian Passage (Z.T.) 1201 Obs. Lat. 31° 10′ S., Long. 166° 3′ E.

Days at Sea. I. Page 217

Compass Error and Deviation by Amplitude.

Course 215° T., log 27.		1200	45° 15′ N.,	12° 17′ W.
			22′ S.	22′ W.
		1633	44° 53′ N.	12° 39′ W.
Dec. S.	22° 34′			
		Amplitude W.	32¾° S.	
Lat.	44° 53′	Bearing	237¼° T.	
			247° C.	
		Error	9¾° W.	
		Variation	11½° W.	
		Deviation	1¾° E.	

P.M. Stars. (A.P. 3720) Page 217

Startime	1709	Co. 215° 31 M.	1200	45° 15′ N.	12° 17′ W.
Long. W.	51			25·4′ S.	25′ W.
G.M.T. startime	1800		1709	44° 50′ N.	12° 42′ W.
Z.T startime	1700				
G.H.A. γ 18h	15° 11′				
Long. W.	12° 42′				
L.H.A. —	2° 29′				

*CAPELLA	ALDEBARAN	Hamal	*Diphda
Hc 38° 43′	28° 21′	58° 0′	26° 27′
Zn 060°	095°	123°	172°
ALTAIR	*VEGA	Kochab	
Hc 23° 43′	30° 21′	32° 16′	
Zn 259°	296°	348°	

Polaris

Lat.	44° 50′	a_0	14·1
a	− 15·4′	a_1	·6
		a_2	·7
	44° 34½′		15·4
	+1°		
	45° 34½′		
8 ft.	+ 4′		
	45° 38½′		
I.E.	− 2′		
Sextant	45° 36½′	Brg. 000½	

M. Stars by A.P. 3270

ay and Date	Friday Jan. 5				
me and Log	1700 Z.T. 31				
one	+1				
. Date	1800 Jan. 5th				
o. and Dist.	215°	31M			
om Lat. ~ Long.	45° 15′ N.	12° 17′ W.			
lat. (dep.) d.long.	25′ S. (20½)	25′ W.			
.R. Sights	44° 50′ N.	12° 42′ W.			
ody	VEGA	ALTAIR	ALDEBARAN	CAPELLA	POLARIS
bs. Alt.	30° 14½′	23° 34′	28° 49½′	39° 12′	45° 35′
or. 8 ft.	− 4½′	5½′	5′	4½′	4′
o T. Alt.	30° 10′	28° 28½′	28° 44½′	39° 7½′	45° 31′
hronometer	5h 59m 17s	6h 0m 31s	6h 0m 56s	6h 2m 10s	6h 5m 0s
rror	+ 2m 5s	2m 5s	2m 5s	2m 5s	2m 5s
.M.T.	18h 1m 22s	18h 2m 36s	18h 3m 1s	18h 4m 15s	18h 7m 5s
.H.A.♈ h.	15° 11·1′	15° 11·1′	15° 11·1′	15° 11·1′	15° 11·1′
c.	20·6′	39·1′	45·4′	1° 4′	1° 46·5′
.H.A.♈	15° 31·7′	15° 50·2′	15° 56·5′	16° 15′	16° 58′
hosen Long. W.	12° 32′	12° 50′	12° 56·5′	13° 15′	12° 42′
.H.A.*	3°	3°	3°	3°	4° 16′
c Zn	30° 2′ 296°	23° 23′ 259°	28° 42′ 095°	39° 2′ 060°	000½°
o	30° 10′	23° 28½′	28° 44½′	39° 7½′	45° 31′
ntercept	8′ T.	5½′ T.	2½′ T.	5½′ T.	−1°

	44° 31′
a_0	13·4′
a_1	·6′
a_2	·7′
Lat.	44° 45·7′ N

700 Fix: 44° 46′ N., 12° 55½′ W.

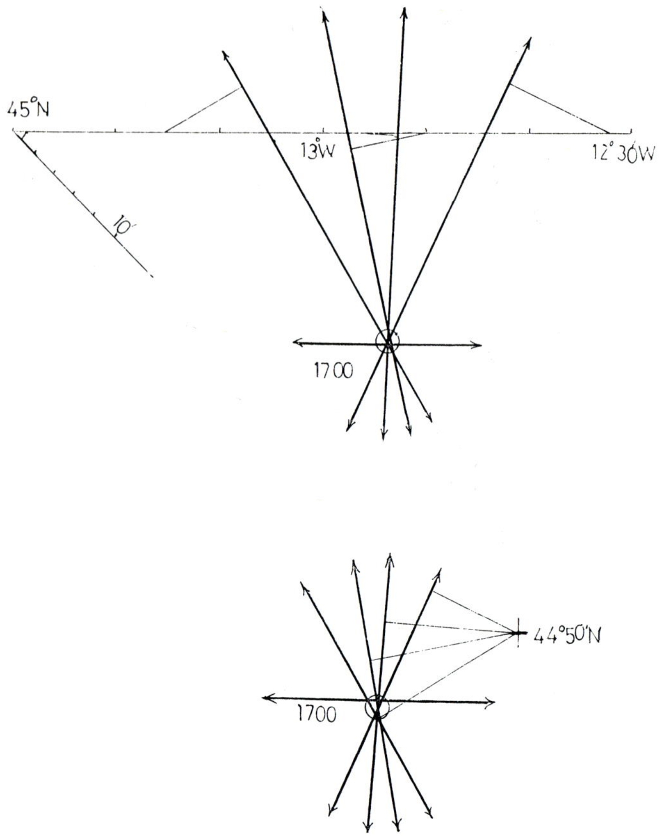

FIGS. 19.1a, 19.1b

.M. Stars by Versine Formula

ay and Date	Friday Jan. 5				
ime and log	1700 Z.T. 31				
one	+1				
. Date	1800 Jan. 5th				
o. and distance	215°	31M			
rom Lat. & Long.	45° 15′ N.	12° 17′ W.			
.lat (dep.) d.long.	25′ S. (20½)	25′ W.			
.R. Sights	44° 50′ N.	12° 42′ W.			
ody	VEGA	ALTAIR	ALDEBARAN	CAPELLA	POLARIS
bs. Alt.	30° 14½′	23° 34′	28° 49½′	39° 12′	45° 35′
or. for 8 ft.	−4½′	5½′	5′	4½′	4′
Ho) T. Alt.	30° 10′	28° 28½′	28° 44½′	39° 7½′	45° 31′
hronometer	5h 59m 17s	6h 0m 31s	6h 0m 56s	6h 2m 10s	6h 5m 0s
rror	+ 2m 5s	2m 5s	2m 5s	2m 5s	2m 5s
.M.T.	18h 1m 22s	18h 2m 36s	18h 3m 1s	18h 4m 15s	18h 7m 5s
.H.A. ♈ h.	15° 11·1′	15° 11·1′	15° 11·1′	15° 11·1′	15° 11·1′
nc.	20·6′	39·1′	45·4′	1° 4′	1° 46·5′
.H.A.	80° 59·9′	62° 38·1′	291° 23·9′	281° 18·6′	—
.H.A.*	96° 31·6′	78° 28·3′	307° 20·4′	297° 33·7′	16° 57·6′
ong. D.R. W.	12° 42′	12° 42′	12° 42′	12° 42′	
.H.A.*	83° 50′	65° 46′	294° 38′	284° 52′	
ec.	N. 38° 45½′	N. 8° 47·8′	N. 16° 27·5′	N. 45° 59½′	
.R. Lat.	N. 44° 50′	N. 44° 50′	N. 44° 50′	N. 44° 50′	
at. ~ Dec.	N. 6° 4½′	N. 36° 2′	N. 28° 22½′	N. 1° 9½′	
at. L.cos.	9·8507	9·8507	9·8507	9·8507	
ec. L.cos.	9·8920	9·9949	9·9818	9·8419	
.H.A. L.ver.	9·9506	9·7705	9·7658	9·8712	
L.ver.	9·6933	9·6161	9·5983	9·5638	
n.ver.	·4935	·4132	·3966	·3663	T. Alt. 45° 31′
at. ~ Dec. n.ver.	·00562	·1913	·1202	·0002	−1°
n.ver.	·4991	·6045	·5168	·3665	44° 31′
en. Dist.	59° 56½′	66° 42′	61° 6′	50° 41½′	a_0 13·4′
c	30° 3½′	23° 18′	28° 54′	39° 18½′	a_1 ·6′
o	30° 10′	23° 28½′	28° 44½′	39° 7½′	a_2 ·7′
ntercept	6½′T	10½′T	9½′A	11′A	Lat. 44° 45·7 N
A	+·114	+·455	+·455	+·249	
B	−·815	−·173	−·326	−1·07	
C	−·701	+·282	+·129	−·821	
earing	296½°	259°	095°	059½°	000½°

A.M. Stars by A.P. 3270.

Day and Date	Sat. Jan. 6th	
Time and Log	0700 115	
Zone	+1	
G. Date	0800 Jan 6th	
Co. and Dist.	215° 84 M.	
from Lat. & Long.	44° 46′ N.	12° 55′ W.
d.lat (dep.) d.long.	1° 9′ S. (48·2)	1° 7′ W.
D.R. Sights	43° 37′ N.	14° 2′ W.

Body	DENEB	VEGA	ANTARES	ARCTURUS	REGULUS
Obs. Alt.	24° 47′	40° 56½′	13° 19′	65° 38½′	28° 57′
Cor. for 8 ft.	− 5′	4′	7′	3½′	4½′
Ho T. Alt.	24° 42′	40° 52½′	13° 12′	65° 35′	28° 52½′
Chronometer	7h 5m 58s	7h 59m 45s	8h 2m 1s	8h 3m 15s	8h 3m 50s
Error	+ 2m 6s	2m 6s	2m 6s	2m 6s	2m 6s
G.M.T.	08h 1m 4s	08h 1m 51s	08h 4m 7s	08h 5m 21s	08h 5m 56s
G.H.A.♈ h.	225° 45·6′	225° 45·6′	225° 45·6′	225° 45·6′	225° 45·6′
inc.	16′	27·8′	1° 1·9′	1° 20·5′	1° 29·2′
G.H.A.♈	226° 1·6′	226° 13·4′	226° 47·5′	227° 6·1′	227° 4·8′
Ch. Long. W.	14° 2′	14° 13½′	13° 47½′	14° 6′	14° 15′
L.H.A.♈	212°	212°	213°	213°	213°
Hc Zn	24° 54′ 050°	40° 50′ 072°	13° 3′ 149°	65° 18′ 179°	28° 57′ 258°
Ho	24° 42′	40° 52½′	13° 12′	65° 35′	28° 52½′
Intercept	12′ A.	2½′T.	9′ T.	17′ T.	4½′ A.

0700 Fix: 43° 43′ N., 14° 3½′ W.

Jan. 5th 1700 44° 46′ N., 12° 55′ W.
Jan. 6th 0700 43° 43′ N., 14° 3½′ W.

1° 3′ N., 1° 8½′ W.
63′ (49·2) 68½′
= Course 218° distance 80 M. Speed 5·7 Kn.

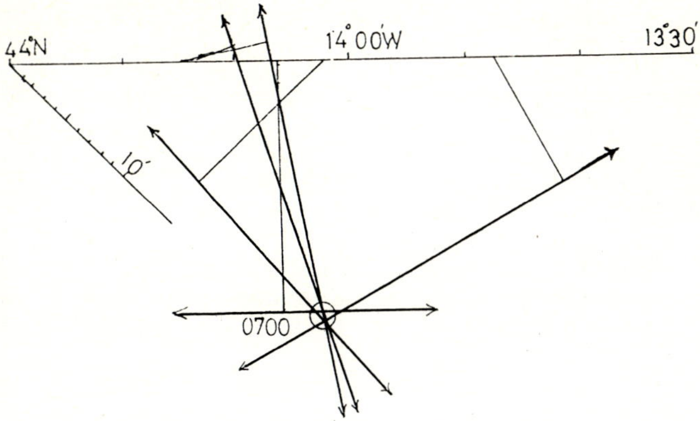

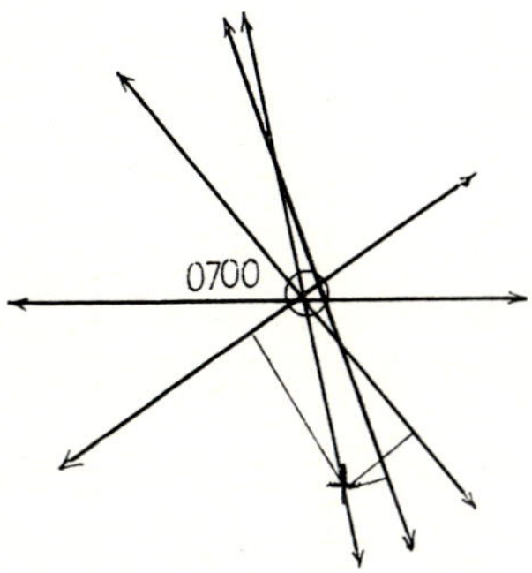

FIGS. 19.2a, 19.2b

A.M. Stars (Versine Formula)

Day and Date	Sat. 6th Jan.				
Time and Log	0700 115				
Zone	+1				
G. Date	0800 6th Jan.				
Co. and Dist.	215° 84 M.				
from Lat. & Long.	44° 46′ N.	12° 55′ N.			
d.lat. (dep.) d.long.	1° 9′ S. (48·2)	1° 7′ W.			
D.R. Sights	43° 37′ N.	14° 2′ W.			
Body	DENEB	VEGA	ANTARES	ARCTURUS	REGULUS
Obs. Alt.	24° 47′	40° 56½′	13° 19′	65° 38½′	28° 57′
Cor. for 8 ft.	— 5′	4′	7′	3½′	4½′
Ho T. Alt.	24° 42′	40° 52½′	13° 12′	65° 35′	28° 52½′
Chronometer	7h 58m 8s	7h 59m 45s	8h 2m 1s	8h 3m 15s	8h 3m 50s
Error	+ 2m 6s	2m 6s	2m 6s	2m 6s	2m 6s
G.M.T.	08h 1m 4s	08h 1m 51s	08h 4m 7s	08h 5m 21s	08h 5m 56s
G.H.A.♈. h.	225° 45·6′	225° 45·6′	225° 45·6′	225° 45·6′	225° 45·6′
inc.	16′	27·8′	1° 1·9′	1° 20·5′	1° 29·2′
S.H.A.	49° 52·6′	80° 59·9′	113° 3·8′	146° 23·4′	208° 15·5′
G.H.A.*	276° 54·2′	307° 13·3′	339° 51·3′	373° 29·5′	435° 30·3′
Long. D.R.	14° 2′	14° 2′	14° 2′	14° 2′	14° 2′
L.H.A.*	261° 52′	293° 11′	325° 49′	359° 27½′	61° 28½′
Dec.	N. 45° 11′	N. 38° 45·4′	S. 26° 22·4′	N. 19° 19′	N. 12° 58′
D.R. Lat.	43° 37′	43° 37′	43° 37′	43° 37′	43° 37′
Lat. and Dec.	1° 34′	4° 51½′	69° 59½′	24° 18′	31° 31′
Lat. L.cos.	9·8597	9·8597	9·8597	9·8597	9·8597
Dec. L.cos.	9·8481	9·8920	9·9523	9·9748	9·9902
L.H.A. L.ver.	0·0575	9·7827	9·2374	5·6500	9·7180
L.ver.	9·7653	9·5344	9·0494	5·4845	9·5679
n.ver.	·5825	·3423	·1121	·00003	·3698
Lat. Dec. n.ver.	·00037	·0036	·6578	·0886	·1475
n.ver.	·58287	·3459	·7699	·08863	·5173
Zen. Dist.	65° 21′	49° 9′	76° 42′	24° 18½′	61° 8′
Hc	24° 39′	40° 51′	13° 18′	65° 41½′	28° 52′
Ho	24° 42′	40° 52½′	13° 12′	65° 35′	28° 52½′
Intercept	3′ T.	1½′ T.	6′ A.	6½′ A.	½′ T.
A	— ·133	+ ·416	+ 1·35	+ 109	+ ·515
B	— 1·01	— ·880	+ ·885	— 40	— ·243
C	— 1·143	— ·464	+ 2·235	+ 69	+ ·272
Bearing	050½°	071°	148½°	176°–180°	259°

A.M. Observation of the Sun.

Day and Date	Sat. Jan. 6th.	
Time and Log	0900 127	
Zone	+1	
Greenwich Date	1000 6th	
Course and Distance	218° 12 M.	
from a.m. Stars	43° 43′ N. 14° 3½′ W.	
	9½′ N. (7·2) 10′ W.	
D.R. 0900	43° 33½′ N. 14° 13½′ W.	
	☉	
Obs. Alt.	11° 57′	
Cor. (8 ft.)	+ 10½′	
Ho T. Alt.	12° 7½′	
Chronometer	10h 1m 12s	
Error	+ 2m 6s	
G.M.T.	10h 3m 18s	
G.H.A. and Dec.	328° 31·7′ S. 22° 29·4′	By A.P. 3270
Inc.	49·5′	
G.H.A.	329° 21·2′	G.H.A. 329° 21·2′
Long. W. D.R.	14° 13·5′ Lat. 43° 33½′ N.	Ch. Long. 14° 21′
L.H.A. ☉	315° 8′ L. ~ D. 66° 3′	L.H.A. ☉ 315°
Lat. L.cos.	9·8601	
Dec. L.cos.	9·9657	
L.H.A. L.ver.	9·4643	
L.ver.	9·2901	
n.ver.	·1950	
Lat. ~ Dec. n.ver.	·5941	From A.P. 3270
		12° 12′ −51
n.ver.	·7891	Cor. for Dec. −25½′
Zen. Dist.	77° 49½′	
Hc Cal. Alt.	12° 10½′	Hc 11° 46½′
Ho True Alt.	12° 7½′	Ho 12° 7½′
Intercept	3′ Away	21′ Towards
A	+·955	
B	+·59	
C	1·545	
Bearing	138°	Zn. 138°

Meridian Altitude of the Sun and Noon Position.

Time of Meridian Passage

Mer. Pass.	1206
Long. W.	58
G.M.T. Mer. Pass.	1304
Z.T. Mer. Pass.	1204

Run to Noon from a.m. Sun Sight

Course and Dist.	218° 18 M.	
0900 D.R.	43° 33½′ N.	14° 13½′ W.
Diffence	14′ S. (11·1)	15′ W.
1200 D.R.	43° 19½′ S.	14° 28½′ W.

Meridian Altitude of Sun.

Obs. Alt.	24° 0′
Cor.	+ 11½′
True Alt.	24° 11½′
Zen. Dist.	65° 48½′
Dec. S.	22° 28½′
Lat.	43° 20′ N.

1200/1204 Obs. 43° 20′ N., 14° 34′ W.

Course, Distance and Speed Made Good.

Jan. 5th 1200	45° 15′ N.,	12° 17′ W.
Jan 6th 1200	43° 20′ N.,	14° 34′ W.
	1° 55′ N.,	2° 17′ W.
	60°	60°
	d.lat. 115′	d.long. 137′ = dep. 98·5′

```
24)152(6·33
   144
   ---
    80
    72
    --
     8
```

Co. 220½°, Distance 152 M., Av. speed 6·3 Kn.

0900 D.R.	43° 33½′ N.	14° 13½′ W.
	14′ S. (11·1)	15′ W.
1200 D.R.	43° 19½′ N.	14° 28½′ W.
		5½′ W.
1200 Obs.	43° 20′ N.	14° 34′ W.

Run to 1200 Co. 218° 18 M.

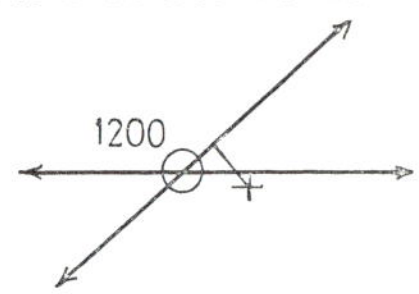

FIG. 19.3a

0900 C.P.	44° 0′ N.	14° 21′ W.
	14′ S.	15′ W.
1200 C.P.	43° 46′ N.	14° 36′ W.
		2′ E.
1200 Obs.	43° 20′ N.	14° 34′ W.

Run to 1200. Co. 218° 18 M.

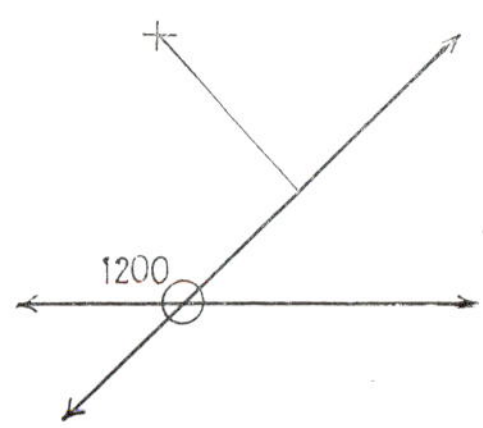

FIG. 19.3b

Days at Sea. II. Page 218

Planet and Sun.

Day and Date	Sunday Sept. 23rd 1973		1504 D.R. Lat.	40° 58′ N.
Time and Log	1504	22	Venus Dec. S.	16° 34′
Zone	2		Zen. Dist.	57° 32′
				90°
Greenwich Date	1304	23rd	True Alt.	32° 28′
Course and Distance	282°	22 M.	Cor.	+ 4′
from 1200 D.R.	40° 53′ N.	18° 31′ E.		
Difference	5′ N. (21·5)	28′ W.	Obs. Alt.	32° 32′
			I.E.	− 2′
D.R. 1504	40° 58′ N.	18° 3′ E.	Set Sextant	32° 30′

	Venus		Sun	
Obs. Alt.	32° 35′		36° 40′	
Cor.	− 4′		+ 12′	
True Altitude	32° 31′		36° 52′	
Chronometer	1h 3m 2s		1h 5m 17s	
Error	+ 4m 36s		+ 4m 36s	
G.M.T.	13h 7m 38s		13h 9m 53s	
G.H.A. h. Dec.	338° 12·8′	S. 16° 34′	16° 54·7′	S. 0° 8½′
inc.	1° 54·3′		2° 28·3′	
G.H.A.	340° 7′	340° 7′	19° 23′	19° 23′
Long. (D.R.)	18° 3′	(Chosen) 17° 53′	18° 3′	17° 37′
L.H.A.	358° 10′	358°	37° 26′	37°
	A.P. 3270			
	32° 58′	−60 178°	37° 04′	−49 131°
	− 34′		− 7′	
Hc	32° 24′		36° 57′	
Ho	32° 31′		36° 52′	
Intercept	7′ Towards	Bearing 178°	5′ Away	Bearing 229°

	Ex-Meridian	*Versine Formula*		
	Factor 1·69	Lat.	L.cos.	9·8780
	1·0 = 1·0	Dec.	L.cos.	0·0
	0·7 = 0·7	H.A.	L.ver.	9·3137
Obs. Alt.	Cor. 1·7′		L.ver.	9·1917
	32° 31′		n.ver.	·1555
Mer. Alt.		Lat. and Dec.		
	32° 33′		n.ver.	·2465
	90°		n.ver.	·4020
Zen. Dist.		Zen. Dist.		53° 16½′
Dec.	57° 27′	Hc		36° 43½′
	S. 16° 34′	Ho		36° 52′
Lat.	40° 53′ N.	Intercept		8½′ Towards
A				+1·13
B				0
C				1·13
Bearing				229½°

1505 Fix: 40° 53′ N., 17° 54′ E.

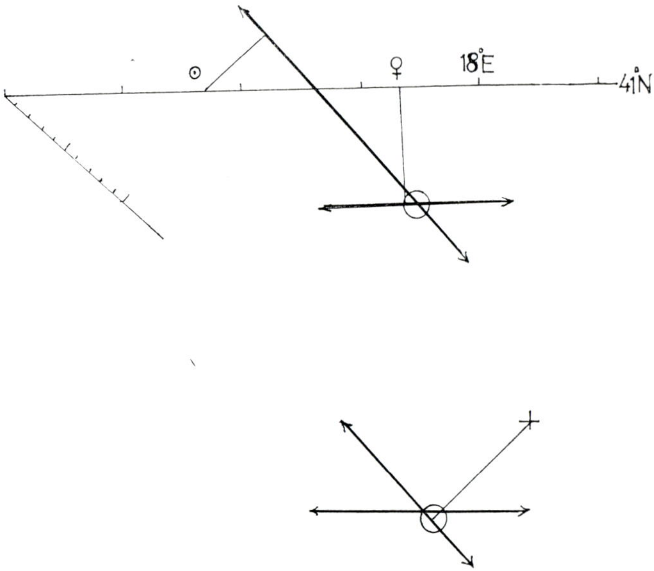

FIG.S 19.4a, 19.4b

Sun and Moon.

Day and Date	Monday 24th Sept. 1973	
Time and Log	0805 148	0858 154
Zone	−1	−1
Greenwich Date	0705 24th	0758 24th
Course and Distance	282° 126 M.	282° 6 M.
from 1505	40° 53′ N. 17° 54′ E.	from 0805 41° 19′ N. 15° 11′ E.
Diff.	26′ N.(23·2) 2° 43′ W.	1′ N. (6) 8′ W.
D.R. 0805	41° 19′ N. 5° 11′ E.	41° 20′ N. 15° 3′ E.
	SUN	MOON
Obs. Alt.	23° 45½′	52° 54½′ (Burton's) 8·8
Cor.	+ 11′	+ 15½′ +6·7
Ho True Altitude	23° 56½′	53° 10′ 15·5
Chronometer	6h 59m 5s	7h 58m 2s
Error	+ 4m 36s	+ 4m 36s
G.M.T.	07h 3m 41s	08h 2m 38s
G.H.A. Dec.	286° 58·6′ S. 0° 26′	330° 13·9′ v. 11·2 N. 6° 37·7′
inc.	55·3′	37·7 H.P. 58·7
G.H.A.	287° 53·9′ 287° 54′	330° 52′ 330° 52′
Long. E. D.R.	15° 11′ Ch. 15° 6′	15° 3′ 15° 8′
L.H.A.	303° 5′ 303°	345° 55′ 346°
	A.P. 3270	*A.P. 3270*
	24° 16′ −43 113°	52° 50′ +57 157
	− 19′	+ 35′
Hc	23° 57′	53° 25′
Ho	23° 56½′	53° 10′
Intercept Bearing	½′ Away 113°	15′ Away 157°
	Versine Formula	*Versine Formula*
Lat. L.cos.	9·8757	9·8756
Dec. L.cos.	0·	9·9971
L.H.A. L.vers.	9·6572	8·4779
L.vers.	9·5329	8·3506
n.vers.	·3411	·0224
Lat. & Dec. n.vers.	·2539	·1780
n.vers.	·5950	·2004
Zen. Dist.	66° 6½′ A +·565	36° 54½′ +3·50
Hc Cal. Alt.	23° 53½′ B —	53° 5½′ − ·47
Ho True Alt.	23° 56½′	53° 10′
	C +·565	+3·03
Intercept Bearing	3′ Towards 113°	4½′ Towards 156½°

0858 (Z−1) Fix: 41° 15′ N., 15° 5′ E.

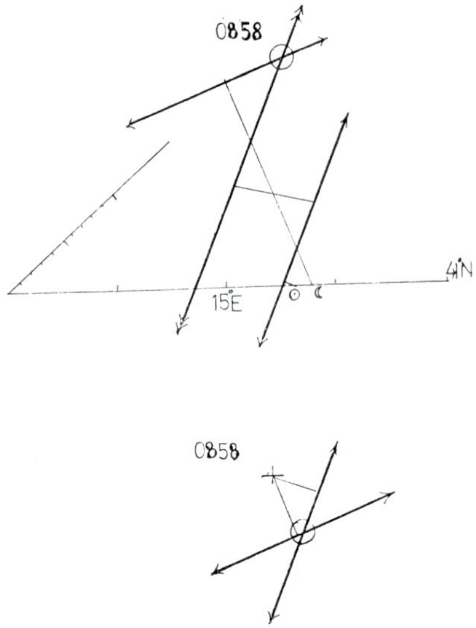

FIGS. 19.5a, 19.5b

Position by Meridian and Equal Altitudes.

Day and Date	Monday 24th Sept.			
Time and Log	1154	175		
Zone	−1			
Greenwich Date	1054	24th		
Course and Distance	282° 21 M.		Mer. Pass.	1152
from 0858	41° 15′ N.	15° 5′ E.	Long. E.	−58
Diff.	4½′ N (20·5)	27′W.		
			G.T. Mer. Pass.	1054
D.R. 1054	41° 19½′ N.	14° 38′ E.	Z.T. Mer. Pass.	1154

Sextant Alt.	47° 24′	Chronometer	10h 28m 10s
	47° 50½′		10h 43m 50s
	47° 54′		10h 46m 15s
	47° 55′		10h 53m 40s—assessed as
	47° 24′		11h 19m 12s Mer. Alt.

			10h 28m 10s
			11h 19m 12s
			2)21h 47m 22s
Obs. Mer. Alt.	47° 57′		
Cor.	12½′	G.M.T. Noon	10h 53m 41s
True Alt.	48° 9½′	G.H.A. 10	331° 59·3′
	90°	inc. 53m 41s	13° 25·3*
Zen. Dist.	41° 50½′		345° 24·6′
Dec.	0° 30′ S.		360°
Lat.	41° 20½′ N.	Long.	14° 35½′ E.

* Increment taken from *Nautical Almanac* as increments over 14 minutes are not included in this book.

EXTRACTS

FROM

The Nautical Almanac
The Admiralty Tide Tables
The Admiralty List of Radio Signals, Vol. 5
Burton's Nautical Tables
Coastguard Addresses and Telephone Numbers

1973 JANUARY 1, 2, 3 (MON., TUES., WED.)

G.M.T. d	h	ARIES G.H.A.	VENUS −3·4 G.H.A.	VENUS Dec.	MARS +1·8 G.H.A.	MARS Dec.	JUPITER −1·4 G.H.A.	JUPITER Dec.	SATURN −0·1 G.H.A.	SATURN Dec.
		° ′	° ′	° ′	° ′	° ′	° ′	° ′	° ′	° ′
1	00	100 30·2	205 04·4	S21 51·7	221 41·2	S20 04·7	171 10·4	S22 25·7	26 16·3	N21 02·2
	01	115 32·7	220 03·5	52·1	236 41·9	05·0	186 12·2	25·6	41 19·0	02·2
	02	130 35·1	235 02·6	52·5	251 42·6	05·4	201 14·1	25·6	56 21·6	02·1
	03	145 37·6	250 01·8	52·9	266 43·3	05·8	216 15·9	25·5	71 24·3	02·1
	04	160 40·0	265 00·9	53·3	281 44·0	06·2	231 17·8	25·4	86 26·9	02·1
	05	175 42·5	280 00·0	53·7	296 44·7	06·5	246 19·6	25·3	101 29·6	02·1
	06	190 45·0	294 59·1	S21 54·2	311 45·4	S20 06·9	261 21·4	S22 25·3	116 32·2	N21 02·1
	07	205 47·4	309 58·3	54·6	326 46·1	07·3	276 23·3	25·2	131 34·9	02·1
	08	220 49·9	324 57·4	55·0	341 46·8	07·7	291 25·1	25·1	146 37·5	02·1
M	09	235 52·4	339 56·5	55·4	356 47·5	08·0	306 27·0	25·0	161 40·2	02·1
O	10	250 54·8	354 55·6	55·8	11 48·2	08·4	321 28·8	25·0	176 42·8	02·0
N	11	265 57·3	9 54·8	56·2	26 48·9	08·8	336 30·6	24·9	191 45·5	02·0
D	12	280 59·8	24 53·9	S21 56·6	41 49·6	S20 09·2	351 32·5	S22 24·8	206 48·1	N21 02·0
A	13	296 02·2	39 53·0	57·0	56 50·3	09·5	6 34·3	24·8	221 50·8	02·0
Y	14	311 04·7	54 52·1	57·4	71 50·9	09·9	21 36·2	24·7	236 53·4	02·0
	15	326 07·2	69 51·2	57·8	86 51·6	10·3	36 38·0	24·6	251 56·1	02·0
	16	341 09·6	84 50·4	58·2	101 52·3	10·6	51 39·8	24·5	266 58·8	02·0
	17	356 12·1	99 49·5	58·6	116 53·0	11·0	66 41·7	24·5	282 01·4	02·0
	18	11 14·5	114 48·6	S21 59·0	131 53·7	S20 11·4	81 43·5	S22 24·4	297 04·1	N21 02·0
	19	26 17·0	129 47·7	59·3	146 54·4	11·8	96 45·4	24·3	312 06·7	01·9
	20	41 19·5	144 46·8	21 59·7	161 55·1	12·1	111 47·2	24·2	327 09·4	01·9
	21	56 21·9	159 46·0	22 00·1	176 55·8	12·5	126 49·0	24·2	342 12·0	01·9
	22	71 24·4	174 45·1	00·5	191 56·5	12·9	141 50·9	24·1	357 14·7	01·9
	23	86 26·9	189 44·2	00·9	206 57·2	13·2	156 52·7	24·0	12 17·3	01·9
2	00	101 29·3	204 43·3	S22 01·3	221 57·9	S20 13·6	171 54·6	S22 23·9	27 20·0	N21 01·9
	01	116 31·8	219 42·4	01·7	236 58·6	14·0	186 56·4	23·9	42 22·6	01·9
	02	131 34·3	234 41·6	02·1	251 59·3	14·3	201 58·2	23·8	57 25·2	01·9
	03	146 36·7	249 40·7	02·4	267 00·0	14·7	217 00·1	23·7	72 27·9	01·9
	04	161 39·2	264 39·8	02·8	282 00·7	15·1	232 01·9	23·6	87 30·5	01·8
	05	176 41·7	279 38·9	03·2	297 01·3	15·4	247 03·8	23·6	102 33·2	01·8
	06	191 44·1	294 38·0	S22 03·6	312 02·0	S20 15·8	262 05·6	S22 23·5	117 35·8	N21 01·8
	07	206 46·6	309 37·2	04·0	327 02·7	16·2	277 07·4	23·4	132 38·5	01·8
T	08	221 49·0	324 36·3	04·3	342 03·4	16·5	292 09·3	23·4	147 41·1	01·8
U	09	236 51·5	339 35·4	04·7	357 04·1	16·9	307 11·1	23·3	162 43·8	01·8
E	10	251 54·0	354 34·5	05·1	12 04·8	17·3	322 12·9	23·2	177 46·4	01·8
S	11	266 56·4	9 33·6	05·5	27 05·5	17·6	337 14·8	23·1	192 49·1	01·8
D	12	281 58·9	24 32·7	S22 05·8	42 06·2	S20 18·0	352 16·6	S22 23·1	207 51·7	N21 01·7
A	13	297 01·4	39 31·9	06·2	57 06·9	18·4	7 18·5	23·0	222 54·4	01·7
Y	14	312 03·8	54 31·0	06·6	72 07·6	18·7	22 20·3	22·9	237 57·0	01·7
	15	327 06·3	69 30·1	07·0	87 08·3	19·1	37 22·1	22·8	252 59·7	01·7
	16	342 08·8	84 29·2	07·3	102 09·0	19·5	52 24·0	22·8	268 02·3	01·7
	17	357 11·2	99 28·3	07·7	117 09·6	19·8	67 25·8	22·7	283 05·0	01·7
	18	12 13·7	114 27·4	S22 08·1	132 10·3	S20 20·2	82 27·7	S22 22·6	298 07·6	N21 01·7
	19	27 16·1	129 26·5	08·4	147 11·0	20·6	97 29·5	22·5	313 10·3	01·7
	20	42 18·6	144 25·6	08·8	162 11·7	20·9	112 31·3	22·5	328 12·9	01·7
	21	57 21·1	159 24·8	09·2	177 12·4	21·3	127 33·2	22·4	343 15·6	01·6
	22	72 23·5	174 23·9	09·5	192 13·1	21·6	142 35·0	22·3	358 18·2	01·6
	23	87 26·0	189 23·0	09·9	207 13·8	22·0	157 36·9	22·2	13 20·9	01·6
3	00	102 28·5	204 22·1	S22 10·2	222 14·5	S20 22·4	172 38·7	S22 22·2	28 23·5	N21 01·6
	01	117 30·9	219 21·2	10·6	237 15·2	22·7	187 40·5	22·1	43 26·1	01·6
	02	132 33·4	234 20·3	10·9	252 15·8	23·1	202 42·4	22·0	58 28·8	01·6
	03	147 35·9	249 19·4	11·3	267 16·5	23·5	217 44·2	21·9	73 31·4	01·6
	04	162 38·3	264 18·5	11·7	282 17·2	23·8	232 46·1	21·9	88 34·1	01·6
	05	177 40·8	279 17·7	12·0	297 17·9	24·2	247 47·9	21·8	103 36·7	01·6
	06	192 43·3	294 16·8	S22 12·4	312 18·6	S20 24·5	262 49·7	S22 21·7	118 39·4	N21 01·5
W	07	207 45·7	309 15·9	12·7	327 19·3	24·9	277 51·6	21·6	133 42·0	01·5
E	08	222 48·2	324 15·0	13·1	342 20·0	25·3	292 53·4	21·6	148 44·7	01·5
D	09	237 50·6	339 14·1	13·4	357 20·7	25·6	307 55·3	21·5	163 47·3	01·5
N	10	252 53·1	354 13·2	13·8	12 21·3	26·0	322 57·1	21·4	178 50·0	01·5
E	11	267 55·6	9 12·3	14·1	27 22·0	26·3	337 58·9	21·3	193 52·6	01·5
S	12	282 58·0	24 11·4	S22 14·5	42 22·7	S20 26·7	353 00·8	S22 21·3	208 55·3	N21 01·5
D	13	298 00·5	39 10·5	14·8	57 23·4	27·1	8 02·6	21·2	223 57·9	01·5
A	14	313 03·0	54 09·6	15·1	72 24·1	27·4	23 04·4	21·1	239 00·5	01·5
Y	15	328 05·4	69 08·7	15·5	87 24·8	27·8	38 06·3	21·0	254 03·2	01·4
	16	343 07·9	84 07·8	15·8	102 25·5	28·1	53 08·1	21·0	269 05·8	01·4
	17	358 10·4	99 06·9	16·2	117 26·1	28·5	68 10·0	20·9	284 08·5	01·4
	18	13 12·8	114 06·1	S22 16·5	132 26·8	S20 28·8	83 11·8	S22 20·8	299 11·1	N21 01·4
	19	28 15·3	129 05·2	16·8	147 27·5	29·2	98 13·6	20·7	314 13·8	01·4
	20	43 17·8	144 04·3	17·2	162 28·2	29·6	113 15·5	20·7	329 16·4	01·4
	21	58 20·2	159 03·4	17·5	177 28·9	29·9	128 17·3	20·6	344 19·1	01·4
	22	73 22·7	174 02·5	17·9	192 29·6	30·3	143 19·2	20·5	359 21·7	01·4
	23	88 25·1	189 01·6	18·2	207 30·3	30·6	158 21·0	20·4	14 24·3	01·4
Mer. Pass.		17 11·2 (h m)	v −0·9	d 0·4	v 0·7	d 0·4	v 1·8	d 0·1	v 2·6	d 0·0

STARS

Name	S.H.A.	Dec.
	° ′	° ′
Acamar	315 41·0	S 40 24·
Achernar	335 48·9	S 57 22·
Acrux	173 43·4	S 62 56·
Adhara	255 36·0	S 28 56·
Aldebaran	291 23·9	N 16 27·
Alioth	166 47·1	N 56 05·
Alkaid	153 22·7	N 49 26·
Al Na'ir	28 21·9	S 47 05·
Alnilam	276 16·8	S 1 13·
Alphard	218 25·6	S 8 32·
Alphecca	126 36·9	N 26 48·
Alpheratz	358 15·0	N 28 56·
Altair	62 38·1	N 8 47·
Ankaa	353 45·5	S 42 27·
Antares	113 03·8	S 26 22·
Arcturus	146 23·5	N 19 19·
Atria	108 33·4	S 68 58·
Avior	234 29·9	S 59 25·
Bellatrix	279 04·2	N 6 19·
Betelgeuse	271 33·8	N 7 24·
Canopus	264 09·0	S 52 40·
Capella	281 18·8	N 45 58·
Deneb	49 52·6	N 45 11·
Denebola	183 04·4	N 14 43·
Diphda	349 26·2	S 18 08·
Dubhe	194 28·2	N 61 53·
Elnath	278 50·6	N 28 35·
Eltanin	91 00·8	N 51 29·
Enif	34 17·1	N 9 45·
Fomalhaut	15 57·5	S 29 46·
Gacrux	172 34·9	S 56 57·
Gienah	176 23·5	S 17 23·
Hadar	149 31·4	S 60 14·
Hamal	328 34·9	N 23 20·
Kaus Aust.	84 24·4	S 34 24·
Kochab	137 19·1	N 74 15·
Markab	14 08·7	N 15 03·
Menkar	314 46·6	N 3 59·
Menkent	148 43·6	S 36 14·
Miaplacidus	221 45·6	S 69 36·
Mirfak	309 23·5	N 49 46·
Nunki	76 36·2	S 26 19·
Peacock	54 07·3	S 56 49·
Pollux	244 04·3	N 28 05·
Procyon	245 31·1	N 5 17·
Rasalhague	96 34·9	N 12 34·
Regulus	208 15·5	N 12 05·
Rigel	281 40·9	S 8 13·
Rigil Kent.	140 33·6	S 60 43·
Sabik	102 47·6	S 15 41·
Schedar	350 15·3	N 56 23·
Shaula	97 03·5	S 37 05·
Sirius	259 00·1	S 16 40·
Spica	159 03·3	S 11 01·
Suhail	223 14·4	S 43 19·
Vega	80 59·9	N 38 45·
Zuben'ubi	137 39·2	S 15 55·

	S.H.A.	Mer. Pass.
	° ′	h m
Venus	103 14·0	10 22
Mars	120 28·6	9 12
Jupiter	70 25·2	12 31
Saturn	285 50·6	22 07

G.M.T. d h	SUN G.H.A. ° ′	SUN Dec. ° ′	MOON G.H.A. ° ′	v ′	MOON Dec. ° ′	d ′	H.P. ′
1 00	179 09·0	S23 02·0	222 58·2	11·5	S24 00·6	4·4	54·0
01	194 08·7	01·8	237 28·7	11·5	24 05·0	4·3	54·0
02	209 08·4	01·6	251 59·2	11·4	24 09·3	4·2	54·0
03	224 08·1	· · 01·4	266 29·6	11·3	24 13·5	4·1	54·0
04	239 07·8	01·2	280 59·9	11·4	24 17·6	3·9	54·0
05	254 07·5	01·0	295 30·3	11·3	24 21·5	3·9	54·0
06	269 07·2	S23 00·8	310 00·6	11·2	S24 25·4	3·7	54·0
07	284 06·9	00·6	324 30·8	11·3	24 29·1	3·7	54·0
08	299 06·6	00·4	339 01·1	11·2	24 32·8	3·5	54·0
09	314 06·3	· · 00·2	353 31·3	11·1	24 36·3	3·4	54·0
10	329 06·0	23 00·0	8 01·4	11·1	24 39·7	3·3	54·0
11	344 05·7	22 59·7	22 31·5	11·1	24 43·0	3·1	54·0
12	359 05·4	S22 59·5	37 01·6	11·1	S24 46·1	3·1	54·0
13	14 05·1	59·3	51 31·7	11·0	24 49·2	2·9	54·0
14	29 04·8	59·1	66 01·7	11·0	24 52·1	2·9	54·0
15	44 04·5	· · 58·9	80 31·7	10·9	24 55·0	2·7	54·0
16	59 04·2	58·7	95 01·6	11·0	24 57·7	2·6	54·1
17	74 03·9	58·5	109 31·6	10·9	25 00·3	2·4	54·1
18	89 03·7	S22 58·3	124 01·5	10·8	S25 02·7	2·4	54·1
19	104 03·4	58·1	138 31·3	10·9	25 05·1	2·2	54·1
20	119 03·1	57·8	153 01·2	10·8	25 07·3	2·1	54·1
21	134 02·8	· · 57·6	167 31·0	10·7	25 09·4	2·0	54·1
22	149 02·5	57·4	182 00·7	10·8	25 11·4	1·9	54·1
23	164 02·2	57·2	196 30·5	10·7	25 13·3	1·8	54·1
2 00	179 01·9	S22 57·0	211 00·2	10·7	S25 15·1	1·6	54·1
01	194 01·6	56·8	225 29·9	10·7	25 16·7	1·5	54·1
02	209 01·3	56·5	239 59·6	10·7	25 18·2	1·4	54·1
03	224 01·0	· · 56·3	254 29·3	10·6	25 19·6	1·3	54·1
04	239 00·7	56·1	268 58·9	10·7	25 20·9	1·2	54·1
05	254 00·4	55·9	283 28·6	10·6	25 22·1	1·0	54·1
06	269 00·1	S22 55·6	297 58·2	10·5	S25 23·1	0·9	54·1
07	283 59·8	55·4	312 27·7	10·6	25 24·0	0·8	54·1
08	298 59·5	55·2	326 57·3	10·5	25 24·8	0·7	54·1
09	313 59·2	· · 55·0	341 26·8	10·6	25 25·5	0·5	54·1
10	328 59·0	54·7	355 56·4	10·5	25 26·0	0·5	54·1
11	343 58·7	54·5	10 25·9	10·5	25 26·5	0·3	54·2
12	358 58·4	S22 54·3	24 55·4	10·5	S25 26·8	0·2	54·2
13	13 58·1	54·1	39 24·9	10·4	25 27·0	0·0	54·2
14	28 57·8	53·8	53 54·3	10·5	25 27·0	0·0	54·2
15	43 57·5	· · 53·6	68 23·8	10·5	25 27·0	0·2	54·2
16	58 57·2	53·4	82 53·3	10·4	25 26·8	0·3	54·2
17	73 56·9	53·1	97 22·7	10·4	25 26·5	0·5	54·2
18	88 56·6	S22 52·9	111 52·1	10·5	S25 26·0	0·5	54·2
19	103 56·3	52·7	126 21·6	10·4	25 25·5	0·7	54·2
20	118 56·0	52·4	140 51·0	10·4	25 24·8	0·8	54·2
21	133 55·7	· · 52·2	155 20·4	10·4	25 24·0	0·9	54·2
22	148 55·5	52·0	169 49·8	10·4	25 23·1	1·1	54·2
23	163 55·2	51·7	184 19·2	10·4	25 22·0	1·2	54·3
3 00	178 54·9	S22 51·5	198 48·6	10·4	S25 20·8	1·3	54·3
01	193 54·6	51·3	213 18·0	10·4	25 19·5	1·4	54·3
02	208 54·3	51·0	227 47·4	10·4	25 18·1	1·5	54·3
03	223 54·0	· · 50·8	242 16·8	10·4	25 16·6	1·7	54·3
04	238 53·7	50·5	256 46·2	10·4	25 14·9	1·8	54·3
05	253 53·4	50·3	271 15·6	10·4	25 13·1	1·9	54·3
06	268 53·1	S22 50·1	285 45·0	10·4	S25 11·2	2·0	54·3
07	283 52·8	49·8	300 14·4	10·4	25 09·2	2·2	54·3
08	298 52·6	49·6	314 43·8	10·5	25 07·0	2·3	54·3
09	313 52·3	· · 49·3	329 13·3	10·4	25 04·7	2·4	54·3
10	328 52·0	49·1	343 42·7	10·4	25 02·3	2·5	54·4
11	343 51·7	48·8	358 12·1	10·5	24 59·8	2·7	54·4
12	358 51·4	S22 48·6	12 41·6	10·4	S24 57·1	2·8	54·4
13	13 51·1	48·3	27 11·0	10·5	24 54·3	2·8	54·4
14	28 50·8	48·1	41 40·5	10·5	24 51·5	3·1	54·4
15	43 50·5	· · 47·8	56 10·0	10·5	24 48·4	3·1	54·4
16	58 50·2	47·6	70 39·5	10·5	24 45·3	3·3	54·4
17	73 50·0	47·3	85 09·0	10·5	24 42·0	3·4	54·4
18	88 49·7	S22 47·1	99 38·5	10·5	S24 38·6	3·5	54·4
19	103 49·4	46·8	114 08·0	10·6	24 35·1	3·6	54·5
20	118 49·1	46·6	128 37·6	10·5	24 31·5	3·7	54·5
21	133 48·8	· · 46·3	143 07·1	10·6	24 27·8	3·9	54·5
22	148 48·5	46·1	157 36·7	10·6	24 23·9	4·0	54·5
23	163 48·2	45·8	172 06·3	10·7	24 19·9	4·1	54·5
	S.D. 16·3	d 0·2	S.D. 14·7		14·8		14·8

Lat. °	Twilight Naut. h m	Twilight Civil h m	Sunrise h m	Moonrise 1 h m	Moonrise 2 h m	Moonrise 3 h m	Moonrise 4 h m
N 72	08 23	10 40	■	■	■	■	■
N 70	08 04	09 48	■	■	■	■	■
68	07 49	09 16	■	■	■	■	■
66	07 37	08 52	10 26	■	■	■	11 05
64	07 26	08 34	09 49	08 05	09 28	10 02	10 07
62	07 17	08 18	09 22	07 20	08 28	09 12	09 34
60	07 09	08 05	09 02	06 51	07 55	08 41	09 09
N 58	07 02	07 54	08 45	06 29	07 30	08 17	08 50
56	06 56	07 44	08 31	06 11	07 11	07 58	08 33
54	06 50	07 35	08 19	05 55	06 54	07 42	08 19
52	06 44	07 28	08 08	05 42	06 40	07 28	08 07
50	06 39	07 20	07 58	05 31	06 28	07 16	07 56
45	06 28	07 05	07 38	05 07	06 02	06 51	07 34
N 40	06 18	06 52	07 22	04 47	05 42	06 32	07 15
35	06 09	06 40	07 08	04 31	05 25	06 15	07 00
30	06 00	06 30	06 56	04 18	05 10	06 01	06 47
20	05 44	06 11	06 35	03 54	04 46	05 36	06 24
N 10	05 28	05 54	06 17	03 34	04 24	05 15	06 04
0	05 12	05 38	06 00	03 15	04 05	04 55	05 46
S 10	04 53	05 20	05 43	02 56	03 45	04 35	05 27
20	04 31	05 00	05 25	02 36	03 24	04 14	05 07
30	04 03	04 36	05 03	02 13	02 59	03 50	04 44
35	03 44	04 21	04 50	01 59	02 45	03 35	04 31
40	03 22	04 03	04 36	01 44	02 28	03 19	04 15
45	02 52	03 41	04 18	01 25	02 08	02 59	03 57
S 50	02 09	03 12	03 56	01 02	01 43	02 34	03 33
52	01 43	02 58	03 46	00 51	01 31	02 22	03 22
54	01 03	02 40	03 34	00 39	01 17	02 08	03 09
56	////	02 19	03 20	00 24	01 01	01 52	02 55
58	////	01 51	03 04	00 07	00 42	01 32	02 37
S 60	////	01 08	02 44	24 19	00 19	01 08	02 16

Lat. °	Sunset h m	Twilight Civil h m	Twilight Naut. h m	Moonset 1 h m	Moonset 2 h m	Moonset 3 h m	Moonset 4 h m
N 72	■	13 29	15 45	■	■	■	■
N 70	■	14 20	16 04	■	■	■	■
68	■	14 52	16 19	■	■	■	■
66	13 42	15 16	16 32	■	■	■	12 59
64	14 20	15 35	16 42	10 44	11 05	12 17	13 56
62	14 46	15 50	16 51	11 29	12 05	13 06	14 29
60	15 06	16 03	16 59	11 58	12 38	13 37	14 53
N 58	15 23	16 14	17 06	12 21	13 03	14 01	15 12
56	15 37	16 24	17 13	12 39	13 23	14 20	15 28
54	15 50	16 33	17 19	12 55	13 39	14 35	15 42
52	16 00	16 41	17 24	13 08	13 53	14 49	15 54
50	16 10	16 48	17 29	13 20	14 06	15 01	16 04
45	16 30	17 04	17 41	13 44	14 31	15 26	16 26
N 40	16 46	17 17	17 51	14 04	14 51	15 45	16 44
35	17 00	17 28	18 00	14 20	15 08	16 02	16 59
30	17 12	17 38	18 08	14 34	15 23	16 16	17 12
20	17 33	17 57	18 24	14 58	15 48	16 40	17 33
N 10	17 51	18 14	18 40	15 19	16 09	17 00	17 52
0	18 08	18 30	18 56	15 39	16 29	17 20	18 10
S 10	18 25	18 48	19 15	15 58	16 49	17 39	18 27
20	18 43	19 08	19 37	16 19	17 10	17 59	18 46
30	19 05	19 32	20 05	16 43	17 35	18 23	19 07
35	19 18	19 47	20 23	16 58	17 49	18 37	19 20
40	19 32	20 05	20 46	17 14	18 06	18 53	19 34
45	19 50	20 27	21 15	17 34	18 26	19 12	19 51
S 50	20 12	20 55	21 59	17 58	18 52	19 36	20 11
52	20 22	21 10	22 24	18 10	19 04	19 47	20 21
54	20 34	21 27	23 03	18 24	19 18	20 00	20 32
56	20 47	21 48	////	18 40	19 34	20 15	20 45
58	21 03	22 15	////	18 59	19 53	20 33	21 00
S 60	21 23	22 57	////	19 22	20 17	20 54	21 17

Day	SUN Eqn. of Time 00ʰ m s	SUN Eqn. of Time 12ʰ m s	SUN Mer. Pass. h m	MOON Mer. Pass. Upper h m	MOON Mer. Pass. Lower h m	MOON Age d	MOON Phase
1	03 23	03 38	12 04	09 27	21 52	27	
2	03 52	04 06	12 04	10 17	22 42	28	
3	04 20	04 34	12 05	11 07	23 33	29	

G.M.T. d h	SUN G.H.A. ° ′	SUN Dec. ° ′	MOON G.H.A. ° ′	v ′	Dec. ° ′	d ′	H.P. ′
4 00	178 48·0	S22 45·6	186 36·0	10·6	S24 15·8	4·2	54·5
01	193 47·7	45·3	201 05·6	10·7	24 11·6	4·4	54·5
02	208 47·4	45·0	215 35·3	10·6	24 07·2	4·4	54·5
03	223 47·1	· · 44·8	230 04·9	10·8	24 02·8	4·6	54·6
04	238 46·8	44·5	244 34·7	10·7	23 58·2	4·7	54·6
05	253 46·5	44·3	259 04·4	10·7	23 53·5	4·8	54·6
06	268 46·2	S22 44·0	273 34·1	10·8	S23 48·7	5·0	54·6
07	283 46·0	43·7	288 03·9	10·8	23 43·7	5·0	54·6
T 08	298 45·7	43·5	302 33·7	10·9	23 38·7	5·2	54·6
H 09	313 45·4	· · 43·2	317 03·6	10·8	23 33·5	5·3	54·6
U 10	328 45·1	43·0	331 33·4	10·9	23 28·2	5·4	54·6
R 11	343 44·8	42·7	346 03·3	10·9	23 22·8	5·5	54·7
S 12	358 44·5	S22 42·4	An Annular Eclipse				
D 13	13 44·2	42·2	of the Sun occurs				
A 14	28 44·0	41·9	to-day. See page 5.				
Y 15	43 43·7	· · 41·6					
16	58 43·4	41·3					
17	73 43·1	41·1					
18	88 42·8	S22 40·8	87 33·4	11·1	S22 41·8	6·3	54·8
19	103 42·5	40·5	102 03·5	11·2	22 35·5	6·4	54·8
20	118 42·3	40·3	116 33·7	11·2	22 29·1	6·6	54·8
21	133 42·0	· · 40·0	131 03·9	11·2	22 22·5	6·6	54·8
22	148 41·7	39·7	145 34·1	11·3	22 15·9	6·7	54·8
23	163 41·4	39·4	160 04·4	11·3	22 09·2	6·9	54·8
5 00	178 41·1	S22 39·2	174 34·7	11·3	S22 02·3	7·0	54·8
01	193 40·8	38·9	189 05·0	11·4	21 55·3	7·0	54·9
02	208 40·6	38·6	203 35·4	11·4	21 48·3	7·2	54·9
03	223 40·3	· · 38·3	218 05·8	11·4	21 41·1	7·3	54·9
04	238 40·0	38·1	232 36·2	11·5	21 33·8	7·4	54·9
05	253 39·7	37·8	247 06·7	11·5	21 26·4	7·5	54·9
06	268 39·4	S22 37·5	261 37·2	11·6	S21 18·9	7·6	54·9
07	283 39·2	37·2	276 07·8	11·6	21 11·3	7·7	55·0
08	298 38·9	36·9	290 38·4	11·6	21 03·6	7·8	55·0
F 09	313 38·6	· · 36·7	305 09·0	11·7	20 55·8	7·9	55·0
R 10	328 38·3	36·4	319 39·7	11·7	20 47·9	8·1	55·0
I 11	343 38·0	36·1	334 10·4	11·7	20 39·8	8·1	55·0
D 12	358 37·8	S22 35·8	348 41·1	11·8	S20 31·7	8·2	55·0
A 13	13 37·5	35·5	3 11·9	11·9	20 23·5	8·3	55·0
Y 14	28 37·2	35·2	17 42·8	11·8	20 15·2	8·4	55·1
15	43 36·9	· · 34·9	32 13·6	11·9	20 06·8	8·5	55·1
16	58 36·6	34·7	46 44·5	12·0	19 58·3	8·6	55·1
17	73 36·4	34·4	61 15·5	12·0	19 49·7	8·7	55·1
18	88 36·1	S22 34·1	75 46·5	12·0	S19 41·0	8·8	55·1
19	103 35·8	33·8	90 17·5	12·1	19 32·2	8·9	55·1
20	118 35·5	33·5	104 48·6	12·1	19 23·3	9·0	55·2
21	133 35·2	· · 33·2	119 19·7	12·1	19 14·3	9·1	55·2
22	148 35·0	32·9	133 50·8	12·2	19 05·2	9·2	55·2
23	163 34·7	32·6	148 22·0	12·2	18 56·0	9·3	55·2
6 00	178 34·4	S22 32·3	162 53·2	12·3	S18 46·7	9·3	55·2
01	193 34·1	32·0	177 24·5	12·3	18 37·4	9·5	55·2
02	208 33·9	31·7	191 55·8	12·3	18 27·9	9·6	55·3
03	223 33·6	· · 31·4	206 27·1	12·4	18 18·3	9·6	55·3
04	238 33·3	31·2	220 58·5	12·4	18 08·7	9·7	55·3
05	253 33·0	30·9	235 29·9	12·5	17 59·0	9·8	55·3
06	268 32·8	S22 30·6	250 01·4	12·5	S17 49·2	9·9	55·3
07	283 32·5	30·3	264 32·9	12·5	17 39·3	10·0	55·3
S 08	298 32·2	30·0	279 04·4	12·6	17 29·3	10·1	55·4
A 09	313 31·9	· · 29·7	293 36·0	12·6	17 19·2	10·1	55·4
T 10	328 31·7	29·4	308 07·6	12·7	17 09·1	10·3	55·4
U 11	343 31·4	29·1	322 39·3	12·7	16 58·8	10·3	55·4
R 12	358 31·1	S22 28·7	337 11·0	12·7	S16 48·5	10·4	55·4
D 13	13 30·8	28·4	351 42·7	12·8	16 38·1	10·5	55·5
A 14	28 30·6	28·1	6 14·5	12·8	16 27·6	10·6	55·5
Y 15	43 30·3	· · 27·8	20 46·3	12·8	16 17·0	10·6	55·5
16	58 30·0	27·5	35 18·1	12·9	16 06·4	10·7	55·5
17	73 29·7	27·2	49 50·0	12·9	15 55·7	10·8	55·5
18	88 29·5	S22 26·9	64 21·9	13·0	S15 44·9	10·9	55·5
19	103 29·2	26·6	78 53·9	13·0	15 34·0	11·0	55·6
20	118 28·9	26·3	93 25·9	13·0	15 23·0	11·0	55·6
21	133 28·6	· · 26·0	107 57·9	13·1	15 12·0	11·1	55·6
22	148 28·4	25·7	122 30·0	13·1	15 00·9	11·2	55·6
23	163 28·1	25·4	137 02·1	13·1	14 49·7	11·2	55·6
	S.D. 16·3	d 0·3	S.D. 14·9		15·0		15·1

Lat. °	Twilight Naut. h m	Twilight Civil h m	Sunrise h m	Moonrise 4 h m	Moonrise 5 h m	Moonrise 6 h m	Moonrise 7 h m
N 72	08 20	10 30	■	■	■	12 08	11 08
N 70	08 02	09 43	■	■	■	11 19	10 46
68	07 47	09 12	11 30	■	11 22	10 48	10 29
66	07 35	08 50	10 20	11 05	10 37	10 24	10 15
64	07 25	08 32	09 45	10 07	10 07	10 06	10 03
62	07 16	08 17	09 20	09 34	09 45	09 51	09 54
60	07 08	08 04	09 00	09 09	09 27	09 38	09 45
N 58	07 01	07 53	08 44	08 50	09 12	09 27	09 38
56	06 55	07 43	08 30	08 33	08 58	09 17	09 31
54	06 49	07 35	08 18	08 19	08 47	09 08	09 25
52	06 44	07 27	08 07	08 07	08 37	09 00	09 20
50	06 39	07 20	07 58	07 56	08 28	08 53	09 15
45	06 28	07 05	07 38	07 34	08 09	08 39	09 04
N 40	06 18	06 52	07 22	07 15	07 53	08 26	08 55
35	06 09	06 41	07 08	07 00	07 40	08 16	08 48
30	06 01	06 30	06 57	06 47	07 28	08 06	08 41
20	05 45	06 12	06 36	06 24	07 09	07 50	08 29
N 10	05 30	05 56	06 18	06 04	06 51	07 36	08 19
0	05 13	05 39	06 02	05 46	06 35	07 23	08 09
S 10	04 55	05 22	05 45	05 27	06 19	07 10	08 00
20	04 33	05 02	05 26	05 07	06 01	06 56	07 49
30	04 05	04 38	05 05	04 44	05 41	06 39	07 37
35	03 47	04 23	04 53	04 31	05 29	06 30	07 30
40	03 25	04 06	04 38	04 15	05 16	06 19	07 22
45	02 56	03 44	04 21	03 57	05 00	06 06	07 13
S 50	02 13	03 16	04 00	03 33	04 40	05 50	07 02
52	01 49	03 02	03 49	03 22	04 30	05 43	06 57
54	01 12	02 45	03 38	03 09	04 20	05 34	06 51
56	////	02 25	03 24	02 55	04 07	05 25	06 45
58	////	01 58	03 08	02 37	03 53	05 15	06 37
S 60	////	01 18	02 49	02 16	03 37	05 02	06 29

Lat. °	Sunset h m	Twilight Civil h m	Twilight Naut. h m	Moonset 4 h m	Moonset 5 h m	Moonset 6 h m	Moonset 7 h m
N 72	■	13 41	15 52	■	■	15 20	17 57
N 70	■	14 28	16 10	■	■	16 07	18 16
68	12 41	14 59	16 24	■	14 25	16 38	18 32
66	13 51	15 22	16 36	12 59	15 08	17 00	18 44
64	14 26	15 40	16 46	13 56	15 37	17 17	18 54
62	14 52	15 55	16 55	14 29	15 59	17 32	19 03
60	15 11	16 07	17 03	14 53	16 17	17 44	19 11
N 58	15 28	16 18	17 10	15 12	16 31	17 54	19 17
56	15 41	16 28	17 16	15 28	16 44	18 03	19 23
54	15 53	16 36	17 22	15 42	16 55	18 11	19 28
52	16 04	16 44	17 27	15 54	17 04	18 18	19 33
50	16 13	16 51	17 32	16 04	17 13	18 24	19 37
45	16 33	17 06	17 43	16 26	17 31	18 38	19 46
N 40	16 49	17 19	17 53	16 44	17 46	18 49	19 53
35	17 02	17 30	18 02	16 59	17 58	18 59	20 00
30	17 14	17 40	18 10	17 12	18 09	19 07	20 05
20	17 35	17 59	18 26	17 33	18 28	19 21	20 15
N 10	17 52	18 15	18 41	17 52	18 43	19 34	20 23
0	18 09	18 32	18 57	18 10	18 58	19 45	20 31
S 10	18 26	18 49	19 16	18 27	19 13	19 57	20 39
20	18 44	19 09	19 38	18 46	19 29	20 09	20 47
30	19 05	19 33	20 05	19 07	19 47	20 23	20 56
35	19 18	19 47	20 23	19 20	19 57	20 31	21 01
40	19 32	20 05	20 45	19 34	20 09	20 40	21 07
45	19 49	20 26	21 14	19 51	20 23	20 50	21 14
S 50	20 11	20 54	21 56	20 11	20 40	21 03	21 23
52	20 21	21 08	22 21	20 21	20 48	21 09	21 26
54	20 33	21 25	22 56	20 32	20 57	21 15	21 31
56	20 46	21 45	////	20 45	21 06	21 22	21 35
58	21 01	22 11	////	21 00	21 18	21 31	21 40
S 60	21 20	22 50	////	21 17	21 31	21 40	21 46

Day	SUN Eqn. of Time 00ʰ m s	SUN Eqn. of Time 12ʰ m s	SUN Mer. Pass. h m	MOON Mer. Pass. Upper h m	MOON Mer. Pass. Lower h m	Age d	Phase
4	04 48	05 01	12 05	11 58	24 22	30	■
5	05 15	05 28	12 05	12 47	00 22	01	
6	05 42	05 55	12 06	13 34	01 11	02	

G.M.T.	SUN G.H.A.	SUN Dec.	MOON G.H.A.	v	MOON Dec.	d	H.P.
h	° ′	° ′	° ′	′	° ′	′	′
19 00	181 30·7	N 1 37·5	281 57·2	4·7	N24 16·0	0·4	59·2
01	196 30·9	36·6	296 20·9	4·8	24 16·4	0·2	59·2
02	211 31·2	35·6	310 44·7	4·7	24 16·6	0·1	59·2
03	226 31·4 ··	34·6	325 08·4	4·8	24 16·7	0·1	59·2
04	241 31·6	33·7	339 32·2	4·7	24 16·6	0·3	59·2
05	256 31·8	32·7	353 55·9	4·7	24 16·3	0·4	59·2
06	271 32·0	N 1 31·7	8 19·6	4·7	N24 15·9	0·6	59·2
W 07	286 32·3	30·8	22 43·3	4·8	24 15·3	0·7	59·2
E 08	301 32·5	29·8	37 07·1	4·7	24 14·6	0·9	59·2
D 09	316 32·7 ··	28·8	51 30·8	4·7	24 13·7	1·1	59·2
N 10	331 32·9	27·8	65 54·5	4·7	24 12·6	1·3	59·2
E 11	346 33·1	26·9	80 18·2	4·8	24 11·3	1·4	59·2
S 12	1 33·4	N 1 25·9	94 42·0	4·7	N24 09·9	1·5	59·2
D 13	16 33·6	24·9	109 05·7	4·8	24 08·4	1·8	59·2
A 14	31 33·8	24·0	123 29·5	4·8	24 06·6	1·9	59·2
Y 15	46 34·0 ··	23·0	137 53·3	4·8	24 04·7	2·0	59·2
16	61 34·3	22·0	152 17·1	4·8	24 02·7	2·3	59·2
17	76 34·5	21·0	166 40·9	4·8	24 00·4	2·3	59·3
18	91 34·7	N 1 20·1	181 04·7	4·8	N23 58·1	2·6	59·3
19	106 34·9	19·1	195 28·5	4·9	23 55·5	2·7	59·3
20	121 35·1	18·1	209 52·4	4·9	23 52·8	2·9	59·3
21	136 35·4 ··	17·2	224 16·3	4·9	23 49·9	3·0	59·3
22	151 35·6	16·2	238 40·2	4·9	23 46·9	3·2	59·3
23	166 35·8	15·2	253 04·1	5·0	23 43·7	3·3	59·3
20 00	181 36·0	N 1 14·3	267 28·1	4·9	N23 40·4	3·5	59·3
01	196 36·3	13·3	281 52·0	5·1	23 36·9	3·7	59·3
02	211 36·5	12·3	296 16·1	5·0	23 33·2	3·8	59·3
03	226 36·7 ··	11·3	310 40·1	5·1	23 29·4	4·0	59·3
04	241 36·9	10·4	325 04·2	5·1	23 25·4	4·1	59·3
05	256 37·1	09·4	339 28·3	5·2	23 21·3	4·3	59·3
06	271 37·4	N 1 08·4	353 52·5	5·2	N23 17·0	4·4	59·3
T 07	286 37·6	07·5	8 16·7	5·2	23 12·6	4·6	59·3
H 08	301 37·8	06·5	22 40·9	5·3	23 08·0	4·8	59·3
U 09	316 38·0 ··	05·5	37 05·2	5·3	23 03·2	4·9	59·3
R 10	331 38·2	04·5	51 29·5	5·4	22 58·3	5·1	59·3
S 11	346 38·5	03·6	65 53·9	5·4	22 53·2	5·2	59·3
D 12	1 38·7	N 1 02·6	80 18·3	5·4	N22 48·0	5·3	59·3
A 13	16 38·9	01·6	94 42·7	5·5	22 42·7	5·5	59·3
Y 14	31 39·1	1 00·7	109 07·2	5·6	22 37·2	5·7	59·3
15	46 39·4	0 59·7	123 31·8	5·5	22 31·5	5·8	59·3
16	61 39·6	58·7	137 56·3	5·7	22 25·7	6·0	59·3
17	76 39·8	57·7	152 21·0	5·7	22 19·7	6·1	59·3
18	91 40·0	N 0 56·8	166 45·7	5·7	N22 13·6	6·2	59·3
19	106 40·2	55·8	181 10·4	5·9	22 07·4	6·4	59·3
20	121 40·5	54·8	195 35·3	5·8	22 01·0	6·5	59·3
21	136 40·7 ··	53·9	210 00·1	5·9	21 54·5	6·7	59·3
22	151 40·9	52·9	224 25·0	6·0	21 47·8	6·8	59·3
23	166 41·1	51·9	238 50·0	6·1	21 41·0	7·0	59·3
21 00	181 41·3	N 0 50·9	253 15·1	6·0	N21 34·0	7·1	59·3
01	196 41·6	50·0	267 40·1	6·2	21 26·9	7·2	59·3
02	211 41·8	49·0	282 05·3	6·2	21 19·7	7·4	59·3
03	226 42·0 ··	48·0	296 30·5	6·3	21 12·3	7·5	59·3
04	241 42·2	47·1	310 55·8	6·3	21 04·8	7·6	59·3
05	256 42·4	46·1	325 21·1	6·4	20 57·2	7·8	59·3
06	271 42·7	N 0 45·1	339 46·5	6·5	N20 49·4	7·9	59·3
07	286 42·9	44·1	354 12·0	6·5	20 41·5	8·1	59·3
08	301 43·1	43·2	8 37·5	6·6	20 33·4	8·1	59·3
F 09	316 43·3 ··	42·2	23 03·1	6·7	20 25·3	8·3	59·3
R 10	331 43·6	41·2	37 28·8	6·7	20 17·0	8·4	59·3
I 11	346 43·8	40·2	51 54·5	6·8	20 08·6	8·6	59·3
D 12	1 44·0	N 0 39·3	66 20·3	6·9	N20 00·0	8·7	59·3
A 13	16 44·2	38·3	80 46·2	6·9	19 51·3	8·8	59·3
Y 14	31 44·4	37·3	95 12·1	7·0	19 42·5	8·9	59·3
15	46 44·7 ··	36·4	109 38·1	7·1	19 33·6	9·1	59·3
16	61 44·9	35·4	124 04·2	7·1	19 24·5	9·1	59·3
17	76 45·1	34·4	138 30·3	7·2	19 15·4	9·3	59·3
18	91 45·3	N 0 33·4	152 56·5	7·3	N19 06·1	9·4	59·3
19	106 45·5	32·5	167 22·8	7·3	18 56·7	9·5	59·3
20	121 45·8	31·5	181 49·1	7·5	18 47·2	9·7	59·3
21	136 46·0 ··	30·5	196 15·6	7·5	18 37·5	9·7	59·3
22	151 46·2	29·5	210 42·1	7·5	18 27·8	9·9	59·3
23	166 46·4	28·6	225 08·6	7·6	18 17·9	9·9	59·3
	S.D. 16·0	d 1·0	S.D. 16·1		16·2		16·2

Lat.	Twilight Naut.	Twilight Civil	Sunrise	Moonrise 19	Moonrise 20	Moonrise 21	Moonrise 22
°	h m	h m	h m	h m	h m	h m	h m
N 72	02 43	04 19	05 28	☐	☐	☐	23 41
N 70	03 06	04 29	05 31	☐	☐	21 32	24 09
68	03 24	04 36	05 33	☐	19 12	22 19	24 30
66	03 38	04 43	05 35	18 29	20 44	22 49	24 47
64	03 49	04 48	05 36	19 42	21 22	23 11	25 00
62	03 58	04 53	05 38	20 19	21 49	23 29	25 11
60	04 06	04 57	05 39	20 45	22 09	23 44	25 20
N 58	04 13	05 00	05 40	21 05	22 26	23 56	25 29
56	04 19	05 03	05 41	21 22	22 41	24 07	00 07
54	04 24	05 06	05 41	21 36	22 53	24 17	00 17
52	04 28	05 08	05 42	21 49	23 04	24 25	00 25
50	04 32	05 10	05 43	22 00	23 13	24 33	00 33
45	04 40	05 15	05 44	22 23	23 34	24 49	00 49
N 40	04 46	05 18	05 45	22 41	23 50	25 02	01 02
35	04 51	05 21	05 46	22 57	24 04	00 04	01 13
30	04 55	05 23	05 47	23 10	24 16	00 16	01 22
20	05 01	05 26	05 48	23 33	24 36	00 36	01 39
N 10	05 04	05 28	05 49	23 53	24 54	00 54	01 54
0	05 06	05 30	05 50	24 11	00 11	01 10	02 07
S 10	05 06	05 30	05 51	24 30	00 30	01 27	02 20
20	05 04	05 30	05 52	24 49	00 49	01 44	02 35
30	05 01	05 29	05 52	00 14	01 12	02 05	02 51
35	04 58	05 28	05 53	00 28	01 26	02 16	03 01
40	04 55	05 26	05 53	00 44	01 41	02 30	03 11
45	04 50	05 24	05 54	01 03	01 59	02 46	03 24
S 50	04 44	05 22	05 54	01 27	02 22	03 05	03 39
52	04 41	05 21	05 54	01 39	02 33	03 15	03 46
54	04 38	05 19	05 54	01 52	02 45	03 25	03 54
56	04 34	05 18	05 55	02 07	02 59	03 37	04 03
58	04 29	05 16	05 55	02 25	03 16	03 50	04 13
S 60	04 24	05 14	05 55	02 47	03 36	04 06	04 24

Lat.	Sunset	Twilight Civil	Twilight Naut.	Moonset 19	Moonset 20	Moonset 21	Moonset 22
°	h m	h m	h m	h m	h m	h m	h m
N 72	18 16	19 25	20 58	☐	☐	☐	18 16
N 70	18 14	19 16	20 36	☐	☐	18 29	17 46
68	18 12	19 08	20 20	☐	18 48	17 41	17 23
66	18 10	19 02	20 06	17 26	17 16	17 10	17 05
64	18 09	18 57	19 55	16 12	16 37	16 47	16 51
62	18 08	18 52	19 47	15 36	16 10	16 28	16 39
60	18 07	18 49	19 39	15 10	15 49	16 13	16 28
N 58	18 06	18 45	19 32	14 49	15 31	16 00	16 19
56	18 05	18 42	19 27	14 32	15 17	15 48	16 11
54	18 04	18 40	19 22	14 17	15 04	15 38	16 04
52	18 04	18 38	19 17	14 05	14 53	15 29	15 57
50	18 03	18 36	19 14	13 54	14 43	15 21	15 51
45	18 02	18 31	19 06	13 30	14 22	15 04	15 38
N 40	18 01	18 28	19 00	13 11	14 05	14 50	15 28
35	18 00	18 25	18 55	12 56	13 50	14 38	15 19
30	17 59	18 23	18 51	12 42	13 38	14 27	15 11
20	17 58	18 20	18 46	12 19	13 16	14 09	14 57
N 10	17 57	18 18	18 43	11 59	12 57	13 53	14 44
0	17 57	18 17	18 41	11 40	12 40	13 38	14 33
S 10	17 56	18 17	18 41	11 21	12 22	13 23	14 21
20	17 56	18 18	18 43	11 00	12 03	13 06	14 09
30	17 55	18 19	18 47	10 37	11 41	12 47	13 54
35	17 55	18 20	18 49	10 23	11 08	12 36	13 46
40	17 54	18 22	18 53	10 07	11 13	12 24	13 36
45	17 54	18 24	18 58	09 48	10 55	12 09	13 25
S 50	17 54	18 26	19 04	09 24	10 33	11 50	13 11
52	17 54	18 28	19 07	09 12	10 22	11 41	13 04
54	17 54	18 29	19 11	08 59	10 10	11 32	12 57
56	17 54	18 31	19 15	08 44	09 57	11 20	12 49
58	17 53	18 33	19 19	08 26	09 40	11 08	12 40
S 60	17 53	18 35	19 25	08 05	09 21	10 52	12 29

Day	SUN Eqn. of Time 00h	SUN Eqn. of Time 12h	SUN Mer. Pass.	MOON Mer. Pass. Upper	MOON Mer. Pass. Lower	Age	Phase
	m s	m s	h m	h m	h m	d	
19	06 02	06 13	11 54	05 25	17 56	22	
20	06 24	06 34	11 53	06 26	18 55	23	
21	06 45	06 56	11 53	07 24	19 52	24	

G.M.T.	SUN G.H.A.	SUN Dec.	MOON G.H.A.	v	MOON Dec.	d	H.P.
d h	° ′	° ′	° ′	′	° ′	′	′
22 00	181 46.6	N 0 27.6	239 35.2	7.7	N18 08.0	10.1	59.3
01	196 46.9	26.6	254 01.9	7.8	17 57.9	10.2	59.3
02	211 47.1	25.7	268 28.7	7.9	17 47.7	10.3	59.3
03	226 47.3 · ·	24.7	282 55.6	7.9	17 37.4	10.4	59.2
04	241 47.5	23.7	297 22.5	8.0	17 27.0	10.5	59.2
05	256 47.7	22.7	311 49.5	8.0	17 16.5	10.6	59.2
06	271 47.9	N 0 21.8	326 16.5	8.2	N17 05.9	10.7	59.2
07	286 48.2	20.8	340 43.7	8.2	16 55.2	10.8	59.2
S 08	301 48.4	19.8	355 10.9	8.2	16 44.4	10.9	59.2
A 09	316 48.6 · ·	18.8	9 38.1	8.4	16 33.5	11.0	59.2
T 10	331 48.8	17.9	24 05.5	8.4	16 22.5	11.1	59.2
U 11	346 49.0	16.9	38 32.9	8.5	16 11.4	11.2	59.2
R 12	1 49.3	N 0 15.9	53 00.4	8.5	N16 00.2	11.2	59.2
D 13	16 49.5	14.9	67 27.9	8.7	15 49.0	11.4	59.2
A 14	31 49.7	14.0	81 55.6	8.7	15 37.6	11.5	59.2
Y 15	46 49.9 · ·	13.0	96 23.3	8.7	15 26.1	11.5	59.2
16	61 50.1	12.0	110 51.0	8.9	15 14.6	11.6	59.2
17	76 50.4	11.0	125 18.9	8.9	15 03.0	11.7	59.2
18	91 50.6	N 0 10.1	139 46.8	9.0	N14 51.3	11.8	59.2
19	106 50.8	09.1	154 14.8	9.0	14 39.5	11.9	59.1
20	121 51.0	08.1	168 42.8	9.1	14 27.6	12.0	59.1
21	136 51.2 · ·	07.2	183 10.9	9.2	14 15.6	12.0	59.1
22	151 51.5	06.2	197 39.1	9.3	14 03.6	12.2	59.1
23	166 51.7	05.2	212 07.4	9.3	13 51.4	12.1	59.1
23 00	181 51.9	N 0 04.2	226 35.7	9.4	N13 39.3	12.3	59.1
01	196 52.1	03.3	241 04.1	9.4	13 27.0	12.4	59.1
02	211 52.3	02.3	255 32.5	9.5	13 14.6	12.4	59.1
03	226 52.5 · ·	01.3	270 01.0	9.6	13 02.2	12.5	59.1
04	241 52.8	N 0 00.3	284 29.6	9.7	12 49.7	12.5	59.1
05	256 53.0	S 0 00.6	298 58.3	9.7	12 37.2	12.6	59.1
06	271 53.2	S 0 01.6	313 27.0	9.7	N12 24.6	12.7	59.0
07	286 53.4	02.6	327 55.7	9.9	12 11.9	12.8	59.0
08	301 53.6	03.6	342 24.6	9.9	11 59.1	12.8	59.0
S 09	316 53.9 · ·	04.5	356 53.5	9.9	11 46.3	12.8	59.0
U 10	331 54.1	05.5	11 22.4	10.1	11 33.5	13.0	59.0
N 11	346 54.3	06.5	25 51.5	10.0	11 20.5	13.0	59.0
D 12	1 54.5	S 0 07.5	40 20.5	10.2	N11 07.5	13.0	59.0
A 13	16 54.7	08.4	54 49.7	10.2	10 54.5	13.1	59.0
Y 14	31 54.9	09.4	69 18.9	10.2	10 41.4	13.2	59.0
15	46 55.2 · ·	10.4	83 48.1	10.4	10 28.2	13.2	58.9
16	61 55.4	11.4	98 17.5	10.3	10 15.0	13.2	58.9
17	76 55.6	12.3	112 46.8	10.5	10 01.8	13.4	58.9
18	91 55.8	S 0 13.3	127 16.3	10.5	N 9 48.4	13.3	58.9
19	106 56.0	14.3	141 45.8	10.5	9 35.1	13.4	58.9
20	121 56.2	15.2	156 15.3	10.6	9 21.7	13.5	58.9
21	136 56.5 · ·	16.2	170 44.9	10.7	9 08.2	13.5	58.9
22	151 56.7	17.2	185 14.6	10.7	8 54.7	13.5	58.9
23	166 56.9	18.2	199 44.3	10.7	8 41.2	13.6	58.8
24 00	181 57.1	S 0 19.1	214 14.0	10.8	N 8 27.6	13.6	58.8
01	196 57.3	20.1	228 43.8	10.9	8 14.0	13.6	58.8
02	211 57.5	21.1	243 13.7	10.9	8 00.4	13.7	58.8
03	226 57.8 · ·	22.1	257 43.6	11.0	7 46.7	13.8	58.8
04	241 58.0	23.0	272 13.6	11.0	7 32.9	13.7	58.8
05	256 58.2	24.0	286 43.6	11.0	7 19.2	13.8	58.8
06	271 58.4	S 0 25.0	301 13.6	11.2	N 7 05.4	13.8	58.7
07	286 58.6	26.0	315 43.8	11.1	6 51.6	13.9	58.7
08	301 58.8	26.9	330 13.9	11.2	6 37.7	13.9	58.7
M 09	316 59.1 · ·	27.9	344 44.1	11.2	6 23.8	13.9	58.7
O 10	331 59.3	28.9	359 14.3	11.3	6 09.9	13.9	58.7
N 11	346 59.5	29.9	13 44.6	11.3	5 56.0	14.0	58.7
D 12	1 59.7	S 0 30.8	28 14.9	11.4	N 5 42.0	13.9	58.6
A 13	16 59.9	31.8	42 45.3	11.4	5 28.1	14.0	58.6
Y 14	32 00.1	32.8	57 15.7	11.5	5 14.1	14.0	58.6
15	47 00.4 · ·	33.8	71 46.2	11.5	5 00.1	14.1	58.6
16	62 00.6	34.7	86 16.7	11.5	4 46.0	14.0	58.6
17	77 00.8	35.7	100 47.2	11.6	4 32.0	14.1	58.6
18	92 01.0	S 0 36.7	115 17.8	11.6	N 4 17.9	14.1	58.5
19	107 01.2	37.7	129 48.4	11.6	4 03.8	14.1	58.5
20	122 01.4	38.6	144 19.0	11.7	3 49.7	14.1	58.5
21	137 01.7 · ·	39.6	158 49.7	11.7	3 35.6	14.1	58.5
22	152 01.9	40.6	173 20.4	11.7	3 21.5	14.1	58.5
23	167 02.1	41.6	187 51.1	11.8	3 07.4	14.1	58.5
	S.D. 16.0	d 1.0	S.D. 16.1		16.1		16.0

Lat.	Twilight Naut.	Twilight Civil	Sun-rise	Moonrise 22	Moonrise 23	Moonrise 24	Moonrise 25
°	h m	h m	h m	h m	h m	h m	h
N 72	03 03	04 33	05 42	23 41	26 07	02 07	04 1
N 70	03 22	04 41	05 43	24 09	00 09	02 20	04 2
68	03 37	04 48	05 44	24 30	00 30	02 29	04 2
66	03 49	04 53	05 44	24 47	00 47	02 38	04 2
64	03 59	04 57	05 45	25 00	01 00	02 44	04 2
62	04 07	05 01	05 45	25 11	01 11	02 50	04 2
60	04 14	05 04	05 46	25 20	01 20	02 55	04 2
N 58	04 20	05 07	05 46	25 29	01 29	03 00	04 2
56	04 25	05 09	05 46	00 07	01 36	03 04	04 2
54	04 30	05 12	05 47	00 17	01 42	03 07	04 3
52	04 33	05 13	05 47	00 25	01 48	03 10	04 3
50	04 37	05 15	05 47	00 33	01 53	03 13	04 3
45	04 44	05 18	05 48	00 49	02 04	03 19	04 3
N 40	04 49	05 21	05 48	01 02	02 14	03 25	04 3
35	04 54	05 23	05 48	01 13	02 22	03 29	04 3
30	04 57	05 25	05 49	01 22	02 29	03 33	04 3
20	05 01	05 27	05 49	01 39	02 40	03 40	04 3
N 10	05 04	05 28	05 49	01 54	02 51	03 46	04 3
0	05 05	05 29	05 49	02 07	03 01	03 51	04 4
S 10	05 04	05 28	05 49	02 20	03 10	03 57	04 4
20	05 01	05 27	05 49	02 35	03 21	04 03	04 4
30	04 57	05 25	05 49	02 51	03 32	04 10	04 4
35	04 54	05 23	05 48	03 01	03 39	04 13	04 4
40	04 50	05 21	05 48	03 11	03 47	04 18	04 4
45	04 44	05 19	05 48	03 24	03 55	04 23	04 4
S 50	04 37	05 15	05 47	03 39	04 06	04 29	04 4
52	04 34	05 13	05 47	03 46	04 11	04 32	04 5
54	04 30	05 12	05 47	03 54	04 16	04 35	04 5
56	04 25	05 10	05 47	04 03	04 22	04 38	04 5
58	04 20	05 07	05 46	04 13	04 29	04 42	04 5
S 60	04 14	05 04	05 46	04 24	04 37	04 46	04 5

Lat.	Sun-set	Twilight Civil	Twilight Naut.	Moonset 22	Moonset 23	Moonset 24	Moonset 25
°	h m	h m	h m	h m	h m	h m	h
N 72	18 01	19 09	20 38	18 16	17 38	17 11	16 4
N 70	18 00	19 01	20 19	17 46	17 23	17 05	16 5
68	17 59	18 55	20 04	17 23	17 11	17 01	16 5
66	17 59	18 50	19 53	17 05	17 01	16 56	16 5
64	17 58	18 46	19 43	16 51	16 52	16 53	16 5
62	17 58	18 42	19 36	16 39	16 45	16 50	16 5
60	17 58	18 39	19 29	16 28	16 39	16 47	16 5
N 58	17 57	18 37	19 23	16 19	16 33	16 45	16 5
56	17 57	18 34	19 18	16 11	16 28	16 43	16 5
54	17 57	18 32	19 14	16 04	16 24	16 41	16 5
52	17 57	18 30	19 10	15 57	16 20	16 39	16 5
50	17 57	18 29	19 07	15 51	16 16	16 37	16 5
45	17 56	18 25	19 00	15 38	16 08	16 34	16 5
N 40	17 56	18 23	18 55	15 28	16 01	16 31	16 5
35	17 56	18 21	18 51	15 19	15 55	16 28	17 0
30	17 56	18 20	18 47	15 11	15 50	16 26	17 0
20	17 56	18 18	18 43	14 57	15 41	16 22	17 0
N 10	17 56	18 17	18 41	14 44	15 33	16 18	17 0
0	17 56	18 16	18 40	14 33	15 25	16 15	17 0
S 10	17 56	18 17	18 41	14 21	15 17	16 11	17 0
20	17 56	18 18	18 44	14 09	15 09	16 08	17 0
30	17 57	18 21	18 48	13 54	15 00	16 03	17 0
35	17 57	18 22	18 52	13 46	14 54	16 01	17 0
40	17 57	18 24	18 56	13 36	14 48	15 58	17 0
45	17 58	18 27	19 02	13 25	14 40	15 55	17 0
S 50	17 58	18 31	19 09	13 11	14 31	15 51	17 0
52	17 59	18 32	19 12	13 04	14 27	15 49	17 0
54	17 59	18 34	19 17	12 57	14 23	15 47	17 0
56	17 59	18 37	19 21	12 49	14 18	15 45	17 1
58	18 00	18 39	19 26	12 40	14 12	15 42	17 1
S 60	18 00	18 42	19 32	12 29	14 05	15 39	17 1

Day	SUN Eqn. of Time 00ʰ	SUN Eqn. of Time 12ʰ	SUN Mer. Pass.	MOON Mer. Pass. Upper	MOON Mer. Pass. Lower	Age	Phase
	m s	m s	h m	h m	h m	d	
22	07 06	07 17	11 53	08 20	20 47	25	☾
23	07 27	07 38	11 52	09 13	21 38	26	
24	07 48	07 58	11 52	10 03	22 28	27	

SUN PLANETS	ARIES	MOON	v or d	Corr^n	v or d	Corr^n	v or d	Corr^n
° ′	° ′	° ′	′	′	′	′	′	′
0 00·0	0 00·0	0 00·0	0·0	0·0	6·0	0·1	12·0	0·1
0 00·3	0 00·3	0 00·2	0·1	0·0	6·1	0·1	12·1	0·1
0 00·5	0 00·5	0 00·5	0·2	0·0	6·2	0·1	12·2	0·1
0 00·8	0 00·8	0 00·7	0·3	0·0	6·3	0·1	12·3	0·1
0 01·0	0 01·0	0 01·0	0·4	0·0	6·4	0·1	12·4	0·1
0 01·3	0 01·3	0 01·2	0·5	0·0	6·5	0·1	12·5	0·1
0 01·5	0 01·5	0 01·4	0·6	0·0	6·6	0·1	12·6	0·1
0 01·8	0 01·8	0 01·7	0·7	0·0	6·7	0·1	12·7	0·1
0 02·0	0 02·0	0 01·9	0·8	0·0	6·8	0·1	12·8	0·1
0 02·3	0 02·3	0 02·1	0·9	0·0	6·9	0·1	12·9	0·1
0 02·5	0 02·5	0 02·4	1·0	0·0	7·0	0·1	13·0	0·1
0 02·8	0 02·8	0 02·6	1·1	0·0	7·1	0·1	13·1	0·1
0 03·0	0 03·0	0 02·9	1·2	0·0	7·2	0·1	13·2	0·1
0 03·3	0 03·3	0 03·1	1·3	0·0	7·3	0·1	13·3	0·1
0 03·5	0 03·5	0 03·3	1·4	0·0	7·4	0·1	13·4	0·1
0 03·8	0 03·8	0 03·6	1·5	0·0	7·5	0·1	13·5	0·1
0 04·0	0 04·0	0 03·8	1·6	0·0	7·6	0·1	13·6	0·1
0 04·3	0 04·3	0 04·1	1·7	0·0	7·7	0·1	13·7	0·1
0 04·5	0 04·5	0 04·3	1·8	0·0	7·8	0·1	13·8	0·1
0 04·8	0 04·8	0 04·5	1·9	0·0	7·9	0·1	13·9	0·1
0 05·0	0 05·0	0 04·8	2·0	0·0	8·0	0·1	14·0	0·1
0 05·3	0 05·3	0 05·0	2·1	0·0	8·1	0·1	14·1	0·1
0 05·5	0 05·5	0 05·2	2·2	0·0	8·2	0·1	14·2	0·1
0 05·8	0 05·8	0 05·5	2·3	0·0	8·3	0·1	14·3	0·1
0 06·0	0 06·0	0 05·7	2·4	0·0	8·4	0·1	14·4	0·1
0 06·3	0 06·3	0 06·0	2·5	0·0	8·5	0·1	14·5	0·1
0 06·5	0 06·5	0 06·2	2·6	0·0	8·6	0·1	14·6	0·1
0 06·8	0 06·8	0 06·4	2·7	0·0	8·7	0·1	14·7	0·1
0 07·0	0 07·0	0 06·7	2·8	0·0	8·8	0·1	14·8	0·1
0 07·3	0 07·3	0 06·9	2·9	0·0	8·9	0·1	14·9	0·1
0 07·5	0 07·5	0 07·2	3·0	0·0	9·0	0·1	15·0	0·1
0 07·8	0 07·8	0 07·4	3·1	0·0	9·1	0·1	15·1	0·1
0 08·0	0 08·0	0 07·6	3·2	0·0	9·2	0·1	15·2	0·1
0 08·3	0 08·3	0 07·9	3·3	0·0	9·3	0·1	15·3	0·1
0 08·5	0 08·5	0 08·1	3·4	0·0	9·4	0·1	15·4	0·1
0 08·8	0 08·8	0 08·4	3·5	0·0	9·5	0·1	15·5	0·1
0 09·0	0 09·0	0 08·6	3·6	0·0	9·6	0·1	15·6	0·1
0 09·3	0 09·3	0 08·8	3·7	0·0	9·7	0·1	15·7	0·1
0 09·5	0 09·5	0 09·1	3·8	0·0	9·8	0·1	15·8	0·1
0 09·8	0 09·8	0 09·3	3·9	0·0	9·9	0·1	15·9	0·1
0 10·0	0 10·0	0 09·5	4·0	0·0	10·0	0·1	16·0	0·1
0 10·3	0 10·3	0 09·8	4·1	0·0	10·1	0·1	16·1	0·1
0 10·5	0 10·5	0 10·0	4·2	0·0	10·2	0·1	16·2	0·1
0 10·8	0 10·8	0 10·3	4·3	0·0	10·3	0·1	16·3	0·1
0 11·0	0 11·0	0 10·5	4·4	0·0	10·4	0·1	16·4	0·1
0 11·3	0 11·3	0 10·7	4·5	0·0	10·5	0·1	16·5	0·1
0 11·5	0 11·5	0 11·0	4·6	0·0	10·6	0·1	16·6	0·1
0 11·8	0 11·8	0 11·2	4·7	0·0	10·7	0·1	16·7	0·1
0 12·0	0 12·0	0 11·5	4·8	0·0	10·8	0·1	16·8	0·1
0 12·3	0 12·3	0 11·7	4·9	0·0	10·9	0·1	16·9	0·1
0 12·5	0 12·5	0 11·9	5·0	0·0	11·0	0·1	17·0	0·1
0 12·8	0 12·8	0 12·2	5·1	0·0	11·1	0·1	17·1	0·1
0 13·0	0 13·0	0 12·4	5·2	0·0	11·2	0·1	17·2	0·1
0 13·3	0 13·3	0 12·6	5·3	0·0	11·3	0·1	17·3	0·1
0 13·5	0 13·5	0 12·9	5·4	0·0	11·4	0·1	17·4	0·1
0 13·8	0 13·8	0 13·1	5·5	0·0	11·5	0·1	17·5	0·1
0 14·0	0 14·0	0 13·4	5·6	0·0	11·6	0·1	17·6	0·1
0 14·3	0 14·3	0 13·6	5·7	0·0	11·7	0·1	17·7	0·1
0 14·5	0 14·5	0 13·8	5·8	0·0	11·8	0·1	17·8	0·1
0 14·8	0 14·8	0 14·1	5·9	0·0	11·9	0·1	17·9	0·1
0 15·0	0 15·0	0 14·3	6·0	0·1	12·0	0·1	18·0	0·2

1^m	SUN PLANETS	ARIES	MOON	v or d	Corr^n	v or d	Corr^n	v or d	Corr^n
s	° ′	° ′	° ′	′	′	′	′	′	′
00	0 15·0	0 15·0	0 14·3	0·0	0·0	6·0	0·2	12·0	0·3
01	0 15·3	0 15·3	0 14·6	0·1	0·0	6·1	0·2	12·1	0·3
02	0 15·5	0 15·5	0 14·8	0·2	0·0	6·2	0·2	12·2	0·3
03	0 15·8	0 15·8	0 15·0	0·3	0·0	6·3	0·2	12·3	0·3
04	0 16·0	0 16·0	0 15·3	0·4	0·0	6·4	0·2	12·4	0·3
05	0 16·3	0 16·3	0 15·5	0·5	0·0	6·5	0·2	12·5	0·3
06	0 16·5	0 16·5	0 15·7	0·6	0·0	6·6	0·2	12·6	0·3
07	0 16·8	0 16·8	0 16·0	0·7	0·0	6·7	0·2	12·7	0·3
08	0 17·0	0 17·0	0 16·2	0·8	0·0	6·8	0·2	12·8	0·3
09	0 17·3	0 17·3	0 16·5	0·9	0·0	6·9	0·2	12·9	0·3
10	0 17·5	0 17·5	0 16·7	1·0	0·0	7·0	0·2	13·0	0·3
11	0 17·8	0 17·8	0 16·9	1·1	0·0	7·1	0·2	13·1	0·3
12	0 18·0	0 18·0	0 17·2	1·2	0·0	7·2	0·2	13·2	0·3
13	0 18·3	0 18·3	0 17·4	1·3	0·0	7·3	0·2	13·3	0·3
14	0 18·5	0 18·6	0 17·7	1·4	0·0	7·4	0·2	13·4	0·3
15	0 18·8	0 18·8	0 17·9	1·5	0·0	7·5	0·2	13·5	0·3
16	0 19·0	0 19·1	0 18·1	1·6	0·0	7·6	0·2	13·6	0·3
17	0 19·3	0 19·3	0 18·4	1·7	0·0	7·7	0·2	13·7	0·3
18	0 19·5	0 19·6	0 18·6	1·8	0·0	7·8	0·2	13·8	0·3
19	0 19·8	0 19·8	0 18·9	1·9	0·0	7·9	0·2	13·9	0·3
20	0 20·0	0 20·1	0 19·1	2·0	0·1	8·0	0·2	14·0	0·4
21	0 20·3	0 20·3	0 19·3	2·1	0·1	8·1	0·2	14·1	0·4
22	0 20·5	0 20·6	0 19·6	2·2	0·1	8·2	0·2	14·2	0·4
23	0 20·8	0 20·8	0 19·8	2·3	0·1	8·3	0·2	14·3	0·4
24	0 21·0	0 21·1	0 20·0	2·4	0·1	8·4	0·2	14·4	0·4
25	0 21·3	0 21·3	0 20·3	2·5	0·1	8·5	0·2	14·5	0·4
26	0 21·5	0 21·6	0 20·5	2·6	0·1	8·6	0·2	14·6	0·4
27	0 21·8	0 21·8	0 20·8	2·7	0·1	8·7	0·2	14·7	0·4
28	0 22·0	0 22·1	0 21·0	2·8	0·1	8·8	0·2	14·8	0·4
29	0 22·3	0 22·3	0 21·2	2·9	0·1	8·9	0·2	14·9	0·4
30	0 22·5	0 22·6	0 21·5	3·0	0·1	9·0	0·2	15·0	0·4
31	0 22·8	0 22·8	0 21·7	3·1	0·1	9·1	0·2	15·1	0·4
32	0 23·0	0 23·1	0 22·0	3·2	0·1	9·2	0·2	15·2	0·4
33	0 23·3	0 23·3	0 22·2	3·3	0·1	9·3	0·2	15·3	0·4
34	0 23·5	0 23·6	0 22·4	3·4	0·1	9·4	0·2	15·4	0·4
35	0 23·8	0 23·8	0 22·7	3·5	0·1	9·5	0·2	15·5	0·4
36	0 24·0	0 24·1	0 22·9	3·6	0·1	9·6	0·2	15·6	0·4
37	0 24·3	0 24·3	0 23·1	3·7	0·1	9·7	0·2	15·7	0·4
38	0 24·5	0 24·6	0 23·4	3·8	0·1	9·8	0·2	15·8	0·4
39	0 24·8	0 24·8	0 23·6	3·9	0·1	9·9	0·2	15·9	0·4
40	0 25·0	0 25·1	0 23·9	4·0	0·1	10·0	0·3	16·0	0·4
41	0 25·3	0 25·3	0 24·1	4·1	0·1	10·1	0·3	16·1	0·4
42	0 25·5	0 25·6	0 24·3	4·2	0·1	10·2	0·3	16·2	0·4
43	0 25·8	0 25·8	0 24·6	4·3	0·1	10·3	0·3	16·3	0·4
44	0 26·0	0 26·1	0 24·8	4·4	0·1	10·4	0·3	16·4	0·4
45	0 26·3	0 26·3	0 25·1	4·5	0·1	10·5	0·3	16·5	0·4
46	0 26·5	0 26·6	0 25·3	4·6	0·1	10·6	0·3	16·6	0·4
47	0 26·8	0 26·8	0 25·5	4·7	0·1	10·7	0·3	16·7	0·4
48	0 27·0	0 27·1	0 25·8	4·8	0·1	10·8	0·3	16·8	0·4
49	0 27·3	0 27·3	0 26·0	4·9	0·1	10·9	0·3	16·9	0·4
50	0 27·5	0 27·6	0 26·2	5·0	0·1	11·0	0·3	17·0	0·4
51	0 27·8	0 27·8	0 26·5	5·1	0·1	11·1	0·3	17·1	0·4
52	0 28·0	0 28·1	0 26·7	5·2	0·1	11·2	0·3	17·2	0·4
53	0 28·3	0 28·3	0 27·0	5·3	0·1	11·3	0·3	17·3	0·4
54	0 28·5	0 28·6	0 27·2	5·4	0·1	11·4	0·3	17·4	0·4
55	0 28·8	0 28·8	0 27·4	5·5	0·1	11·5	0·3	17·5	0·4
56	0 29·0	0 29·1	0 27·7	5·6	0·1	11·6	0·3	17·6	0·4
57	0 29·3	0 29·3	0 27·9	5·7	0·1	11·7	0·3	17·7	0·4
58	0 29·5	0 29·6	0 28·2	5·8	0·1	11·8	0·3	17·8	0·4
59	0 29·8	0 29·8	0 28·4	5·9	0·1	11·9	0·3	17·9	0·4
60	0 30·0	0 30·1	0 28·6	6·0	0·2	12·0	0·3	18·0	0·5

2m	SUN PLANETS	ARIES	MOON	v or d	Corrn	v or d	Corrn	v or d	Corrn
s	° ′	° ′	° ′	′	′	′	′	′	′
00	0 30·0	0 30·1	0 28·6	0·0	0·0	6·0	0·3	12·0	0·5
01	0 30·3	0 30·3	0 28·9	0·1	0·0	6·1	0·3	12·1	0·5
02	0 30·5	0 30·6	0 29·1	0·2	0·0	6·2	0·3	12·2	0·5
03	0 30·8	0 30·8	0 29·3	0·3	0·0	6·3	0·3	12·3	0·5
04	0 31·0	0 31·1	0 29·6	0·4	0·0	6·4	0·3	12·4	0·5
05	0 31·3	0 31·3	0 29·8	0·5	0·0	6·5	0·3	12·5	0·5
06	0 31·5	0 31·6	0 30·1	0·6	0·0	6·6	0·3	12·6	0·5
07	0 31·8	0 31·8	0 30·3	0·7	0·0	6·7	0·3	12·7	0·5
08	0 32·0	0 32·1	0 30·5	0·8	0·0	6·8	0·3	12·8	0·5
09	0 32·3	0 32·3	0 30·8	0·9	0·0	6·9	0·3	12·9	0·5
10	0 32·5	0 32·6	0 31·0	1·0	0·0	7·0	0·3	13·0	0·5
11	0 32·8	0 32·8	0 31·3	1·1	0·0	7·1	0·3	13·1	0·5
12	0 33·0	0 33·1	0 31·5	1·2	0·1	7·2	0·3	13·2	0·6
13	0 33·3	0 33·3	0 31·7	1·3	0·1	7·3	0·3	13·3	0·6
14	0 33·5	0 33·6	0 32·0	1·4	0·1	7·4	0·3	13·4	0·6
15	0 33·8	0 33·8	0 32·2	1·5	0·1	7·5	0·3	13·5	0·6
16	0 34·0	0 34·1	0 32·5	1·6	0·1	7·6	0·3	13·6	0·6
17	0 34·3	0 34·3	0 32·7	1·7	0·1	7·7	0·3	13·7	0·6
18	0 34·5	0 34·6	0 32·9	1·8	0·1	7·8	0·3	13·8	0·6
19	0 34·8	0 34·8	0 33·2	1·9	0·1	7·9	0·3	13·9	0·6
20	0 35·0	0 35·1	0 33·4	2·0	0·1	8·0	0·3	14·0	0·6
21	0 35·3	0 35·3	0 33·6	2·1	0·1	8·1	0·3	14·1	0·6
22	0 35·5	0 35·6	0 33·9	2·2	0·1	8·2	0·3	14·2	0·6
23	0 35·8	0 35·8	0 34·1	2·3	0·1	8·3	0·3	14·3	0·6
24	0 36·0	0 36·1	0 34·4	2·4	0·1	8·4	0·4	14·4	0·6
25	0 36·3	0 36·3	0 34·6	2·5	0·1	8·5	0·4	14·5	0·6
26	0 36·5	0 36·6	0 34·8	2·6	0·1	8·6	0·4	14·6	0·6
27	0 36·8	0 36·9	0 35·1	2·7	0·1	8·7	0·4	14·7	0·6
28	0 37·0	0 37·1	0 35·3	2·8	0·1	8·8	0·4	14·8	0·6
29	0 37·3	0 37·4	0 35·6	2·9	0·1	8·9	0·4	14·9	0·6
30	0 37·5	0 37·6	0 35·8	3·0	0·1	9·0	0·4	15·0	0·6
31	0 37·8	0 37·9	0 36·0	3·1	0·1	9·1	0·4	15·1	0·6
32	0 38·0	0 38·1	0 36·3	3·2	0·1	9·2	0·4	15·2	0·6
33	0 38·3	0 38·4	0 36·5	3·3	0·1	9·3	0·4	15·3	0·6
34	0 38·5	0 38·6	0 36·7	3·4	0·1	9·4	0·4	15·4	0·6
35	0 38·8	0 38·9	0 37·0	3·5	0·1	9·5	0·4	15·5	0·6
36	0 39·0	0 39·1	0 37·2	3·6	0·2	9·6	0·4	15·6	0·7
37	0 39·3	0 39·4	0 37·5	3·7	0·2	9·7	0·4	15·7	0·7
38	0 39·5	0 39·6	0 37·7	3·8	0·2	9·8	0·4	15·8	0·7
39	0 39·8	0 39·9	0 37·9	3·9	0·2	9·9	0·4	15·9	0·7
40	0 40·0	0 40·1	0 38·2	4·0	0·2	10·0	0·4	16·0	0·7
41	0 40·3	0 40·4	0 38·4	4·1	0·2	10·1	0·4	16·1	0·7
42	0 40·5	0 40·6	0 38·7	4·2	0·2	10·2	0·4	16·2	0·7
43	0 40·8	0 40·9	0 38·9	4·3	0·2	10·3	0·4	16·3	0·7
44	0 41·0	0 41·1	0 39·1	4·4	0·2	10·4	0·4	16·4	0·7
45	0 41·3	0 41·4	0 39·4	4·5	0·2	10·5	0·4	16·5	0·7
46	0 41·5	0 41·6	0 39·6	4·6	0·2	10·6	0·4	16·6	0·7
47	0 41·8	0 41·9	0 39·8	4·7	0·2	10·7	0·4	16·7	0·7
48	0 42·0	0 42·1	0 40·1	4·8	0·2	10·8	0·5	16·8	0·7
49	0 42·3	0 42·4	0 40·3	4·9	0·2	10·9	0·5	16·9	0·7
50	0 42·5	0 42·6	0 40·6	5·0	0·2	11·0	0·5	17·0	0·7
51	0 42·8	0 42·9	0 40·8	5·1	0·2	11·1	0·5	17·1	0·7
52	0 43·0	0 43·1	0 41·0	5·2	0·2	11·2	0·5	17·2	0·7
53	0 43·3	0 43·4	0 41·3	5·3	0·2	11·3	0·5	17·3	0·7
54	0 43·5	0 43·6	0 41·5	5·4	0·2	11·4	0·5	17·4	0·7
55	0 43·8	0 43·9	0 41·8	5·5	0·2	11·5	0·5	17·5	0·7
56	0 44·0	0 44·1	0 42·0	5·6	0·2	11·6	0·5	17·6	0·7
57	0 44·3	0 44·4	0 42·2	5·7	0·2	11·7	0·5	17·7	0·7
58	0 44·5	0 44·6	0 42·5	5·8	0·2	11·8	0·5	17·8	0·7
59	0 44·8	0 44·9	0 42·7	5·9	0·2	11·9	0·5	17·9	0·7
60	0 45·0	0 45·1	0 43·0	6·0	0·3	12·0	0·5	18·0	0·8

3m	SUN PLANETS	ARIES	MOON	v or d	Corrn	v or d	Corrn	v or d	Corrn
s	° ′	° ′	° ′	′	′	′	′	′	′
00	0 45·0	0 45·1	0 43·0	0·0	0·0	6·0	0·4	12·0	0·
01	0 45·3	0 45·4	0 43·2	0·1	0·0	6·1	0·4	12·1	0·
02	0 45·5	0 45·6	0 43·4	0·2	0·0	6·2	0·4	12·2	0·
03	0 45·8	0 45·9	0 43·7	0·3	0·0	6·3	0·4	12·3	0·
04	0 46·0	0 46·1	0 43·9	0·4	0·0	6·4	0·4	12·4	0·
05	0 46·3	0 46·4	0 44·1	0·5	0·0	6·5	0·4	12·5	0·
06	0 46·5	0 46·6	0 44·4	0·6	0·0	6·6	0·4	12·6	0·
07	0 46·8	0 46·9	0 44·6	0·7	0·0	6·7	0·4	12·7	0·
08	0 47·0	0 47·1	0 44·9	0·8	0·0	6·8	0·4	12·8	0·
09	0 47·3	0 47·4	0 45·1	0·9	0·1	6·9	0·4	12·9	0·
10	0 47·5	0 47·6	0 45·3	1·0	0·1	7·0	0·4	13·0	0·
11	0 47·8	0 47·9	0 45·6	1·1	0·1	7·1	0·4	13·1	0·
12	0 48·0	0 48·1	0 45·8	1·2	0·1	7·2	0·4	13·2	0·
13	0 48·3	0 48·4	0 46·1	1·3	0·1	7·3	0·4	13·3	0·
14	0 48·5	0 48·6	0 46·3	1·4	0·1	7·4	0·4	13·4	0·
15	0 48·8	0 48·9	0 46·5	1·5	0·1	7·5	0·4	13·5	0·
16	0 49·0	0 49·1	0 46·8	1·6	0·1	7·6	0·4	13·6	0·
17	0 49·3	0 49·4	0 47·0	1·7	0·1	7·7	0·4	13·7	0·
18	0 49·5	0 49·6	0 47·2	1·8	0·1	7·8	0·5	13·8	0·
19	0 49·8	0 49·9	0 47·5	1·9	0·1	7·9	0·5	13·9	0·
20	0 50·0	0 50·1	0 47·7	2·0	0·1	8·0	0·5	14·0	0·
21	0 50·3	0 50·4	0 48·0	2·1	0·1	8·1	0·5	14·1	0·
22	0 50·5	0 50·6	0 48·2	2·2	0·1	8·2	0·5	14·2	0·
23	0 50·8	0 50·9	0 48·4	2·3	0·1	8·3	0·5	14·3	0·
24	0 51·0	0 51·1	0 48·7	2·4	0·1	8·4	0·5	14·4	0·
25	0 51·3	0 51·4	0 48·9	2·5	0·1	8·5	0·5	14·5	0·
26	0 51·5	0 51·6	0 49·2	2·6	0·2	8·6	0·5	14·6	0
27	0 51·8	0 51·9	0 49·4	2·7	0·2	8·7	0·5	14·7	0·
28	0 52·0	0 52·1	0 49·6	2·8	0·2	8·8	0·5	14·8	0·
29	0 52·3	0 52·4	0 49·9	2·9	0·2	8·9	0·5	14·9	0·
30	0 52·5	0 52·6	0 50·1	3·0	0·2	9·0	0·5	15·0	0·
31	0 52·8	0 52·9	0 50·3	3·1	0·2	9·1	0·5	15·1	0·
32	0 53·0	0 53·1	0 50·6	3·2	0·2	9·2	0·5	15·2	0·
33	0 53·3	0 53·4	0 50·8	3·3	0·2	9·3	0·5	15·3	0·
34	0 53·5	0 53·6	0 51·1	3·4	0·2	9·4	0·5	15·4	0·
35	0 53·8	0 53·9	0 51·3	3·5	0·2	9·5	0·6	15·5	0·
36	0 54·0	0 54·1	0 51·5	3·6	0·2	9·6	0·6	15·6	0
37	0 54·3	0 54·4	0 51·8	3·7	0·2	9·7	0·6	15·7	0·
38	0 54·5	0 54·6	0 52·0	3·8	0·2	9·8	0·6	15·8	0
39	0 54·8	0 54·9	0 52·3	3·9	0·2	9·9	0·6	15·9	0·
40	0 55·0	0 55·2	0 52·5	4·0	0·2	10·0	0·6	16·0	0·
41	0 55·3	0 55·4	0 52·7	4·1	0·2	10·1	0·6	16·1	0·
42	0 55·5	0 55·7	0 53·0	4·2	0·2	10·2	0·6	16·2	0
43	0 55·8	0 55·9	0 53·2	4·3	0·3	10·3	0·6	16·3	1
44	0 56·0	0 56·2	0 53·4	4·4	0·3	10·4	0·6	16·4	1
45	0 56·3	0 56·4	0 53·7	4·5	0·3	10·5	0·6	16·5	1
46	0 56·5	0 56·7	0 53·9	4·6	0·3	10·6	0·6	16·6	1·
47	0 56·8	0 56·9	0 54·2	4·7	0·3	10·7	0·6	16·7	1
48	0 57·0	0 57·2	0 54·4	4·8	0·3	10·8	0·6	16·8	1
49	0 57·3	0 57·4	0 54·6	4·9	0·3	10·9	0·6	16·9	1
50	0 57·5	0 57·7	0 54·9	5·0	0·3	11·0	0·6	17·0	1
51	0 57·8	0 57·9	0 55·1	5·1	0·3	11·1	0·6	17·1	1
52	0 58·0	0 58·2	0 55·4	5·2	0·3	11·2	0·7	17·2	1
53	0 58·3	0 58·4	0 55·6	5·3	0·3	11·3	0·7	17·3	1
54	0 58·5	0 58·7	0 55·8	5·4	0·3	11·4	0·7	17·4	1
55	0 58·8	0 58·9	0 56·1	5·5	0·3	11·5	0·7	17·5	1
56	0 59·0	0 59·2	0 56·3	5·6	0·3	11·6	0·7	17·6	1·
57	0 59·3	0 59·4	0 56·6	5·7	0·3	11·7	0·7	17·7	1
58	0 59·5	0 59·7	0 56·8	5·8	0·3	11·8	0·7	17·8	1
59	0 59·8	0 59·9	0 57·0	5·9	0·3	11·9	0·7	17·9	1
60	1 00·0	1 00·2	0 57·3	6·0	0·4	12·0	0·7	18·0	1

4m	SUN PLANETS	ARIES	MOON	v or d	Corrn	v or d	Corrn	v or d	Corrn
s	° ′	° ′	° ′	′	′	′	′	′	′
00	1 00·0	1 00·2	0 57·3	0·0	0·0	6·0	0·5	12·0	0·9
01	1 00·3	1 00·4	0 57·5	0·1	0·0	6·1	0·5	12·1	0·9
02	1 00·5	1 00·7	0 57·7	0·2	0·0	6·2	0·5	12·2	0·9
03	1 00·8	1 00·9	0 58·0	0·3	0·0	6·3	0·5	12·3	0·9
04	1 01·0	1 01·2	0 58·2	0·4	0·0	6·4	0·5	12·4	0·9
05	1 01·3	1 01·4	0 58·5	0·5	0·0	6·5	0·5	12·5	0·9
06	1 01·5	1 01·7	0 58·7	0·6	0·0	6·6	0·5	12·6	0·9
07	1 01·8	1 01·9	0 58·9	0·7	0·1	6·7	0·5	12·7	1·0
08	1 02·0	1 02·2	0 59·2	0·8	0·1	6·8	0·5	12·8	1·0
09	1 02·3	1 02·4	0 59·4	0·9	0·1	6·9	0·5	12·9	1·0
10	1 02·5	1 02·7	0 59·7	1·0	0·1	7·0	0·5	13·0	1·0
11	1 02·8	1 02·9	0 59·9	1·1	0·1	7·1	0·5	13·1	1·0
12	1 03·0	1 03·2	1 00·1	1·2	0·1	7·2	0·5	13·2	1·0
13	1 03·3	1 03·4	1 00·4	1·3	0·1	7·3	0·5	13·3	1·0
14	1 03·5	1 03·7	1 00·6	1·4	0·1	7·4	0·6	13·4	1·0
15	1 03·8	1 03·9	1 00·8	1·5	0·1	7·5	0·6	13·5	1·0
16	1 04·0	1 04·2	1 01·1	1·6	0·1	7·6	0·6	13·6	1·0
17	1 04·3	1 04·4	1 01·3	1·7	0·1	7·7	0·6	13·7	1·0
18	1 04·5	1 04·7	1 01·6	1·8	0·1	7·8	0·6	13·8	1·0
19	1 04·8	1 04·9	1 01·8	1·9	0·1	7·9	0·6	13·9	1·0
20	1 05·0	1 05·2	1 02·0	2·0	0·2	8·0	0·6	14·0	1·1
21	1 05·3	1 05·4	1 02·3	2·1	0·2	8·1	0·6	14·1	1·1
22	1 05·5	1 05·7	1 02·5	2·2	0·2	8·2	0·6	14·2	1·1
23	1 05·8	1 05·9	1 02·8	2·3	0·2	8·3	0·6	14·3	1·1
24	1 06·0	1 06·2	1 03·0	2·4	0·2	8·4	0·6	14·4	1·1
25	1 06·3	1 06·4	1 03·2	2·5	0·2	8·5	0·6	14·5	1·1
26	1 06·5	1 06·7	1 03·5	2·6	0·2	8·6	0·6	14·6	1·1
27	1 06·8	1 06·9	1 03·7	2·7	0·2	8·7	0·7	14·7	1·1
28	1 07·0	1 07·2	1 03·9	2·8	0·2	8·8	0·7	14·8	1·1
29	1 07·3	1 07·4	1 04·2	2·9	0·2	8·9	0·7	14·9	1·1
30	1 07·5	1 07·7	1 04·4	3·0	0·2	9·0	0·7	15·0	1·1
31	1 07·8	1 07·9	1 04·7	3·1	0·2	9·1	0·7	15·1	1·1
32	1 08·0	1 08·2	1 04·9	3·2	0·2	9·2	0·7	15·2	1·1
33	1 08·3	1 08·4	1 05·1	3·3	0·2	9·3	0·7	15·3	1·1
34	1 08·5	1 08·7	1 05·4	3·4	0·3	9·4	0·7	15·4	1·2
35	1 08·8	1 08·9	1 05·6	3·5	0·3	9·5	0·7	15·5	1·2
36	1 09·0	1 09·2	1 05·9	3·6	0·3	9·6	0·7	15·6	1·2
37	1 09·3	1 09·4	1 06·1	3·7	0·3	9·7	0·7	15·7	1·2
38	1 09·5	1 09·7	1 06·3	3·8	0·3	9·8	0·7	15·8	1·2
39	1 09·8	1 09·9	1 06·6	3·9	0·3	9·9	0·7	15·9	1·2
40	1 10·0	1 10·2	1 06·8	4·0	0·3	10·0	0·8	16·0	1·2
41	1 10·3	1 10·4	1 07·0	4·1	0·3	10·1	0·8	16·1	1·2
42	1 10·5	1 10·7	1 07·3	4·2	0·3	10·2	0·8	16·2	1·2
43	1 10·8	1 10·9	1 07·5	4·3	0·3	10·3	0·8	16·3	1·2
44	1 11·0	1 11·2	1 07·8	4·4	0·3	10·4	0·8	16·4	1·2
45	1 11·3	1 11·4	1 08·0	4·5	0·3	10·5	0·8	16·5	1·2
46	1 11·5	1 11·7	1 08·2	4·6	0·3	10·6	0·8	16·6	1·2
47	1 11·8	1 11·9	1 08·5	4·7	0·4	10·7	0·8	16·7	1·3
48	1 12·0	1 12·2	1 08·7	4·8	0·4	10·8	0·8	16·8	1·3
49	1 12·3	1 12·4	1 09·0	4·9	0·4	10·9	0·8	16·9	1·3
50	1 12·5	1 12·7	1 09·2	5·0	0·4	11·0	0·8	17·0	1·3
51	1 12·8	1 12·9	1 09·4	5·1	0·4	11·1	0·8	17·1	1·3
52	1 13·0	1 13·2	1 09·7	5·2	0·4	11·2	0·8	17·2	1·3
53	1 13·3	1 13·5	1 09·9	5·3	0·4	11·3	0·8	17·3	1·3
54	1 13·5	1 13·7	1 10·2	5·4	0·4	11·4	0·9	17·4	1·3
55	1 13·8	1 14·0	1 10·4	5·5	0·4	11·5	0·9	17·5	1·3
56	1 14·0	1 14·2	1 10·6	5·6	0·4	11·6	0·9	17·6	1·3
57	1 14·3	1 14·5	1 10·9	5·7	0·4	11·7	0·9	17·7	1·3
58	1 14·5	1 14·7	1 11·1	5·8	0·4	11·8	0·9	17·8	1·3
59	1 14·8	1 15·0	1 11·3	5·9	0·4	11·9	0·9	17·9	1·3
60	1 15·0	1 15·2	1 11·6	6·0	0·5	12·0	0·9	18·0	1·4

5m	SUN PLANETS	ARIES	MOON	v or d	Corrn	v or d	Corrn	v or d	Corrn
s	° ′	° ′	° ′	′	′	′	′	′	′
00	1 15·0	1 15·2	1 11·6	0·0	0·0	6·0	0·6	12·0	1·1
01	1 15·3	1 15·5	1 11·8	0·1	0·0	6·1	0·6	12·1	1·1
02	1 15·5	1 15·7	1 12·1	0·2	0·0	6·2	0·6	12·2	1·1
03	1 15·8	1 16·0	1 12·3	0·3	0·0	6·3	0·6	12·3	1·1
04	1 16·0	1 16·2	1 12·5	0·4	0·0	6·4	0·6	12·4	1·1
05	1 16·3	1 16·5	1 12·8	0·5	0·0	6·5	0·6	12·5	1·1
06	1 16·5	1 16·7	1 13·0	0·6	0·1	6·6	0·6	12·6	1·2
07	1 16·8	1 17·0	1 13·3	0·7	0·1	6·7	0·6	12·7	1·2
08	1 17·0	1 17·2	1 13·5	0·8	0·1	6·8	0·6	12·8	1·2
09	1 17·3	1 17·5	1 13·7	0·9	0·1	6·9	0·6	12·9	1·2
10	1 17·5	1 17·7	1 14·0	1·0	0·1	7·0	0·6	13·0	1·2
11	1 17·8	1 18·0	1 14·2	1·1	0·1	7·1	0·7	13·1	1·2
12	1 18·0	1 18·2	1 14·4	1·2	0·1	7·2	0·7	13·2	1·2
13	1 18·3	1 18·5	1 14·7	1·3	0·1	7·3	0·7	13·3	1·2
14	1 18·5	1 18·7	1 14·9	1·4	0·1	7·4	0·7	13·4	1·2
15	1 18·8	1 19·0	1 15·2	1·5	0·1	7·5	0·7	13·5	1·2
16	1 19·0	1 19·2	1 15·4	1·6	0·1	7·6	0·7	13·6	1·2
17	1 19·3	1 19·5	1 15·6	1·7	0·2	7·7	0·7	13·7	1·3
18	1 19·5	1 19·7	1 15·9	1·8	0·2	7·8	0·7	13·8	1·3
19	1 19·8	1 20·0	1 16·1	1·9	0·2	7·9	0·7	13·9	1·3
20	1 20·0	1 20·2	1 16·4	2·0	0·2	8·0	0·7	14·0	1·3
21	1 20·3	1 20·5	1 16·6	2·1	0·2	8·1	0·7	14·1	1·3
22	1 20·5	1 20·7	1 16·8	2·2	0·2	8·2	0·8	14·2	1·3
23	1 20·8	1 21·0	1 17·1	2·3	0·2	8·3	0·8	14·3	1·3
24	1 21·0	1 21·2	1 17·3	2·4	0·2	8·4	0·8	14·4	1·3
25	1 21·3	1 21·5	1 17·5	2·5	0·2	8·5	0·8	14·5	1·3
26	1 21·5	1 21·7	1 17·8	2·6	0·2	8·6	0·8	14·6	1·3
27	1 21·8	1 22·0	1 18·0	2·7	0·2	8·7	0·8	14·7	1·3
28	1 22·0	1 22·2	1 18·3	2·8	0·3	6·8	0·8	14·8	1·4
29	1 22·3	1 22·5	1 18·5	2·9	0·3	8·9	0·8	14·9	1·4
30	1 22·5	1 22·7	1 18·7	3·0	0·3	9·0	0·8	15·0	1·4
31	1 22·8	1 23·0	1 19·0	3·1	0·3	9·1	0·8	15·1	1·4
32	1 23·0	1 23·2	1 19·2	3·2	0·3	9·2	0·8	15·2	1·4
33	1 23·3	1 23·5	1 19·5	3·3	0·3	9·3	0·9	15·3	1·4
34	1 23·5	1 23·7	1 19·7	3·4	0·3	9·4	0·9	15·4	1·4
35	1 23·8	1 24·0	1 19·9	3·5	0·3	9·5	0·9	15·5	1·4
36	1 24·0	1 24·2	1 20·2	3·6	0·3	9·6	0·9	15·6	1·4
37	1 24·3	1 24·5	1 20·4	3·7	0·3	9·7	0·9	15·7	1·4
38	1 24·5	1 24·7	1 20·7	3·8	0·3	9·8	0·9	15·8	1·4
39	1 24·8	1 25·0	1 20·9	3·9	0·4	9·9	0·9	15·9	1·5
40	1 25·0	1 25·2	1 21·1	4·0	0·4	10·0	0·9	16·0	1·5
41	1 25·3	1 25·5	1 21·4	4·1	0·4	10·1	0·9	16·1	1·5
42	1 25·5	1 25·7	1 21·6	4·2	0·4	10·2	0·9	16·2	1·5
43	1 25·8	1 26·0	1 21·8	4·3	0·4	10·3	0·9	16·3	1·5
44	1 26·0	1 26·2	1 22·1	4·4	0·4	10·4	1·0	16·4	1·5
45	1 26·3	1 26·5	1 22·3	4·5	0·4	10·5	1·0	16·5	1·5
46	1 26·5	1 26·7	1 22·6	4·6	0·4	10·6	1·0	16·6	1·5
47	1 26·8	1 27·0	1 22·8	4·7	0·4	10·7	1·0	16·7	1·5
48	1 27·0	1 27·2	1 23·0	4·8	0·4	10·8	1·0	16·8	1·5
49	1 27·3	1 27·5	1 23·3	4·9	0·4	10·9	1·0	16·9	1·5
50	1 27·5	1 27·7	1 23·5	5·0	0·5	11·0	1·0	17·0	1·6
51	1 27·8	1 28·0	1 23·8	5·1	0·5	11·1	1·0	17·1	1·6
52	1 28·0	1 28·2	1 24·0	5·2	0·5	11·2	1·0	17·2	1·6
53	1 28·3	1 28·5	1 24·2	5·3	0·5	11·3	1·0	17·3	1·6
54	1 28·5	1 28·7	1 24·5	5·4	0·5	11·4	1·0	17·4	1·6
55	1 28·8	1 29·0	1 24·7	5·5	0·5	11·5	1·1	17·5	1·6
56	1 29·0	1 29·2	1 24·9	5·6	0·5	11·6	1·1	17·6	1·6
57	1 29·3	1 29·5	1 25·2	5·7	0·5	11·7	1·1	17·7	1·6
58	1 29·5	1 29·7	1 25·4	5·8	0·5	11·8	1·1	17·8	1·6
59	1 29·8	1 30·0	1 25·7	5·9	0·5	11·9	1·1	17·9	1·6
60	1 30·0	1 30·2	1 25·9	6·0	0·6	12·0	1·1	18·0	1·7

6^m	SUN PLANETS	ARIES	MOON	*v* or *d*	Corr^n	*v* or *d*	Corr^n	*v* or *d*	Corr^n
s	° ′	° ′	° ′	′	′	′	′	′	′
00	1 30·0	1 30·2	1 25·9	0·0	0·0	6·0	0·7	12·0	1·3
01	1 30·3	1 30·5	1 26·1	0·1	0·0	6·1	0·7	12·1	1·3
02	1 30·5	1 30·7	1 26·4	0·2	0·0	6·2	0·7	12·2	1·3
03	1 30·8	1 31·0	1 26·6	0·3	0·0	6·3	0·7	12·3	1·3
04	1 31·0	1 31·2	1 26·9	0·4	0·0	6·4	0·7	12·4	1·3
05	1 31·3	1 31·5	1 27·1	0·5	0·1	6·5	0·7	12·5	1·4
06	1 31·5	1 31·8	1 27·3	0·6	0·1	6·6	0·7	12·6	1·4
07	1 31·8	1 32·0	1 27·6	0·7	0·1	6·7	0·7	12·7	1·4
08	1 32·0	1 32·3	1 27·8	0·8	0·1	6·8	0·7	12·8	1·4
09	1 32·3	1 32·5	1 28·0	0·9	0·1	6·9	0·7	12·9	1·4
10	1 32·5	1 32·8	1 28·3	1·0	0·1	7·0	0·8	13·0	1·4
11	1 32·8	1 33·0	1 28·5	1·1	0·1	7·1	0·8	13·1	1·4
12	1 33·0	1 33·3	1 28·8	1·2	0·1	7·2	0·8	13·2	1·4
13	1 33·3	1 33·5	1 29·0	1·3	0·1	7·3	0·8	13·3	1·4
14	1 33·5	1 33·8	1 29·2	1·4	0·2	7·4	0·8	13·4	1·5
15	1 33·8	1 34·0	1 29·5	1·5	0·2	7·5	0·8	13·5	1·5
16	1 34·0	1 34·3	1 29·7	1·6	0·2	7·6	0·8	13·6	1·5
17	1 34·3	1 34·5	1 30·0	1·7	0·2	7·7	0·8	13·7	1·5
18	1 34·5	1 34·8	1 30·2	1·8	0·2	7·8	0·8	13·8	1·5
19	1 34·8	1 35·0	1 30·4	1·9	0·2	7·9	0·9	13·9	1·5
20	1 35·0	1 35·3	1 30·7	2·0	0·2	8·0	0·9	14·0	1·5
21	1 35·3	1 35·5	1 30·9	2·1	0·2	8·1	0·9	14·1	1·5
22	1 35·5	1 35·8	1 31·1	2·2	0·2	8·2	0·9	14·2	1·5
23	1 35·8	1 36·0	1 31·4	2·3	0·2	8·3	0·9	14·3	1·5
24	1 36·0	1 36·3	1 31·6	2·4	0·3	8·4	0·9	14·4	1·6
25	1 36·3	1 36·5	1 31·9	2·5	0·3	8·5	0·9	14·5	1·6
26	1 36·5	1 36·8	1 32·1	2·6	0·3	8·6	0·9	14·6	1·6
27	1 36·8	1 37·0	1 32·3	2·7	0·3	8·7	0·9	14·7	1·6
28	1 37·0	1 37·3	1 32·6	2·8	0·3	8·8	1·0	14·8	1·6
29	1 37·3	1 37·5	1 32·8	2·9	0·3	8·9	1·0	14·9	1·6
30	1 37·5	1 37·8	1 33·1	3·0	0·3	9·0	1·0	15·0	1·6
31	1 37·8	1 38·0	1 33·3	3·1	0·3	9·1	1·0	15·1	1·6
32	1 38·0	1 38·3	1 33·5	3·2	0·3	9·2	1·0	15·2	1·6
33	1 38·3	1 38·5	1 33·8	3·3	0·4	9·3	1·0	15·3	1·7
34	1 38·5	1 38·8	1 34·0	3·4	0·4	9·4	1·0	15·4	1·7
35	1 38·8	1 39·0	1 34·3	3·5	0·4	9·5	1·0	15·5	1·7
36	1 39·0	1 39·3	1 34·5	3·6	0·4	9·6	1·0	15·6	1·7
37	1 39·3	1 39·5	1 34·7	3·7	0·4	9·7	1·1	15·7	1·7
38	1 39·5	1 39·8	1 35·0	3·8	0·4	9·8	1·1	15·8	1·7
39	1 39·8	1 40·0	1 35·2	3·9	0·4	9·9	1·1	15·9	1·7
40	1 40·0	1 40·3	1 35·4	4·0	0·4	10·0	1·1	16·0	1·7
41	1 40·3	1 40·5	1 35·7	4·1	0·4	10·1	1·1	16·1	1·7
42	1 40·5	1 40·8	1 35·9	4·2	0·5	10·2	1·1	16·2	1·8
43	1 40·8	1 41·0	1 36·2	4·3	0·5	10·3	1·1	16·3	1·8
44	1 41·0	1 41·3	1 36·4	4·4	0·5	10·4	1·1	16·4	1·8
45	1 41·3	1 41·5	1 36·6	4·5	0·5	10·5	1·1	16·5	1·8
46	1 41·5	1 41·8	1 36·9	4·6	0·5	10·6	1·1	16·6	1·8
47	1 41·8	1 42·0	1 37·1	4·7	0·5	10·7	1·2	16·7	1·8
48	1 42·0	1 42·3	1 37·4	4·8	0·5	10·8	1·2	16·8	1·8
49	1 42·3	1 42·5	1 37·6	4·9	0·5	10·9	1·2	16·9	1·8
50	1 42·5	1 42·8	1 37·8	5·0	0·5	11·0	1·2	17·0	1·8
51	1 42·8	1 43·0	1 38·1	5·1	0·6	11·1	1·2	17·1	1·9
52	1 43·0	1 43·3	1 38·3	5·2	0·6	11·2	1·2	17·2	1·9
53	1 43·3	1 43·5	1 38·5	5·3	0·6	11·3	1·2	17·3	1·9
54	1 43·5	1 43·8	1 38·8	5·4	0·6	11·4	1·2	17·4	1·9
55	1 43·8	1 44·0	1 39·0	5·5	0·6	11·5	1·2	17·5	1·9
56	1 44·0	1 44·3	1 39·3	5·6	0·6	11·6	1·3	17·6	1·9
57	1 44·3	1 44·5	1 39·5	5·7	0·6	11·7	1·3	17·7	1·9
58	1 44·5	1 44·8	1 39·7	5·8	0·6	11·8	1·3	17·8	1·9
59	1 44·8	1 45·0	1 40·0	5·9	0·6	11·9	1·3	17·9	1·9
60	1 45·0	1 45·3	1 40·2	6·0	0·7	12·0	1·3	18·0	2·0

7^m	SUN PLANETS	ARIES	MOON	*v* or *d*	Corr^n	*v* or *d*	Corr^n	*v* or *d*	Corr^n
s	° ′	° ′	° ′	′	′	′	′	′	′
00	1 45·0	1 45·3	1 40·2	0·0	0·0	6·0	0·8	12·0	1·5
01	1 45·3	1 45·5	1 40·5	0·1	0·0	6·1	0·8	12·1	1·5
02	1 45·5	1 45·8	1 40·7	0·2	0·0	6·2	0·8	12·2	1·5
03	1 45·8	1 46·0	1 40·9	0·3	0·0	6·3	0·8	12·3	1·5
04	1 46·0	1 46·3	1 41·2	0·4	0·1	6·4	0·8	12·4	1·6
05	1 46·3	1 46·5	1 41·4	0·5	0·1	6·5	0·8	12·5	1·6
06	1 46·5	1 46·8	1 41·6	0·6	0·1	6·6	0·8	12·6	1·6
07	1 46·8	1 47·0	1 41·9	0·7	0·1	6·7	0·8	12·7	1·6
08	1 47·0	1 47·3	1 42·1	0·8	0·1	6·8	0·9	12·8	1·6
09	1 47·3	1 47·5	1 42·4	0·9	0·1	6·9	0·9	12·9	1·6
10	1 47·5	1 47·8	1 42·6	1·0	0·1	7·0	0·9	13·0	1·6
11	1 47·8	1 48·0	1 42·8	1·1	0·1	7·1	0·9	13·1	1·6
12	1 48·0	1 48·3	1 43·1	1·2	0·2	7·2	0·9	13·2	1·7
13	1 48·3	1 48·5	1 43·3	1·3	0·2	7·3	0·9	13·3	1·7
14	1 48·5	1 48·8	1 43·6	1·4	0·2	7·4	0·9	13·4	1·7
15	1 48·8	1 49·0	1 43·8	1·5	0·2	7·5	0·9	13·5	1·7
16	1 49·0	1 49·3	1 44·0	1·6	0·2	7·6	1·0	13·6	1·7
17	1 49·3	1 49·5	1 44·3	1·7	0·2	7·7	1·0	13·7	1·7
18	1 49·5	1 49·8	1 44·5	1·8	0·2	7·8	1·0	13·8	1·7
19	1 49·8	1 50·1	1 44·8	1·9	0·2	7·9	1·0	13·9	1·7
20	1 50·0	1 50·3	1 45·0	2·0	0·3	8·0	1·0	14·0	1·8
21	1 50·3	1 50·6	1 45·2	2·1	0·3	8·1	1·0	14·1	1·8
22	1 50·5	1 50·8	1 45·5	2·2	0·3	8·2	1·0	14·2	1·8
23	1 50·8	1 51·1	1 45·7	2·3	0·3	8·3	1·0	14·3	1·8
24	1 51·0	1 51·3	1 45·9	2·4	0·3	8·4	1·1	14·4	1·8
25	1 51·3	1 51·6	1 46·2	2·5	0·3	8·5	1·1	14·5	1·8
26	1 51·5	1 51·8	1 46·4	2·6	0·3	8·6	1·1	14·6	1·8
27	1 51·8	1 52·1	1 46·7	2·7	0·3	8·7	1·1	14·7	1·8
28	1 52·0	1 52·3	1 46·9	2·8	0·4	8·8	1·1	14·8	1·9
29	1 52·3	1 52·6	1 47·1	2·9	0·4	8·9	1·1	14·9	1·9
30	1 52·5	1 52·8	1 47·4	3·0	0·4	9·0	1·1	15·0	1·9
31	1 52·8	1 53·1	1 47·6	3·1	0·4	9·1	1·1	15·1	1·9
32	1 53·0	1 53·3	1 47·9	3·2	0·4	9·2	1·2	15·2	1·9
33	1 53·3	1 53·6	1 48·1	3·3	0·4	9·3	1·2	15·3	1·9
34	1 53·5	1 53·8	1 48·3	3·4	0·4	9·4	1·2	15·4	1·9
35	1 53·8	1 54·1	1 48·6	3·5	0·4	9·5	1·2	15·5	1·9
36	1 54·0	1 54·3	1 48·8	3·6	0·5	9·6	1·2	15·6	2·0
37	1 54·3	1 54·6	1 49·0	3·7	0·5	9·7	1·2	15·7	2·0
38	1 54·5	1 54·8	1 49·3	3·8	0·5	9·8	1·2	15·8	2·0
39	1 54·8	1 55·1	1 49·5	3·9	0·5	9·9	1·2	15·9	2·0
40	1 55·0	1 55·3	1 49·8	4·0	0·5	10·0	1·3	16·0	2·0
41	1 55·3	1 55·6	1 50·0	4·1	0·5	10·1	1·3	16·1	2·0
42	1 55·5	1 55·8	1 50·2	4·2	0·5	10·2	1·3	16·2	2·0
43	1 55·8	1 56·1	1 50·5	4·3	0·5	10·3	1·3	16·3	2·0
44	1 56·0	1 56·3	1 50·7	4·4	0·6	10·4	1·3	16·4	2·1
45	1 56·3	1 56·6	1 51·0	4·5	0·6	10·5	1·3	16·5	2·1
46	1 56·5	1 56·8	1 51·2	4·6	0·6	10·6	1·3	16·6	2·1
47	1 56·8	1 57·1	1 51·4	4·7	0·6	10·7	1·3	16·7	2·1
48	1 57·0	1 57·3	1 51·7	4·8	0·6	10·8	1·4	16·8	2·1
49	1 57·3	1 57·6	1 51·9	4·9	0·6	10·9	1·4	16·9	2·1
50	1 57·5	1 57·8	1 52·1	5·0	0·6	11·0	1·4	17·0	2·1
51	1 57·8	1 58·1	1 52·4	5·1	0·6	11·1	1·4	17·1	2·1
52	1 58·0	1 58·3	1 52·6	5·2	0·7	11·2	1·4	17·2	2·2
53	1 58·3	1 58·6	1 52·9	5·3	0·7	11·3	1·4	17·3	2·2
54	1 58·5	1 58·8	1 53·1	5·4	0·7	11·4	1·4	17·4	2·2
55	1 58·8	1 59·1	1 53·3	5·5	0·7	11·5	1·4	17·5	2·2
56	1 59·0	1 59·3	1 53·6	5·6	0·7	11·6	1·5	17·6	2·2
57	1 59·3	1 59·6	1 53·8	5·7	0·7	11·7	1·5	17·7	2·2
58	1 59·5	1 59·8	1 54·1	5·8	0·7	11·8	1·5	17·8	2·2
59	1 59·8	2 00·1	1 54·3	5·9	0·7	11·9	1·5	17·9	2·2
60	2 00·0	2 00·3	1 54·5	6·0	0·8	12·0	1·5	18·0	2·3

8ᵐ	SUN PLANETS	ARIES	MOON	v or d	Corrⁿ	v or d	Corrⁿ	v or d	Corrⁿ
s	° ′	° ′	° ′	′	′	′	′	′	′
00	2 00·0	2 00·3	1 54·5	0·0	0·0	6·0	0·9	12·0	1·7
01	2 00·3	2 00·6	1 54·8	0·1	0·0	6·1	0·9	12·1	1·7
02	2 00·5	2 00·8	1 55·0	0·2	0·0	6·2	0·9	12·2	1·7
03	2 00·8	2 01·1	1 55·2	0·3	0·0	6·3	0·9	12·3	1·7
04	2 01·0	2 01·3	1 55·5	0·4	0·1	6·4	0·9	12·4	1·8
05	2 01·3	2 01·6	1 55·7	0·5	0·1	6·5	0·9	12·5	1·8
06	2 01·5	2 01·8	1 56·0	0·6	0·1	6·6	0·9	12·6	1·8
07	2 01·8	2 02·1	1 56·2	0·7	0·1	6·7	0·9	12·7	1·8
08	2 02·0	2 02·3	1 56·4	0·8	0·1	6·8	1·0	12·8	1·8
09	2 02·3	2 02·6	1 56·7	0·9	0·1	6·9	1·0	12·9	1·8
10	2 02·5	2 02·8	1 56·9	1·0	0·1	7·0	1·0	13·0	1·8
11	2 02·8	2 03·1	1 57·2	1·1	0·2	7·1	1·0	13·1	1·9
12	2 03·0	2 03·3	1 57·4	1·2	0·2	7·2	1·0	13·2	1·9
13	2 03·3	2 03·6	1 57·6	1·3	0·2	7·3	1·0	13·3	1·9
14	2 03·5	2 03·8	1 57·9	1·4	0·2	7·4	1·0	13·4	1·9
15	2 03·8	2 04·1	1 58·1	1·5	0·2	7·5	1·1	13·5	1·9
16	2 04·0	2 04·3	1 58·4	1·6	0·2	7·6	1·1	13·6	1·9
17	2 04·3	2 04·6	1 58·6	1·7	0·2	7·7	1·1	13·7	1·9
18	2 04·5	2 04·8	1 58·8	1·8	0·3	7·8	1·1	13·8	2·0
19	2 04·8	2 05·1	1 59·1	1·9	0·3	7·9	1·1	13·9	2·0
20	2 05·0	2 05·3	1 59·3	2·0	0·3	8·0	1·1	14·0	2·0
21	2 05·3	2 05·6	1 59·5	2·1	0·3	8·1	1·1	14·1	2·0
22	2 05·5	2 05·8	1 59·8	2·2	0·3	8·2	1·2	14·2	2·0
23	2 05·8	2 06·1	2 00·0	2·3	0·3	8·3	1·2	14·3	2·0
24	2 06·0	2 06·3	2 00·3	2·4	0·3	8·4	1·2	14·4	2·0
25	2 06·3	2 06·6	2 00·5	2·5	0·4	8·5	1·2	14·5	2·1
26	2 06·5	2 06·8	2 00·7	2·6	0·4	8·6	1·2	14·6	2·1
27	2 06·8	2 07·1	2 01·0	2·7	0·4	8·7	1·2	14·7	2·1
28	2 07·0	2 07·3	2 01·2	2·8	0·4	8·8	1·2	14·8	2·1
29	2 07·3	2 07·6	2 01·5	2·9	0·4	8·9	1·3	14·9	2·1
30	2 07·5	2 07·8	2 01·7	3·0	0·4	9·0	1·3	15·0	2·1
31	2 07·8	2 08·1	2 01·9	3·1	0·4	9·1	1·3	15·1	2·1
32	2 08·0	2 08·4	2 02·2	3·2	0·5	9·2	1·3	15·2	2·2
33	2 08·3	2 08·6	2 02·4	3·3	0·5	9·3	1·3	15·3	2·2
34	2 08·5	2 08·9	2 02·6	3·4	0·5	9·4	1·3	15·4	2·2
35	2 08·8	2 09·1	2 02·9	3·5	0·5	9·5	1·3	15·5	2·2
36	2 09·0	2 09·4	2 03·1	3·6	0·5	9·6	1·4	15·6	2·2
37	2 09·3	2 09·6	2 03·4	3·7	0·5	9·7	1·4	15·7	2·2
38	2 09·5	2 09·9	2 03·6	3·8	0·5	9·8	1·4	15·8	2·2
39	2 09·8	2 10·1	2 03·8	3·9	0·6	9·9	1·4	15·9	2·3
40	2 10·0	2 10·4	2 04·1	4·0	0·6	10·0	1·4	16·0	2·3
41	2 10·3	2 10·6	2 04·3	4·1	0·6	10·1	1·4	16·1	2·3
42	2 10·5	2 10·9	2 04·6	4·2	0·6	10·2	1·4	16·2	2·3
43	2 10·8	2 11·1	2 04·8	4·3	0·6	10·3	1·5	16·3	2·3
44	2 11·0	2 11·4	2 05·0	4·4	0·6	10·4	1·5	16·4	2·3
45	2 11·3	2 11·6	2 05·3	4·5	0·6	10·5	1·5	16·5	2·3
46	2 11·5	2 11·9	2 05·5	4·6	0·7	10·6	1·5	16·6	2·4
47	2 11·8	2 12·1	2 05·7	4·7	0·7	10·7	1·5	16·7	2·4
48	2 12·0	2 12·4	2 06·0	4·8	0·7	10·8	1·5	16·8	2·4
49	2 12·3	2 12·6	2 06·2	4·9	0·7	10·9	1·5	16·9	2·4
50	2 12·5	2 12·9	2 06·5	5·0	0·7	11·0	1·6	17·0	2·4
51	2 12·8	2 13·1	2 06·7	5·1	0·7	11·1	1·6	17·1	2·4
52	2 13·0	2 13·4	2 06·9	5·2	0·7	11·2	1·6	17·2	2·4
53	2 13·3	2 13·6	2 07·2	5·3	0·8	11·3	1·6	17·3	2·5
54	2 13·5	2 13·9	2 07·4	5·4	0·8	11·4	1·6	17·4	2·5
55	2 13·8	2 14·1	2 07·7	5·5	0·8	11·5	1·6	17·5	2·5
56	2 14·0	2 14·4	2 07·9	5·6	0·8	11·6	1·6	17·6	2·5
57	2 14·3	2 14·6	2 08·1	5·7	0·8	11·7	1·7	17·7	2·5
58	2 14·5	2 14·9	2 08·4	5·8	0·8	11·8	1·7	17·8	2·5
59	2 14·8	2 15·1	2 08·6	5·9	0·8	11·9	1·7	17·9	2·5
60	2 15·0	2 15·4	2 08·9	6·0	0·9	12·0	1·7	18·0	2·6

9ᵐ	SUN PLANETS	ARIES	MOON	v or d	Corrⁿ	v or d	Corrⁿ	v or d	Corrⁿ
s	° ′	° ′	° ′	′	′	′	′	′	′
00	2 15·0	2 15·4	2 08·9	0·0	0·0	6·0	1·0	12·0	1·9
01	2 15·3	2 15·6	2 09·1	0·1	0·0	6·1	1·0	12·1	1·9
02	2 15·5	2 15·9	2 09·3	0·2	0·0	6·2	1·0	12·2	1·9
03	2 15·8	2 16·1	2 09·6	0·3	0·0	6·3	1·0	12·3	1·9
04	2 16·0	2 16·4	2 09·8	0·4	0·1	6·4	1·0	12·4	2·0
05	2 16·3	2 16·6	2 10·0	0·5	0·1	6·5	1·0	12·5	2·0
06	2 16·5	2 16·9	2 10·3	0·6	0·1	6·6	1·0	12·6	2·0
07	2 16·8	2 17·1	2 10·5	0·7	0·1	6·7	1·1	12·7	2·0
08	2 17·0	2 17·4	2 10·8	0·8	0·1	6·8	1·1	12·8	2·0
09	2 17·3	2 17·6	2 11·0	0·9	0·1	6·9	1·1	12·9	2·0
10	2 17·5	2 17·9	2 11·2	1·0	0·2	7·0	1·1	13·0	2·1
11	2 17·8	2 18·1	2 11·5	1·1	0·2	7·1	1·1	13·1	2·1
12	2 18·0	2 18·4	2 11·7	1·2	0·2	7·2	1·1	13·2	2·1
13	2 18·3	2 18·6	2 12·0	1·3	0·2	7·3	1·2	13·3	2·1
14	2 18·5	2 18·9	2 12·2	1·4	0·2	7·4	1·2	13·4	2·1
15	2 18·8	2 19·1	2 12·4	1·5	0·2	7·5	1·2	13·5	2·1
16	2 19·0	2 19·4	2 12·7	1·6	0·3	7·6	1·2	13·6	2·2
17	2 19·3	2 19·6	2 12·9	1·7	0·3	7·7	1·2	13·7	2·2
18	2 19·5	2 19·9	2 13·1	1·8	0·3	7·8	1·2	13·8	2·2
19	2 19·8	2 20·1	2 13·4	1·9	0·3	7·9	1·3	13·9	2·2
20	2 20·0	2 20·4	2 13·6	2·0	0·3	8·0	1·3	14·0	2·2
21	2 20·3	2 20·6	2 13·9	2·1	0·3	8·1	1·3	14·1	2·2
22	2 20·5	2 20·9	2 14·1	2·2	0·3	8·2	1·3	14·2	2·2
23	2 20·8	2 21·1	2 14·3	2·3	0·4	8·3	1·3	14·3	2·3
24	2 21·0	2 21·4	2 14·6	2·4	0·4	8·4	1·3	14·4	2·3
25	2 21·3	2 21·6	2 14·8	2·5	0·4	8·5	1·3	14·5	2·3
26	2 21·5	2 21·9	2 15·1	2·6	0·4	8·6	1·4	14·6	2·3
27	2 21·8	2 22·1	2 15·3	2·7	0·4	8·7	1·4	14·7	2·3
28	2 22·0	2 22·4	2 15·5	2·8	0·4	8·8	1·4	14·8	2·3
29	2 22·3	2 22·6	2 15·8	2·9	0·5	8·9	1·4	14·9	2·4
30	2 22·5	2 22·9	2 16·0	3·0	0·5	9·0	1·4	15·0	2·4
31	2 22·8	2 23·1	2 16·2	3·1	0·5	9·1	1·4	15·1	2·4
32	2 23·0	2 23·4	2 16·5	3·2	0·5	9·2	1·5	15·2	2·4
33	2 23·3	2 23·6	2 16·7	3·3	0·5	9·3	1·5	15·3	2·4
34	2 23·5	2 23·9	2 17·0	3·4	0·5	9·4	1·5	15·4	2·4
35	2 23·8	2 24·1	2 17·2	3·5	0·6	9·5	1·5	15·5	2·5
36	2 24·0	2 24·4	2 17·4	3·6	0·6	9·6	1·5	15·6	2·5
37	2 24·3	2 24·6	2 17·7	3·7	0·6	9·7	1·5	15·7	2·5
38	2 24·5	2 24·9	2 17·9	3·8	0·6	9·8	1·6	15·8	2·5
39	2 24·8	2 25·1	2 18·2	3·9	0·6	9·9	1·6	15·9	2·5
40	2 25·0	2 25·4	2 18·4	4·0	0·6	10·0	1·6	16·0	2·5
41	2 25·3	2 25·6	2 18·6	4·1	0·6	10·1	1·6	16·1	2·5
42	2 25·5	2 25·9	2 18·9	4·2	0·7	10·2	1·6	16·2	2·6
43	2 25·8	2 26·1	2 19·1	4·3	0·7	10·3	1·6	16·3	2·6
44	2 26·0	2 26·4	2 19·3	4·4	0·7	10·4	1·6	16·4	2·6
45	2 26·3	2 26·7	2 19·6	4·5	0·7	10·5	1·7	16·5	2·6
46	2 26·5	2 26·9	2 19·8	4·6	0·7	10·6	1·7	16·6	2·6
47	2 26·8	2 27·2	2 20·1	4·7	0·7	10·7	1·7	16·7	2·6
48	2 27·0	2 27·4	2 20·3	4·8	0·8	10·8	1·7	16·8	2·7
49	2 27·3	2 27·7	2 20·5	4·9	0·8	10·9	1·7	16·9	2·7
50	2 27·5	2 27·9	2 20·8	5·0	0·8	11·0	1·7	17·0	2·7
51	2 27·8	2 28·2	2 21·0	5·1	0·8	11·1	1·8	17·1	2·7
52	2 28·0	2 28·4	2 21·3	5·2	0·8	11·2	1·8	17·2	2·7
53	2 28·3	2 28·7	2 21·5	5·3	0·8	11·3	1·8	17·3	2·7
54	2 28·5	2 28·9	2 21·7	5·4	0·9	11·4	1·8	17·4	2·8
55	2 28·8	2 29·2	2 22·0	5·5	0·9	11·5	1·8	17·5	2·8
56	2 29·0	2 29·4	2 22·2	5·6	0·9	11·6	1·8	17·6	2·8
57	2 29·3	2 29·7	2 22·5	5·7	0·9	11·7	1·9	17·7	2·8
58	2 29·5	2 29·9	2 22·7	5·8	0·9	11·8	1·9	17·8	2·8
59	2 29·8	2 30·2	2 22·9	5·9	0·9	11·9	1·9	17·9	2·8
60	2 30·0	2 30·4	2 23·2	6·0	1·0	12·0	1·9	18·0	2·9

10ᵐ	SUN PLANETS	ARIES	MOON	v or d	Corrⁿ	v or d	Corrⁿ	v or d	Corrⁿ
s	° ′	° ′	° ′	′	′	′	′	′	′
00	2 30·0	2 30·4	2 23·2	0·0	0·0	6·0	1·1	12·0	2·1
01	2 30·3	2 30·7	2 23·4	0·1	0·0	6·1	1·1	12·1	2·1
02	2 30·5	2 30·9	2 23·6	0·2	0·0	6·2	1·1	12·2	2·1
03	2 30·8	2 31·2	2 23·9	0·3	0·1	6·3	1·1	12·3	2·2
04	2 31·0	2 31·4	2 24·1	0·4	0·1	6·4	1·1	12·4	2·2
05	2 31·3	2 31·7	2 24·4	0·5	0·1	6·5	1·1	12·5	2·2
06	2 31·5	2 31·9	2 24·6	0·6	0·1	6·6	1·2	12·6	2·2
07	2 31·8	2 32·2	2 24·8	0·7	0·1	6·7	1·2	12·7	2·2
08	2 32·0	2 32·4	2 25·1	0·8	0·1	6·8	1·2	12·8	2·2
09	2 32·3	2 32·7	2 25·3	0·9	0·2	6·9	1·2	12·9	2·3
10	2 32·5	2 32·9	2 25·6	1·0	0·2	7·0	1·2	13·0	2·3
11	2 32·8	2 33·2	2 25·8	1·1	0·2	7·1	1·2	13·1	2·3
12	2 33·0	2 33·4	2 26·0	1·2	0·2	7·2	1·3	13·2	2·3
13	2 33·3	2 33·7	2 26·3	1·3	0·2	7·3	1·3	13·3	2·3
14	2 33·5	2 33·9	2 26·5	1·4	0·2	7·4	1·3	13·4	2·3
15	2 33·8	2 34·2	2 26·7	1·5	0·3	7·5	1·3	13·5	2·4
16	2 34·0	2 34·4	2 27·0	1·6	0·3	7·6	1·3	13·6	2·4
17	2 34·3	2 34·7	2 27·2	1·7	0·3	7·7	1·3	13·7	2·4
18	2 34·5	2 34·9	2 27·5	1·8	0·3	7·8	1·4	13·8	2·4
19	2 34·8	2 35·2	2 27·7	1·9	0·3	7·9	1·4	13·9	2·4
20	2 35·0	2 35·4	2 27·9	2·0	0·4	8·0	1·4	14·0	2·5
21	2 35·3	2 35·7	2 28·2	2·1	0·4	8·1	1·4	14·1	2·5
22	2 35·5	2 35·9	2 28·4	2·2	0·4	8·2	1·4	14·2	2·5
23	2 35·8	2 36·2	2 28·7	2·3	0·4	8·3	1·5	14·3	2·5
24	2 36·0	2 36·4	2 28·9	2·4	0·4	8·4	1·5	14·4	2·5
25	2 36·3	2 36·7	2 29·1	2·5	0·4	8·5	1·5	14·5	2·5
26	2 36·5	2 36·9	2 29·4	2·6	0·5	8·6	1·5	14·6	2·6
27	2 36·8	2 37·2	2 29·6	2·7	0·5	8·7	1·5	14·7	2·6
28	2 37·0	2 37·4	2 29·8	2·8	0·5	8·8	1·5	14·8	2·6
29	2 37·3	2 37·7	2 30·1	2·9	0·5	8·9	1·6	14·9	2·6
30	2 37·5	2 37·9	2 30·3	3·0	0·5	9·0	1·6	15·0	2·6
31	2 37·8	2 38·2	2 30·6	3·1	0·5	9·1	1·6	15·1	2·6
32	2 38·0	2 38·4	2 30·8	3·2	0·6	9·2	1·6	15·2	2·7
33	2 38·3	2 38·7	2 31·0	3·3	0·6	9·3	1·6	15·3	2·7
34	2 38·5	2 38·9	2 31·3	3·4	0·6	9·4	1·6	15·4	2·7
35	2 38·8	2 39·2	2 31·5	3·5	0·6	9·5	1·7	15·5	2·7
36	2 39·0	2 39·4	2 31·8	3·6	0·6	9·6	1·7	15·6	2·7
37	2 39·3	2 39·7	2 32·0	3·7	0·6	9·7	1·7	15·7	2·7
38	2 39·5	2 39·9	2 32·2	3·8	0·7	9·8	1·7	15·8	2·8
39	2 39·8	2 40·2	2 32·5	3·9	0·7	9·9	1·7	15·9	2·8
40	2 40·0	2 40·4	2 32·7	4·0	0·7	10·0	1·8	16·0	2·8
41	2 40·3	2 40·7	2 32·9	4·1	0·7	10·1	1·8	16·1	2·8
42	2 40·5	2 40·9	2 33·2	4·2	0·7	10·2	1·8	16·2	2·8
43	2 40·8	2 41·2	2 33·4	4·3	0·8	10·3	1·8	16·3	2·9
44	2 41·0	2 41·4	2 33·7	4·4	0·8	10·4	1·8	16·4	2·9
45	2 41·3	2 41·7	2 33·9	4·5	0·8	10·5	1·8	16·5	2·9
46	2 41·5	2 41·9	2 34·1	4·6	0·8	10·6	1·9	16·6	2·9
47	2 41·8	2 42·2	2 34·4	4·7	0·8	10·7	1·9	16·7	2·9
48	2 42·0	2 42·4	2 34·6	4·8	0·8	10·8	1·9	16·8	2·9
49	2 42·3	2 42·7	2 34·9	4·9	0·9	10·9	1·9	16·9	3·0
50	2 42·5	2 42·9	2 35·1	5·0	0·9	11·0	1·9	17·0	3·0
51	2 42·8	2 43·2	2 35·3	5·1	0·9	11·1	1·9	17·1	3·0
52	2 43·0	2 43·4	2 35·6	5·2	0·9	11·2	2·0	17·2	3·0
53	2 43·3	2 43·7	2 35·8	5·3	0·9	11·3	2·0	17·3	3·0
54	2 43·5	2 43·9	2 36·1	5·4	0·9	11·4	2·0	17·4	3·0
55	2 43·8	2 44·2	2 36·3	5·5	1·0	11·5	2·0	17·5	3·1
56	2 44·0	2 44·4	2 36·5	5·6	1·0	11·6	2·0	17·6	3·1
57	2 44·3	2 44·7	2 36·8	5·7	1·0	11·7	2·0	17·7	3·1
58	2 44·5	2 45·0	2 37·0	5·8	1·0	11·8	2·1	17·8	3·1
59	2 44·8	2 45·2	2 37·2	5·9	1·0	11·9	2·1	17·9	3·1
60	2 45·0	2 45·5	2 37·5	6·0	1·1	12·0	2·1	18·0	3·2

11ᵐ	SUN PLANETS	ARIES	MOON	v or d	Corrⁿ	v or d	Corrⁿ	v or d	Corrⁿ
s	° ′	° ′	° ′	′	′	′	′	′	′
00	2 45·0	2 45·5	2 37·5	0·0	0·0	6·0	1·2	12·0	2·3
01	2 45·3	2 45·7	2 37·7	0·1	0·0	6·1	1·2	12·1	2·3
02	2 45·5	2 46·0	2 38·0	0·2	0·0	6·2	1·2	12·2	2·3
03	2 45·8	2 46·2	2 38·2	0·3	0·1	6·3	1·2	12·3	2·4
04	2 46·0	2 46·5	2 38·4	0·4	0·1	6·4	1·2	12·4	2·4
05	2 46·3	2 46·7	2 38·7	0·5	0·1	6·5	1·2	12·5	2·4
06	2 46·5	2 47·0	2 38·9	0·6	0·1	6·6	1·3	12·6	2·4
07	2 46·8	2 47·2	2 39·2	0·7	0·1	6·7	1·3	12·7	2·4
08	2 47·0	2 47·5	2 39·4	0·8	0·2	6·8	1·3	12·8	2·5
09	2 47·3	2 47·7	2 39·6	0·9	0·2	6·9	1·3	12·9	2·5
10	2 47·5	2 48·0	2 39·9	1·0	0·2	7·0	1·3	13·0	2·5
11	2 47·8	2 48·2	2 40·1	1·1	0·2	7·1	1·4	13·1	2·5
12	2 48·0	2 48·5	2 40·3	1·2	0·2	7·2	1·4	13·2	2·5
13	2 48·3	2 48·7	2 40·6	1·3	0·2	7·3	1·4	13·3	2·5
14	2 48·5	2 49·0	2 40·8	1·4	0·3	7·4	1·4	13·4	2·6
15	2 48·8	2 49·2	2 41·1	1·5	0·3	7·5	1·4	13·5	2·6
16	2 49·0	2 49·5	2 41·3	1·6	0·3	7·6	1·5	13·6	2·6
17	2 49·3	2 49·7	2 41·5	1·7	0·3	7·7	1·5	13·7	2·6
18	2 49·5	2 50·0	2 41·8	1·8	0·3	7·8	1·5	13·8	2·6
19	2 49·8	2 50·2	2 42·0	1·9	0·4	7·9	1·5	13·9	2·7
20	2 50·0	2 50·5	2 42·3	2·0	0·4	8·0	1·5	14·0	2·7
21	2 50·3	2 50·7	2 42·5	2·1	0·4	8·1	1·6	14·1	2·7
22	2 50·5	2 51·0	2 42·7	2·2	0·4	8·2	1·6	14·2	2·7
23	2 50·8	2 51·2	2 43·0	2·3	0·4	8·3	1·6	14·3	2·7
24	2 51·0	2 51·5	2 43·2	2·4	0·5	8·4	1·6	14·4	2·8
25	2 51·3	2 51·7	2 43·4	2·5	0·5	8·5	1·6	14·5	2·8
26	2 51·5	2 52·0	2 43·7	2·6	0·5	8·6	1·6	14·6	2·8
27	2 51·8	2 52·2	2 43·9	2·7	0·5	8·7	1·7	14·7	2·8
28	2 52·0	2 52·5	2 44·2	2·8	0·5	8·8	1·7	14·8	2·8
29	2 52·3	2 52·7	2 44·4	2·9	0·6	8·9	1·7	14·9	2·9
30	2 52·5	2 53·0	2 44·6	3·0	0·6	9·0	1·7	15·0	2·9
31	2 52·8	2 53·2	2 44·9	3·1	0·6	9·1	1·7	15·1	2·9
32	2 53·0	2 53·5	2 45·1	3·2	0·6	9·2	1·8	15·2	2·9
33	2 53·3	2 53·7	2 45·4	3·3	0·6	9·3	1·8	15·3	2·9
34	2 53·5	2 54·0	2 45·6	3·4	0·7	9·4	1·8	15·4	3·0
35	2 53·8	2 54·2	2 45·8	3·5	0·7	9·5	1·8	15·5	3·0
36	2 54·0	2 54·5	2 46·1	3·6	0·7	9·6	1·8	15·6	3·0
37	2 54·3	2 54·7	2 46·3	3·7	0·7	9·7	1·9	15·7	3·0
38	2 54·5	2 55·0	2 46·6	3·8	0·7	9·8	1·9	15·8	3·0
39	2 54·8	2 55·2	2 46·8	3·9	0·7	9·9	1·9	15·9	3·0
40	2 55·0	2 55·5	2 47·0	4·0	0·8	10·0	1·9	16·0	3·1
41	2 55·3	2 55·7	2 47·3	4·1	0·8	10·1	1·9	16·1	3·1
42	2 55·5	2 56·0	2 47·5	4·2	0·8	10·2	2·0	16·2	3·1
43	2 55·8	2 56·2	2 47·7	4·3	0·8	10·3	2·0	16·3	3·1
44	2 56·0	2 56·5	2 48·0	4·4	0·8	10·4	2·0	16·4	3·1
45	2 56·3	2 56·7	2 48·2	4·5	0·9	10·5	2·0	16·5	3·2
46	2 56·5	2 57·0	2 48·5	4·6	0·9	10·6	2·0	16·6	3·2
47	2 56·8	2 57·2	2 48·7	4·7	0·9	10·7	2·1	16·7	3·2
48	2 57·0	2 57·5	2 48·9	4·8	0·9	10·8	2·1	16·8	3·2
49	2 57·3	2 57·7	2 49·2	4·9	0·9	10·9	2·1	16·9	3·2
50	2 57·5	2 58·0	2 49·4	5·0	1·0	11·0	2·1	17·0	3·3
51	2 57·8	2 58·2	2 49·7	5·1	1·0	11·1	2·1	17·1	3·3
52	2 58·0	2 58·5	2 49·9	5·2	1·0	11·2	2·1	17·2	3·3
53	2 58·3	2 58·7	2 50·1	5·3	1·0	11·3	2·2	17·3	3·3
54	2 58·5	2 59·0	2 50·4	5·4	1·0	11·4	2·2	17·4	3·3
55	2 58·8	2 59·2	2 50·6	5·5	1·1	11·5	2·2	17·5	3·4
56	2 59·0	2 59·5	2 50·8	5·6	1·1	11·6	2·2	17·6	3·4
57	2 59·3	2 59·7	2 51·1	5·7	1·1	11·7	2·2	17·7	3·4
58	2 59·5	3 00·0	2 51·3	5·8	1·1	11·8	2·3	17·8	3·4
59	2 59·8	3 00·2	2 51·6	5·9	1·1	11·9	2·3	17·9	3·4
60	3 00·0	3 00·5	2 51·8	6·0	1·2	12·0	2·3	18·0	3·5

12m s	SUN PLANETS	ARIES	MOON	v or d	Corrⁿ	v or d	Corrⁿ	v or d	Corrⁿ
s	° ′	° ′	° ′	′	′	′	′	′	′
00	3 00·0	3 00·5	2 51·8	0·0	0·0	6·0	1·3	12·0	2·5
01	3 00·3	3 00·7	2 52·0	0·1	0·0	6·1	1·3	12·1	2·5
02	3 00·5	3 01·0	2 52·3	0·2	0·0	6·2	1·3	12·2	2·5
03	3 00·8	3 01·2	2 52·5	0·3	0·1	6·3	1·3	12·3	2·6
04	3 01·0	3 01·5	2 52·8	0·4	0·1	6·4	1·3	12·4	2·6
05	3 01·3	3 01·7	2 53·0	0·5	0·1	6·5	1·4	12·5	2·6
06	3 01·5	3 02·0	2 53·2	0·6	0·1	6·6	1·4	12·6	2·6
07	3 01·8	3 02·2	2 53·5	0·7	0·1	6·7	1·4	12·7	2·6
08	3 02·0	3 02·5	2 53·7	0·8	0·2	6·8	1·4	12·8	2·7
09	3 02·3	3 02·7	2 53·9	0·9	0·2	6·9	1·4	12·9	2·7
10	3 02·5	3 03·0	2 54·2	1·0	0·2	7·0	1·5	13·0	2·7
11	3 02·8	3 03·3	2 54·4	1·1	0·2	7·1	1·5	13·1	2·7
12	3 03·0	3 03·5	2 54·7	1·2	0·3	7·2	1·5	13·2	2·8
13	3 03·3	3 03·8	2 54·9	1·3	0·3	7·3	1·5	13·3	2·8
14	3 03·5	3 04·0	2 55·1	1·4	0·3	7·4	1·5	13·4	2·8
15	3 03·8	3 04·3	2 55·4	1·5	0·3	7·5	1·6	13·5	2·8
16	3 04·0	3 04·5	2 55·6	1·6	0·3	7·6	1·6	13·6	2·8
17	3 04·3	3 04·8	2 55·9	1·7	0·4	7·7	1·6	13·7	2·9
18	3 04·5	3 05·0	2 56·1	1·8	0·4	7·8	1·6	13·8	2·9
19	3 04·8	3 05·3	2 56·3	1·9	0·4	7·9	1·6	13·9	2·9
20	3 05·0	3 05·5	2 56·6	2·0	0·4	8·0	1·7	14·0	2·9
21	3 05·3	3 05·8	2 56·8	2·1	0·4	8·1	1·7	14·1	2·9
22	3 05·5	3 06·0	2 57·0	2·2	0·5	8·2	1·7	14·2	3·0
23	3 05·8	3 06·3	2 57·3	2·3	0·5	8·3	1·7	14·3	3·0
24	3 06·0	3 06·5	2 57·5	2·4	0·5	8·4	1·8	14·4	3·0
25	3 06·3	3 06·8	2 57·8	2·5	0·5	8·5	1·8	14·5	3·0
26	3 06·5	3 07·0	2 58·0	2·6	0·5	8·6	1·8	14·6	3·0
27	3 06·8	3 07·3	2 58·2	2·7	0·6	8·7	1·8	14·7	3·1
28	3 07·0	3 07·5	2 58·5	2·8	0·6	8·8	1·8	14·8	3·1
29	3 07·3	3 07·8	2 58·7	2·9	0·6	8·9	1·9	14·9	3·1
30	3 07·5	3 08·0	2 59·0	3·0	0·6	9·0	1·9	15·0	3·1
31	3 07·8	3 08·3	2 59·2	3·1	0·6	9·1	1·9	15·1	3·1
32	3 08·0	3 08·5	2 59·4	3·2	0·7	9·2	1·9	15·2	3·2
33	3 08·3	3 08·8	2 59·7	3·3	0·7	9·3	1·9	15·3	3·2
34	3 08·5	3 09·0	2 59·9	3·4	0·7	9·4	2·0	15·4	3·2
35	3 08·8	3 09·3	3 00·2	3·5	0·7	9·5	2·0	15·5	3·2
36	3 09·0	3 09·5	3 00·4	3·6	0·8	9·6	2·0	15·6	3·3
37	3 09·3	3 09·8	3 00·6	3·7	0·8	9·7	2·0	15·7	3·3
38	3 09·5	3 10·0	3 00·9	3·8	0·8	9·8	2·0	15·8	3·3
39	3 09·8	3 10·3	3 01·1	3·9	0·8	9·9	2·1	15·9	3·3
40	3 10·0	3 10·5	3 01·3	4·0	0·8	10·0	2·1	16·0	3·3
41	3 10·3	3 10·8	3 01·6	4·1	0·9	10·1	2·1	16·1	3·4
42	3 10·5	3 11·0	3 01·8	4·2	0·9	10·2	2·1	16·2	3·4
43	3 10·8	3 11·3	3 02·1	4·3	0·9	10·3	2·1	16·3	3·4
44	3 11·0	3 11·5	3 02·3	4·4	0·9	10·4	2·2	16·4	3·4
45	3 11·3	3 11·8	3 02·5	4·5	0·9	10·5	2·2	16·5	3·4
46	3 11·5	3 12·0	3 02·8	4·6	1·0	10·6	2·2	16·6	3·5
47	3 11·8	3 12·3	3 03·0	4·7	1·0	10·7	2·2	16·7	3·5
48	3 12·0	3 12·5	3 03·3	4·8	1·0	10·8	2·3	16·8	3·5
49	3 12·3	3 12·8	3 03·5	4·9	1·0	10·9	2·3	16·9	3·5
50	3 12·5	3 13·0	3 03·7	5·0	1·0	11·0	2·3	17·0	3·5
51	3 12·8	3 13·3	3 04·0	5·1	1·1	11·1	2·3	17·1	3·6
52	3 13·0	3 13·5	3 04·2	5·2	1·1	11·2	2·3	17·2	3·6
53	3 13·3	3 13·8	3 04·4	5·3	1·1	11·3	2·4	17·3	3·6
54	3 13·5	3 14·0	3 04·7	5·4	1·1	11·4	2·4	17·4	3·6
55	3 13·8	3 14·3	3 04·9	5·5	1·1	11·5	2·4	17·5	3·6
56	3 14·0	3 14·5	3 05·2	5·6	1·2	11·6	2·4	17·6	3·7
57	3 14·3	3 14·8	3 05·4	5·7	1·2	11·7	2·4	17·7	3·7
58	3 14·5	3 15·0	3 05·6	5·8	1·2	11·8	2·5	17·8	3·7
59	3 14·8	3 15·3	3 05·9	5·9	1·2	11·9	2·5	17·9	3·7
60	3 15·0	3 15·5	3 06·1	6·0	1·3	12·0	2·5	18·0	3·8

13m s	SUN PLANETS	ARIES	MOON	v or d	Corrⁿ	v or d	Corrⁿ	v or d	Corrⁿ
s	° ′	° ′	° ′	′	′	′	′	′	′
00	3 15·0	3 15·5	3 06·1	0·0	0·0	6·0	1·4	12·0	2·7
01	3 15·3	3 15·8	3 06·4	0·1	0·0	6·1	1·4	12·1	2·7
02	3 15·5	3 16·0	3 06·6	0·2	0·0	6·2	1·4	12·2	2·7
03	3 15·8	3 16·3	3 06·8	0·3	0·1	6·3	1·4	12·3	2·8
04	3 16·0	3 16·5	3 07·1	0·4	0·1	6·4	1·4	12·4	2·8
05	3 16·3	3 16·8	3 07·3	0·5	0·1	6·5	1·5	12·5	2·8
06	3 16·5	3 17·0	3 07·5	0·6	0·1	6·6	1·5	12·6	2·8
07	3 16·8	3 17·3	3 07·8	0·7	0·2	6·7	1·5	12·7	2·9
08	3 17·0	3 17·5	3 08·0	0·8	0·2	6·8	1·5	12·8	2·9
09	3 17·3	3 17·8	3 08·3	0·9	0·2	6·9	1·6	12·9	2·9
10	3 17·5	3 18·0	3 08·5	1·0	0·2	7·0	1·6	13·0	2·9
11	3 17·8	3 18·3	3 08·7	1·1	0·2	7·1	1·6	13·1	2·9
12	3 18·0	3 18·5	3 09·0	1·2	0·3	7·2	1·6	13·2	3·0
13	3 18·3	3 18·8	3 09·2	1·3	0·3	7·3	1·6	13·3	3·0
14	3 18·5	3 19·0	3 09·5	1·4	0·3	7·4	1·7	13·4	3·0
15	3 18·8	3 19·3	3 09·7	1·5	0·3	7·5	1·7	13·5	3·0
16	3 19·0	3 19·5	3 09·9	1·6	0·4	7·6	1·7	13·6	3·1
17	3 19·3	3 19·8	3 10·2	1·7	0·4	7·7	1·7	13·7	3·1
18	3 19·5	3 20·0	3 10·4	1·8	0·4	7·8	1·8	13·8	3·1
19	3 19·8	3 20·3	3 10·7	1·9	0·4	7·9	1·8	13·9	3·1
20	3 20·0	3 20·5	3 10·9	2·0	0·5	8·0	1·8	14·0	3·2
21	3 20·3	3 20·8	3 11·1	2·1	0·5	8·1	1·8	14·1	3·2
22	3 20·5	3 21·0	3 11·4	2·2	0·5	8·2	1·8	14·2	3·2
23	3 20·8	3 21·3	3 11·6	2·3	0·5	8·3	1·9	14·3	3·2
24	3 21·0	3 21·6	3 11·8	2·4	0·5	8·4	1·9	14·4	3·2
25	3 21·3	3 21·8	3 12·1	2·5	0·6	8·5	1·9	14·5	3·3
26	3 21·5	3 22·1	3 12·3	2·6	0·6	8·6	1·9	14·6	3·3
27	3 21·8	3 22·3	3 12·6	2·7	0·6	8·7	2·0	14·7	3·3
28	3 22·0	3 22·6	3 12·8	2·8	0·6	8·8	2·0	14·8	3·3
29	3 22·3	3 22·8	3 13·0	2·9	0·7	8·9	2·0	14·9	3·4
30	3 22·5	3 23·1	3 13·3	3·0	0·7	9·0	2·0	15·0	3·4
31	3 22·8	3 23·3	3 13·5	3·1	0·7	9·1	2·0	15·1	3·4
32	3 23·0	3 23·6	3 13·8	3·2	0·7	9·2	2·1	15·2	3·4
33	3 23·3	3 23·8	3 14·0	3·3	0·7	9·3	2·1	15·3	3·4
34	3 23·5	3 24·1	3 14·2	3·4	0·8	9·4	2·1	15·4	3·5
35	3 23·8	3 24·3	3 14·5	3·5	0·8	9·5	2·1	15·5	3·5
36	3 24·0	3 24·6	3 14·7	3·6	0·8	9·6	2·2	15·6	3·5
37	3 24·3	3 24·8	3 14·9	3·7	0·8	9·7	2·2	15·7	3·5
38	3 24·5	3 25·1	3 15·2	3·8	0·9	9·8	2·2	15·8	3·6
39	3 24·8	3 25·3	3 15·4	3·9	0·9	9·9	2·2	15·9	3·6
40	3 25·0	3 25·6	3 15·7	4·0	0·9	10·0	2·3	16·0	3·6
41	3 25·3	3 25·8	3 15·9	4·1	0·9	10·1	2·3	16·1	3·6
42	3 25·5	3 26·1	3 16·1	4·2	0·9	10·2	2·3	16·2	3·6
43	3 25·8	3 26·3	3 16·4	4·3	1·0	10·3	2·3	16·3	3·7
44	3 26·0	3 26·6	3 16·6	4·4	1·0	10·4	2·3	16·4	3·7
45	3 26·3	3 26·8	3 16·9	4·5	1·0	10·5	2·4	16·5	3·7
46	3 26·5	3 27·1	3 17·1	4·6	1·0	10·6	2·4	16·6	3·7
47	3 26·8	3 27·3	3 17·3	4·7	1·1	10·7	2·4	16·7	3·8
48	3 27·0	3 27·6	3 17·6	4·8	1·1	10·8	2·4	16·8	3·8
49	3 27·3	3 27·8	3 17·8	4·9	1·1	10·9	2·5	16·9	3·8
50	3 27·5	3 28·1	3 18·0	5·0	1·1	11·0	2·5	17·0	3·8
51	3 27·8	3 28·3	3 18·3	5·1	1·1	11·1	2·5	17·1	3·8
52	3 28·0	3 28·6	3 18·5	5·2	1·2	11·2	2·5	17·2	3·9
53	3 28·3	3 28·8	3 18·8	5·3	1·2	11·3	2·5	17·3	3·9
54	3 28·5	3 29·1	3 19·0	5·4	1·2	11·4	2·6	17·4	3·9
55	3 28·8	3 29·3	3 19·2	5·5	1·2	11·5	2·6	17·5	3·9
56	3 29·0	3 29·6	3 19·5	5·6	1·3	11·6	2·6	17·6	4·0
57	3 29·3	3 29·8	3 19·7	5·7	1·3	11·7	2·6	17·7	4·0
58	3 29·5	3 30·1	3 20·0	5·8	1·3	11·8	2·7	17·8	4·0
59	3 29·8	3 30·3	3 20·2	5·9	1·3	11·9	2·7	17·9	4·0
60	3 30·0	3 30·6	3 20·4	6·0	1·4	12·0	2·7	18·0	4·1

TABLES FOR INTERPOLATING SUNRISE, MOONRISE, ETC.

TABLE I—FOR LATITUDE

Tabular Interval 10°	5°	2°	Difference between the times for consecutive latitudes 5m	10m	15m	20m	25m	30m	35m	40m	45m	50m	55m	60m	1h 05m	1h 10m	1h 15m	1h 20m
° ′	° ′	° ′	m	m	m	m	m	m	m	m	m	m	m	m	h m	h m	h m	h m
0 30	0 15	0 06	0	0	1	1	1	1	1	2	2	2	2	2	0 02	0 02	0 02	0 02
1 00	0 30	0 12	0	1	1	2	2	3	3	3	4	4	4	5	05	05	05	05
1 30	0 45	0 18	1	1	2	3	3	4	4	5	5	6	7	7	07	07	07	07
2 00	1 00	0 24	1	2	3	4	5	5	6	7	7	8	9	10	10	10	10	10
2 30	1 15	0 30	1	2	4	5	6	7	8	9	9	10	11	12	12	13	13	13
3 00	1 30	0 36	1	3	4	6	7	8	9	10	11	12	13	14	0 15	0 15	0 16	0 1
3 30	1 45	0 42	2	3	5	7	8	10	11	12	13	14	16	17	18	18	19	19
4 00	2 00	0 48	2	4	6	8	9	11	13	14	15	16	18	19	20	21	22	22
4 30	2 15	0 54	2	4	7	9	11	13	15	16	18	19	21	22	23	24	25	26
5 00	2 30	1 00	2	5	7	10	12	14	16	18	20	22	23	25	26	27	28	29
5 30	2 45	1 06	3	5	8	11	13	16	18	20	22	24	26	28	0 29	0 30	0 31	0 32
6 00	3 00	1 12	3	6	9	12	14	17	20	22	24	26	29	31	32	33	34	36
6 30	3 15	1 18	3	6	10	13	16	19	22	24	26	29	31	34	36	37	38	40
7 00	3 30	1 24	3	7	10	14	17	20	23	26	29	31	34	37	39	41	42	44
7 30	3 45	1 30	4	7	11	15	18	22	25	28	31	34	37	40	43	44	46	48
8 00	4 00	1 36	4	8	12	16	20	23	27	30	34	37	41	44	0 47	0 48	0 51	0 53
8 30	4 15	1 42	4	8	13	17	21	25	29	33	36	40	44	48	0 51	0 53	0 56	0 58
9 00	4 30	1 48	4	9	13	18	22	27	31	35	39	43	47	52	0 55	0 58	1 01	1 04
9 30	4 45	1 54	5	9	14	19	24	28	33	38	42	47	51	56	1 00	1 04	1 08	1 12
10 00	5 00	2 00	5	10	15	20	25	30	35	40	45	50	55	60	1 05	1 10	1 15	1 20

Table I is for interpolating the L.M.T. of sunrise, twilight, moonrise, etc. for latitude. It is to be noted that the interpolation is not linear, so that when using this table it is essential to take out the required phenomenon for the latitude *less* than the true latitude. The table is entered with the nearest value of the difference between the times for the tabular latitude and the next higher one, and, in the appropriate column, with the difference between true latitude and tabular latitude; the correction so obtained is applied to the time for the tabular latitude; the sign of the correction can be seen by inspection.

TABLE II—FOR LONGITUDE

Long. East or West	Difference between the times for given date and preceding date (for east longitude) or for given date and following date (for west longitude) 10m	20m	30m	40m	50m	60m	1h + 10m	1h + 20m	1h + 30m	1h + 40m	1h + 50m	1h + 60m	2h 10m	2h 20m	2h 30m	2h 40m	2h 50m	3h 00m
°	m	m	m	m	m	m	m	m	m	m	m	m	h m	h m	h m	h m	h m	h m
0	0	0	0	0	0	0	0	0	0	0	0	0	0 00	0 00	0 00	0 00	0 00	0 00
10	0	1	1	1	1	2	2	2	2	3	3	3	04	04	04	04	05	05
20	1	1	2	2	3	3	4	4	5	6	6	7	07	08	08	09	09	10
30	1	2	2	3	4	5	6	7	7	8	9	10	11	12	12	13	14	15
40	1	2	3	4	6	7	8	9	10	11	12	13	14	16	17	18	19	20
50	1	3	4	6	7	8	10	11	12	14	15	17	0 18	0 19	0 21	0 22	0 24	0 25
60	2	3	5	7	8	10	12	13	15	17	18	20	22	23	25	27	28	30
70	2	4	6	8	10	12	14	16	17	19	21	23	25	27	29	31	33	35
80	2	4	7	9	11	13	16	18	20	22	24	27	29	31	33	36	38	40
90	2	5	7	10	12	15	17	20	22	25	27	30	32	35	37	40	42	45
100	3	6	8	11	14	17	19	22	25	28	31	33	0 36	0 39	0 42	0 44	0 47	0 50
110	3	6	9	12	15	18	21	24	27	31	34	37	40	43	46	49	0 52	0 55
120	3	7	10	13	17	20	23	27	30	33	37	40	43	47	50	53	0 57	1 00
130	4	7	11	14	18	22	25	29	32	36	40	43	47	51	54	0 58	1 01	1 05
140	4	8	12	16	19	23	27	31	35	39	43	47	51	54	0 58	1 02	1 06	1 10
150	4	8	13	17	21	25	29	33	38	42	46	50	0 54	0 58	1 03	1 07	1 11	1 15
160	4	9	13	18	22	27	31	36	40	44	49	53	0 58	1 02	1 07	1 11	1 16	1 20
170	5	9	14	19	24	28	33	38	42	47	52	57	1 01	1 06	1 11	1 16	1 20	1 25
180	5	10	15	20	25	30	35	40	45	50	55	60	1 05	1 10	1 15	1 20	1 25	1 30

Table II is for interpolating the L.M.T. of moonrise, moonset and the Moon's meridian passage for longitude. It is entered with longitude and with the difference between the times for the given date and for the preceding date (in east longitudes) or following date (in west longitudes). The correction is normally *added* for west longitudes and *subtracted* for east longitudes, but if, as occasionally happens, the times become earlier each day instead of later, the signs of the corrections must be reversed.

STANDARD TIMES (Corrected to September 1970)

LIST I—PLACES FAST ON G.M.T. (mainly those EAST OF GREENWICH)

The times given below should be *added* to G.M.T. to give Standard Time. / *subtracted* from Standard Time to give G.M.T.

	h	m
Aden (Southern Yemen)	03	
Admiralty Islands	10	
Afghanistan	04	30
Albania*	01	
Amirante Islands	04	
Andaman Islands	05	30
Angola (Portuguese West Africa)	01	
Annobon Island‡	01	
Arabian Emirates, Federation of	04	
Australia		
Australian Capital Territory	10	
New South Wales[1]	10	
Northern Territory	09	30
Queensland	10	
South Australia	09	30
Tasmania*	10	
Victoria	10	
Western Australia	08	
Austria	01	
Balearic Islands‡	01	
Belgium	01	
Botswana, Republic of	02	
British New Guinea	10	
Brunei	08	
Bulgaria	02	
Burma	06	30
Cambodia	07	
Cameroun Republic	01	
Caroline Islands, east of long. E. 160°	12	
west of long. E. 160°	10	
Truk, Ponape	11	
Central African Republic‡	01	
Ceylon	05	30
Chad	01	
Chagos Archipelago	05	
Chatham Islands‡	12	45
China[2]	08	
Christmas Island, Indian Ocean	07	
Cocos Keeling Islands	06	30
Comoro Islands	03	
Congo Republic	01	
Congolese Republic, eastern part	02	
western part	01	
Corsica‡	01	
Crete	02	
Cyprus	02	
Cyrenaica‡	02	
Czechoslovakia	01	
Dahomey, Republic of	01	
Denmark	01	
Egypt* (United Arab Republic)	02	
Ellice Islands	12	
Equatorial Guinea, Republic of	01	
Estonia	03	
Ethiopia	03	
Fernando Póo‡	01	
Fiji	12	
Finland	02	
Formosa* (Taiwan)	08	
France‡	01	
French Territory of the Afars and Issas	03	
Friendly Islands	13	
Gabon	01	
Germany	01	
Gibraltar‡	01	
Gilbert and Ellice Islands	12	
Greece	02	
Guam	10	
Holland (The Netherlands)	01	
Hong Kong*	08	
Hungary	01	
India	05	30
Indonesia, Republic of		
Bali, Bangka, Billiton, Java, Lombok, Madura, Sumatra	07	
Borneo, Celebes, Flores, Sumba, Sumbawa, Timor	08	
Aru, Kei, Moluccas, Tanimbar, West Irian	09	
Iran	03	30
Iraq	03	
Irish Republic	01	
Israel	02	
Italy*	01	
Japan	09	
Jordan	02	
Kamchatka Peninsula	12	
Kenya	03	
Korea	09	
Kuril Islands	11	
Kuwait	03	
Laccadive Islands	05	30
Ladrone Islands	10	
Laos	07	
Latvia	03	

* Summer time may be kept in these countries.

‡ This time is used throughout the year, but may differ from the legal time.

1 Except Broken Hill Area, which keeps $09^h\ 30^m$.

2 All the coast, but some areas may keep summer time.

POLARIS (POLE STAR) TABLES, 1973

FOR DETERMINING LATITUDE FROM SEXTANT ALTITUDE AND FOR AZIMUTH

L.H.A. ARIES	0°–9°	10°–19°	20°–29°	30°–39°	40°–49°	50°–59°	60°–69°	70°–79°	80°–89°	90°–99°	100°–109°	110°–119°
	a_0	a_0	a_0	a_0	a_0	a_0	a_0	a_0	a_0	a_0	a_0	a_0
°	° ′	° ′	° ′	° ′	° ′	° ′	° ′	° ′	° ′	° ′	° ′	° ′
0	0 15·2	0 11·2	0 08·6	0 07·5	0 08·1	0 10·2	0 13·8	0 18·8	0 25·0	0 32·3	0 40·3	0 48·9
1	14·8	10·8	08·4	07·5	08·2	10·5	14·2	19·3	25·7	33·0	41·2	49·8
2	14·3	10·5	08·2	07·5	08·4	10·8	14·7	19·9	26·4	33·8	42·0	50·7
3	13·9	10·2	08·1	07·5	08·5	11·1	15·1	20·5	27·1	34·6	42·8	51·6
4	13·4	09·9	08·0	07·5	08·7	11·4	15·6	21·1	27·8	35·4	43·7	52·5
5	0 13·0	0 09·7	0 07·8	0 07·6	0 08·9	0 11·8	0 16·1	0 21·7	0 28·5	0 36·2	0 44·6	0 53·4
6	12·6	09·4	07·7	07·7	09·1	12·2	16·6	22·4	29·2	37·0	45·4	54·2
7	12·2	09·2	07·7	07·7	09·4	12·5	17·1	23·0	30·0	37·8	46·3	55·1
8	11·9	09·0	07·6	07·8	09·6	12·9	17·7	23·7	30·7	38·6	47·2	56·0
9	11·5	08·8	07·6	07·9	09·9	13·4	18·2	24·3	31·5	39·5	48·0	56·9
10	0 11·2	0 08·6	0 07·5	0 08·1	0 10·2	0 13·8	0 18·8	0 25·0	0 32·3	0 40·3	0 48·9	0 57·8

Lat.	a_1	a_1	a_1	a_1	a_1	a_1	a_1	a_1	a_1	a_1	a_1	a_1
°	′	′	′	′	′	′	′	′	′	′	′	′
0	0·5	0·6	0·6	0·6	0·6	0·5	0·5	0·4	0·3	0·2	0·2	0·1
10	·5	·6	·6	·6	·6	·5	·5	·4	·3	·3	·2	·2
20	·5	·6	·6	·6	·6	·5	·5	·5	·4	·3	·3	·3
30	·6	·6	·6	·6	·6	·6	·5	·5	·4	·4	·4	·4
40	0·6	0·6	0·6	0·6	0·6	0·6	0·6	0·5	0·5	0·5	0·5	0·5
45	·6	·6	·6	·6	·6	·6	·6	·6	·6	·5	·5	·5
50	·6	·6	·6	·6	·6	·6	·6	·6	·6	·6	·6	·6
55	·6	·6	·6	·6	·6	·6	·6	·6	·7	·7	·7	·7
60	·6	·6	·6	·6	·6	·6	·7	·7	·7	·8	·8	·8
62	0·7	0·6	0·6	0·6	0·6	0·6	0·7	0·7	0·8	0·8	0·8	0·9
64	·7	·6	·6	·6	·6	·7	·7	·8	·8	·9	0·9	0·9
66	·7	·6	·6	·6	·6	·7	·7	·8	·9	0·9	1·0	1·0
68	0·7	0·6	0·6	0·6	0·6	0·7	0·7	0·8	0·9	1·0	1·1	1·1

Month	a_2	a_2	a_2	a_2	a_2	a_2	a_2	a_2	a_2	a_2	a_2	a_2
	′	′	′	′	′	′	′	′	′	′	′	′
Jan.	0·7	0·8	0·8	0·8	0·8	0·8	0·8	0·8	0·8	0·7	0·7	0·7
Feb.	·7	·7	·7	·8	·8	·8	·9	·9	·9	·9	·9	·8
Mar.	·5	·6	·6	·7	·8	·8	·8	·9	·9	·9	·9	0·9
Apr.	0·4	0·4	0·5	0·6	0·6	0·7	0·8	0·8	0·9	0·9	0·9	1·0
May	·3	·3	·4	·4	·5	·5	·6	·7	·7	·8	·8	0·9
June	·2	·2	·3	·3	·3	·4	·5	·5	·6	·6	·7	·8
July	0·3	0·3	0·3	0·3	0·3	0·3	0·3	0·4	0·4	0·5	0·5	0·6
Aug.	·4	·4	·3	·3	·3	·3	·3	·3	·3	·4	·4	·4
Sept.	·6	·5	·5	·4	·4	·4	·3	·3	·3	·3	·3	·3
Oct.	0·8	0·7	0·7	0·6	0·5	0·5	0·4	0·4	0·3	0·3	0·3	0·2
Nov.	1·0	0·9	0·8	·8	·7	·6	·6	·5	·4	·4	·3	·3
Dec.	1·1	1·0	1·0	0·9	0·9	0·8	0·7	0·7	0·6	0·5	0·4	0·3

Lat.	AZIMUTH											
°	°	°	°	°	°	°	°	°	°	°	°	°
0	0·4	0·2	0·1	359·9	359·8	359·7	359·5	359·4	359·3	359·2	359·2	359·2
20	0·4	0·3	0·1	359·9	359·8	359·6	359·5	359·4	359·3	359·2	359·1	359·1
40	0·5	0·3	0·1	359·9	359·7	359·6	359·4	359·2	359·1	359·0	358·9	358·9
50	0·6	0·4	0·2	359·9	359·7	359·5	359·3	359·1	358·9	358·8	358·7	358·7
55	0·7	0·4	0·2	359·9	359·6	359·4	359·2	359·0	358·8	358·7	358·6	358·5
60	0·8	0·5	0·2	359·9	359·6	359·3	359·0	358·8	358·6	358·5	358·3	358·3
65	0·9	0·6	0·2	359·9	359·5	359·2	358·9	358·6	358·3	358·2	358·0	358·0

Latitude = Apparent altitude (corrected for refraction) $-$ 1° $+ a_0 + a_1 + a_2$

The table is entered with L.H.A. Aries to determine the column to be used; each column refers to a range of 10°. a_0 is taken, with mental interpolation, from the upper table with the units of L.H.A. Aries in degrees as argument; a_1, a_2 are taken, without interpolation, from the second and third tables with arguments latitude and month respectively. a_0, a_1, a_2 are always positive. The final table gives the azimuth of *Polaris*.

FRANCE, WEST COAST — BREST

Lat. 48° 23′ N. Long. 4° 29′ W.

IME ZONE: -0100 TIMES AND HEIGHTS OF HIGH AND LOW WATERS YEAR 1973

JANUARY

Day	Time	M	Ft.
1 M	0242	6.0	19.8
	0903	2.8	9.3
	1508	6.1	20.0
	2127	2.8	9.1
2 Tu	0331	6.3	20.6
	0950	2.6	8.5
	1553	6.3	20.6
	2211	2.5	8.3
3 W	0413	6.5	21.4
	1031	2.3	7.7
	1632	6.5	21.3
	2249	2.3	7.7
4 Th	0451	6.8	22.3
	1108	2.1	6.8
	1708	6.6	21.8
	2325	2.1	7.0
5 F	0526	6.9	22.8
	1143	1.9	6.4
	1743	6.8	22.3
6 Sa	0000	2.0	6.5
	0600	7.1	23.2
	1217	1.8	6.0
	1818	6.9	22.6
7 Su	0034	1.9	6.4
	0635	7.1	23.4
	1252	1.7	5.7
	1851	6.9	22.6
8 M	0109	1.9	6.2
	0709	7.1	23.4
	1327	1.8	5.9
	1927	6.8	22.4
9 Tu	0146	1.9	6.4
	0745	7.0	23.1
	1405	1.9	6.2
	2004	6.7	22.1
10 W	0225	2.0	6.7
	0824	6.9	22.6
	1446	2.0	6.7
	2044	6.6	21.6
11 Th	0308	2.2	7.3
	0906	6.6	21.8
	1531	2.2	7.3
	2130	6.4	21.0
12 F	0356	2.4	8.0
	0957	6.4	21.0
	1624	2.5	8.2
	2225	6.2	20.3
13 Sa	0454	2.6	8.6
	1058	6.2	20.3
	1728	2.7	8.8
	2335	6.1	20.0
14 Su	0605	2.7	9.0
	1215	6.1	20.0
	1843	2.7	9.0
15 M	0055	6.1	20.1
	0723	2.6	8.6
	1334	6.2	20.3
	2001	2.6	8.5
16 Tu	0212	6.4	21.1
	0838	2.3	7.7
	1447	6.5	21.3
	2111	2.2	7.3
17 W	0319	6.8	22.4
	0942	1.9	6.4
	1549	6.8	22.4
	2211	1.8	6.0
18 Th	0417	7.2	23.7
	1038	1.5	5.0
	1643	7.2	23.6
	2303	1.5	5.0
19 F	0507	7.5	24.7
	1127	1.2	4.1
	1731	7.4	24.2
	2350	1.3	4.2
20 Sa	0553	7.7	25.4
	1212	1.1	3.6
	1815	7.4	24.4
21 Su	0034	1.2	4.1
	0636	7.7	25.4
	1254	1.1	3.6
	1856	7.4	24.2
22 M	0115	1.3	4.4
	0716	7.6	24.9
	1334	1.3	4.2
	1934	7.2	23.6
23 Tu	0154	1.6	5.2
	0752	7.3	23.9
	1413	1.6	5.4
	2011	6.8	22.4
24 W	0232	1.9	6.4
	0830	6.9	22.6
	1451	2.0	6.7
	2049	6.5	21.3
25 Th	0311	2.3	7.7
	0908	6.4	21.1
	1532	2.5	8.2
	2129	6.1	20.0
26 F	0354	2.7	9.0
	0953	6.0	19.6
	1618	2.9	9.6
	2218	5.7	18.8
27 Sa	0446	3.1	10.3
	1049	5.6	18.5
	1718	3.3	10.8
	2324	5.5	18.2
28 Su	0555	3.3	10.9
	1205	5.4	17.8
	1836	3.4	11.3
29 M	0048	5.5	18.0
	0718	3.3	10.9
	1332	5.4	17.8
	1959	3.3	10.8
30 Tu	0210	5.7	18.7
	0834	3.1	10.1
	1442	5.7	18.7
	2104	3.0	10.0
31 W	0310	6.0	19.6
	0930	2.7	9.0
	1534	6.0	19.6
	2153	2.7	8.8

FEBRUARY

Day	Time	M	Ft.
1 Th	0356	6.3	20.8
	1014	2.4	7.8
	1616	6.3	20.8
	2233	2.3	7.7
2 F	0435	6.7	21.9
	1051	2.0	6.7
	1652	6.6	21.8
	2309	2.0	6.5
3 Sa	0510	7.0	23.1
	1126	1.7	5.5
	1726	6.9	22.8
	2343	1.7	5.5
4 Su	0543	7.3	23.9
	1159	1.4	4.7
	1759	7.1	23.4
5 M	0016	1.5	4.9
	0616	7.4	24.4
	1232	1.3	4.2
	1833	7.2	23.7
6 Tu	0049	1.4	4.5
	0649	7.5	24.6
	1306	1.3	4.2
	1906	7.2	23.7
7 W	0124	1.4	4.7
	0722	7.4	24.2
	1341	1.4	4.5
	1941	7.1	23.4
8 Th	0200	1.6	5.2
	0758	7.2	23.6
	1419	1.6	5.4
	2018	6.9	22.6
9 F	0239	1.8	6.0
	0838	6.8	22.4
	1501	2.0	6.5
	2059	6.6	21.6
10 Sa	0325	2.2	7.3
	0924	6.4	21.1
	1551	2.4	8.0
	2152	6.2	20.3
11 Su	0421	2.6	8.6
	1025	6.0	19.8
	1656	2.8	9.3
	2303	5.9	19.3
12 M	0537	2.9	9.5
	1149	5.8	19.0
	1823	3.0	9.8
13 Tu	0037	5.9	19.3
	0710	2.8	9.3
	1325	5.9	19.3
	1955	2.8	9.1
14 W	0208	6.2	20.3
	0835	2.5	8.2
	1445	6.2	20.5
	2110	2.3	7.7
15 Th	0318	6.6	21.8
	0940	2.0	6.5
	1546	6.7	21.9
	2207	1.8	6.0
16 F	0411	7.1	23.4
	1031	1.5	4.9
	1635	7.1	23.2
	2254	1.4	4.7
17 Sa	0457	7.5	24.6
	1115	1.1	3.7
	1717	7.4	24.2
	2335	1.2	3.9
18 Su	0537	7.7	25.2
	1154	1.0	3.2
	1756	7.5	24.6
19 M	0013	1.1	3.6
	0614	7.7	25.4
	1231	1.0	3.2
	1832	7.4	24.4
20 Tu	0048	1.2	3.9
	0648	7.6	24.9
	1305	1.2	3.9
	1905	7.3	23.9
21 W	0122	1.4	4.5
	0721	7.3	24.1
	1339	1.5	4.9
	1937	7.0	22.9
22 Th	0155	1.7	5.7
	0753	6.9	22.8
	1412	1.9	6.4
	2009	6.6	21.8
23 F	0229	2.2	7.2
	0826	6.5	21.3
	1447	2.4	7.8
	2044	6.2	20.3
24 Sa	0306	2.6	8.5
	0903	6.0	19.6
	1527	2.8	9.3
	2125	5.8	19.0
25 Su	0350	3.0	10.0
	0950	5.5	18.2
	1618	3.3	10.8
	2221	5.4	17.8
26 M	0453	3.4	11.1
	1100	5.2	17.2
	1735	3.5	11.6
	2349	5.3	17.3
27 Tu	0624	3.5	11.4
	1241	5.2	17.0
	1914	3.5	11.4
28 W	0130	5.4	17.8
	0758	3.2	10.6
	1410	5.4	17.8
	2034	3.1	10.3

MARCH

Day	Time	M	Ft.
1 Th	0242	5.8	19.0
	0903	2.8	9.3
	1508	5.8	19.1
	2127	2.7	8.8
2 F	0330	6.2	20.5
	0949	2.3	7.7
	1551	6.3	20.6
	2208	2.2	7.3
3 Sa	0409	6.7	21.9
	1026	1.9	6.2
	1627	6.7	22.1
	2243	1.8	5.9
4 Su	0444	7.1	23.2
	1100	1.5	4.9
	1701	7.1	23.2
	2317	1.4	4.5
5 M	0517	7.4	24.4
	1133	1.1	3.7
	1734	7.4	24.2
	2350	1.1	3.7
6 Tu	0550	7.6	25.1
	1207	1.0	3.2
	1807	7.5	24.7
7 W	0024	1.0	3.4
	0624	7.7	25.2
	1241	1.0	3.2
	1842	7.5	24.7
8 Th	0059	1.1	3.6
	0659	7.6	24.9
	1317	1.1	3.7
	1917	7.4	24.2
9 F	0136	1.3	4.4
	0736	7.3	23.9
	1356	1.5	4.9
	1955	7.0	23.1
10 Sa	0217	1.7	5.5
	0817	6.8	22.4
	1440	2.0	6.5
	2040	6.6	21.8
11 Su	0305	2.2	7.2
	0905	6.3	20.8
	1533	2.5	8.2
	2135	6.1	20.1
12 M	0405	2.7	8.8
	1010	5.8	19.1
	1644	2.9	9.6
	2253	5.8	19.1
13 Tu	0529	2.9	9.6
	1143	5.6	18.5
	1818	3.0	10.0
14 W	0035	5.8	19.1
	0708	2.8	9.3
	1324	5.8	19.0
	1952	2.7	9.0
15 Th	0205	6.2	20.3
	0830	2.4	7.8
	1439	6.2	20.3
	2102	2.3	7.5
16 F	0308	6.6	21.8
	0929	1.9	6.4
	1534	6.6	21.8
	2153	1.8	6.0
17 Sa	0357	7.0	23.1
	1015	1.5	4.9
	1618	7.0	23.1
	2235	1.4	4.7
18 Su	0437	7.4	24.2
	1054	1.2	4.1
	1656	7.3	23.9
	2313	1.2	4.1
19 M	0513	7.5	24.7
	1130	1.1	3.6
	1730	7.4	24.2
	2347	1.2	3.9
20 Tu	0547	7.5	24.7
	1203	1.1	3.7
	1803	7.4	24.2
21 W	0019	1.2	4.1
	0619	7.4	24.4
	1235	1.3	4.4
	1835	7.2	23.7
22 Th	0051	1.4	4.7
	0650	7.2	23.6
	1307	1.6	5.2
	1906	7.0	22.9
23 F	0123	1.8	5.9
	0721	6.8	22.4
	1339	2.0	6.5
	1937	6.6	21.8
24 Sa	0155	2.1	7.0
	0752	6.4	21.1
	1412	2.4	7.8
	2010	6.3	20.6
25 Su	0230	2.5	8.3
	0828	6.0	19.6
	1450	2.8	9.1
	2049	5.9	19.3
26 M	0312	2.9	9.6
	0911	5.6	18.3
	1539	3.2	10.5
	2140	5.5	18.2
27 Tu	0410	3.3	10.8
	1015	5.2	17.2
	1649	3.4	11.3
	2258	5.3	17.5
28 W	0535	3.4	11.1
	1149	5.2	17.0
	1824	3.4	11.3
29 Th	0039	5.4	17.8
	0710	3.2	10.6
	1323	5.4	17.7
	1949	3.1	10.3
30 F	0158	5.7	18.8
	0821	2.8	9.1
	1427	5.8	19.1
	2047	2.6	8.6
31 Sa	0251	6.2	20.5
	0910	2.3	7.7
	1513	6.3	20.6
	2131	2.1	7.0

BREST

MEAN SPRING AND NEAP CURVES

For instructions see page xiv

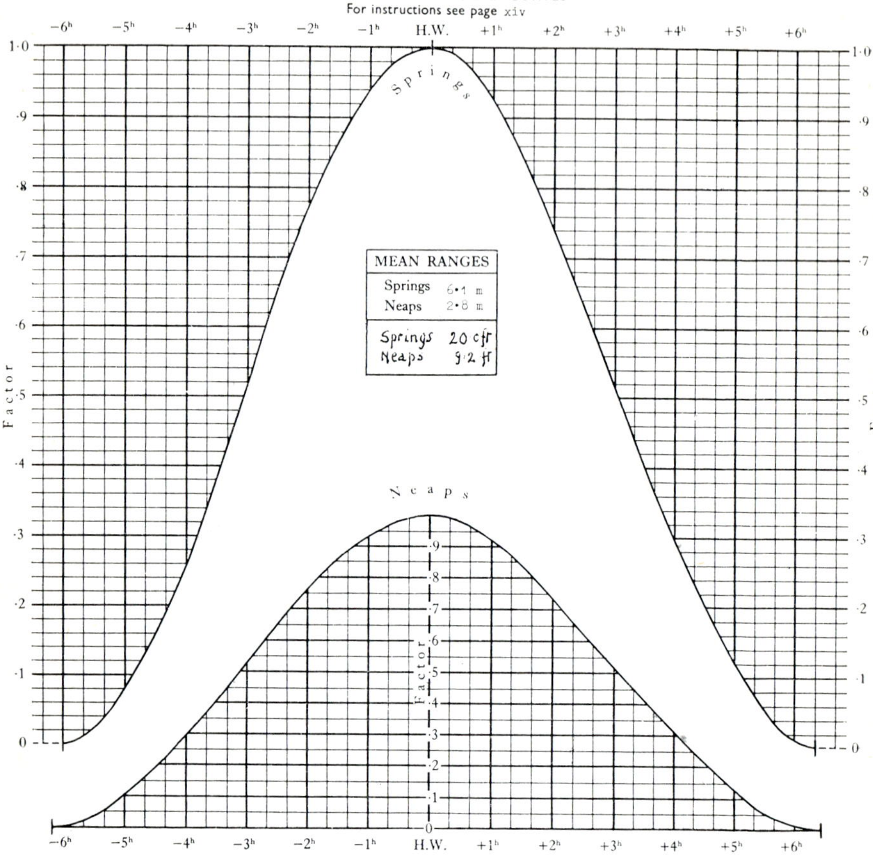

No.	PLACE	POSITION		TIMES AT STANDARD PORT				HEIGHTS (IN FEET) AT STANDARD PORT			
	STANDARD PORT	Lat. N.	Long. W.	High Water at 0500 and 1700	High Water at 1100 and 2300	Low Water at 1100 and 2300	Low Water at 0500 and 1700	MHWS	MHWN	MLWN	MLWS
1638	**BREST**	(see page 277)						24·3	18·9	9·9	4·4
	SECONDARY PORTS			TIME DIFFERENCES (Zone −0100)				HEIGHT DIFFERENCES			
	Odet river										
1646	Benodet	47 53	4 07	−0020	−0015	−0025	−0025	−8·2	−6·4	−4·0	−2·0
1647	Lanroz	47 58	4 07	+0015	+0010	−0015	−0015	−5·9	−4·3	−2·5	−1·4
1648	Concarneau . . .	47 52	3 54	−0020	−0010	−0035	−0015	−8·3	−6·8	−4·0	−2·7
	Iles de Glenan										
1649	Penfret	47 44	3 57	−0025	−0015	−0030	−0025	−8·2	−6·4	−4·0	−2·0
1650	Port Louis . . . F	47 42	3 21	−0025	−0015	−0030	−0020	−8·0	−6·6	−4·3	−2·7
1651	Lorient	47 45	3 21	−0030	+0005	−0035	−0020	−8·0	−6·6	−4·3	−2·7
1653	Port-Haliguen . . .	47 29	3 06	−0020	+0005	−0025	−0015	−7·2	−5·9	−3·6	−2·5
	Belle-Ile										
1654	Le Palais	47 21	3 09	−0030	−0005	−0030	−0025	−7·5	−6·3	−3·8	−2·4
	Crac'h river										
1655	La Trinite	47 35	3 01	−0020	−0010	−0025	−0015	−6·7	−5·4	−3·3	−2·4
	Morbihan										
1656	Port-Navalo	47 33	2 55	−0010	+0005	−0010	0000	−8·2	−6·6	−4·3	−2·7
1657	Auray	47 40	2 59	+0015	+0035	◇	◇	−7·0	−5·8	−3·6	−2·2
1658	Vannes	47 40	2 46	+0200	+0200	◇	◇	−9·1	−6·1	−3·0	−0·4
1659	Ile de Hoedic . . .	47 20	2 52	−0030	−0015	−0027	−0014	−7·3	−5·9	−3·6	−2·7
1660	Penerf	47 31	2 37	−0020	0000	−0020	−0015	−6·5	−5·5	−3·8	−2·7
1661	Le Croisic	47 18	2 31	−0030	−0015	−0025	−0025	−6·0	−5·0	−3·2	−1·5
	La Loire										
1662	Le Pouliguen . . .	47 17	2 25	−0035	+0020	0000	−0020	−6·9	−5·9	−4·0	−2·6
1663	Le Grand-Charpentier .	47 13	2 19	−0035	+0020	0000	−0015	−6·9	−5·9	−4·2	−2·6
1664	St. Nazaire . . . F	47 16	2 12	−0025	+0030	+0015	−0005	−6·7	−5·7	−4·5	−2·8
1665	Paimboeuf	47 17	2 02	−0005	+0055	+0115	+0040	−6·4	−5·2	−4·7	−2·0
1666	Le Pellerin . . .	47 12	1 45	+0055	+0125	+0300	+0220	−5·6	−4·4	−4·0	+0·5
1667	Nantes (Chantenay) .	47 12	1 35	+0125	+0140	+0330	+0300	−5·6	−4·2	−3·7	+0·6
1668	Pornic	47 06	2 07	−0035	−0010	+0005	+0005	−7·0	−5·5	−4·5	−2·0
	Ile de Noirmoutier										
1669	Bois de la Chaise . .	47 01	2 13	−0030	−0015	−0005	0000	−7·0	−6·0	−4·7	−3·0
1670	Fromentine	46 54	2 10	−0025	−0015	+0010	−0005	−7·0	−6·0	−4·2	−2·3
	Ile d'Yeu										
1671	Port Joinville . . .	46 42	2 20	−0050	−0020	−0035	−0025	−8·3	−6·7	−4·5	−2·8
1672	St. Gilles-sur-Vie . .	46 41	1 56	−0030	−0015	−0035	−0025	−7·0	−5·9	−3·8	−2·0
1673	Les Sables-d'Olonne . .	46 30	1 47	−0020	+0015	−0015	−0020	−6·5	−5·9	−3·5	−2·0
				0600 and 1800	0000 and 1200	1000 and 2200	0400 and 1600				
1680	**POINTE DE GRAVE**	(see page 282)						17·6	13·8	6·8	3·0
	Ile de Re										
1674	St. Martin	46 12	1 22	−0005	+0010	−0005	0000	+2·6	+0·2	+0·3	−0·4
1675	La Pallice	46 10	1 13	−0045	−0005	−0020	−0015	+0·6	+0·4	−0·2	−1·4
1676	La Rochelle . . . F	46 09	1 09	−0045	−0010	−0010	−0015	+2·1	+2·0	+1·1	−0·4
1677	Ile d'Aix	46 01	1 10	−0040	+0005	−0010	−0010	+2·6	+1·9	+0·9	−0·1
	La Charente										
1678	Rochefort	45 57	0 58	−0030	+0015	+0120	+0050	+3·6	+2·9	+0·3	+1·8
1679	La Cayenne	45 47	1 08	−0040	−0005	−0005	−0015	+1·0	+1·1	+1·1	+1·3

F Tides predicted in French Tide Tables. ◇ No data.

TABLE II

MULTIPLICATION TABLE

FACTOR	RANGE AT STANDARD PORT																						FACTOR
1·00	3·0	4·0	5·0	6·0	7·0	8·0	9·0	10·0	11·0	12·0	13·0	14·0	15·0	16·0	17·0	18·0	19·0	20·0	21·0	22·0	23·0	24·0	25·0	26·0	1·00
·98	2·9	3·9	4·9	5·9	6·9	7·8	8·8	9·8	10·8	11·8	12·7	13·7	14·7	15·7	16·7	17·6	18·6	19·6	20·6	21·6	22·5	23·5	24·5	25·5	·98
·96	2·9	3·8	4·8	5·8	6·7	7·7	8·6	9·6	10·6	11·5	12·5	13·4	14·4	15·4	16·3	17·3	18·2	19·2	20·2	21·1	22·1	23·0	24·0	25·0	·96
·94	2·8	3·8	4·7	5·6	6·6	7·5	8·5	9·4	10·3	11·3	12·2	13·2	14·1	15·0	16·0	16·9	17·9	18·8	19·7	20·7	21·6	22·6	23·5	24·4	·94
·92	2·8	3·7	4·6	5·5	6·4	7·4	8·3	9·2	10·1	11·0	12·0	12·9	13·8	14·7	15·6	16·6	17·5	18·4	19·3	20·2	21·2	21·1	23·0	23·9	·92
·90	2·7	3·6	4·5	5·4	6·3	7·2	8·1	9·0	9·9	10·8	11·7	12·6	13·5	14·4	15·3	16·2	17·1	18·0	18·9	19·8	20·7	21·6	22·5	23·4	·90
·88	2·6	3·5	4·4	5·3	6·2	7·0	7·9	8·8	9·7	10·6	11·4	12·3	13·2	14·1	15·0	15·8	16·7	17·6	18·5	19·4	20·2	21·1	22·0	22·9	·88
·86	2·6	3·4	4·3	5·2	6·0	6·9	7·7	8·6	9·5	10·3	11·2	12·0	12·9	13·8	14·6	15·5	16·3	17·2	18·1	18·9	19·8	20·6	21·5	22·4	·86
·84	2·5	3·4	4·2	5·0	5·9	6·7	7·6	8·4	9·2	10·1	10·9	11·8	12·6	13·4	14·3	15·1	16·0	16·8	17·6	18·5	19·3	20·2	21·0	21·8	·84
·82	2·5	3·3	4·1	4·9	5·7	6·6	7·4	8·2	9·0	9·8	10·7	11·5	12·3	13·1	13·9	14·8	15·6	16·4	17·2	18·0	18·9	19·7	20·5	21·3	·82
·80	2·4	3·2	4·0	4·8	5·6	6·4	7·2	8·0	8·8	9·6	10·4	11·2	12·0	12·8	13·6	14·4	15·2	16·0	16·8	17·6	18·4	19·2	20·0	20·8	·80
·78	2·3	3·1	3·9	4·7	5·5	6·2	7·0	7·8	8·6	9·4	10·1	10·9	11·7	12·5	13·3	14·0	14·8	15·6	16·4	17·2	17·9	18·7	19·5	20·3	·78
·76	2·3	3·0	3·8	4·6	5·3	6·1	6·8	7·6	8·4	9·1	9·9	10·6	11·4	12·2	12·9	13·7	14·4	15·2	16·0	16·7	17·5	18·2	19·0	19·8	·76
·74	2·2	3·0	3·7	4·4	5·2	5·9	6·7	7·4	8·1	8·9	9·6	10·4	11·1	11·8	12·6	13·3	14·1	14·8	15·5	16·3	17·0	17·8	18·5	19·2	·74
·72	2·2	2·9	3·6	4·3	5·0	5·8	6·5	7·2	7·9	8·6	9·4	10·1	10·8	11·5	12·2	13·0	13·7	14·4	15·1	15·8	16·6	17·3	18·0	18·7	·72
·70	2·1	2·8	3·5	4·2	4·9	5·6	6·3	7·0	7·7	8·4	9·1	9·8	10·5	11·2	11·9	12·6	13·3	14·0	14·7	15·4	16·1	16·8	17·5	18·2	·70
·68	2·0	2·7	3·4	4·1	4·8	5·4	6·1	6·8	7·5	8·2	8·8	9·5	10·2	10·9	11·6	12·2	12·9	13·6	14·3	15·0	15·6	16·3	17·0	17·7	·68
·66	2·0	2·6	3·3	4·0	4·6	5·3	5·9	6·6	7·3	7·9	8·6	9·2	9·9	10·6	11·2	11·9	12·5	13·2	13·9	14·5	15·2	15·8	16·5	17·2	·66
·64	1·9	2·6	3·2	3·8	4·5	5·1	5·8	6·4	7·0	7·7	8·3	9·0	9·6	10·2	10·9	11·5	12·2	12·8	13·4	14·1	14·7	15·4	16·0	16·6	·64
·62	1·9	2·5	3·1	3·7	4·3	5·0	5·6	6·2	6·8	7·4	8·1	8·7	9·3	9·9	10·5	11·2	11·8	12·4	13·0	13·6	14·3	14·9	15·5	16·1	·62
·60	1·8	2·4	3·0	3·6	4·2	4·8	5·4	6·0	6·6	7·2	7·8	8·4	9·0	9·6	10·2	10·8	11·4	12·0	12·6	13·2	13·8	14·4	15·0	15·6	·60
·58	1·7	2·3	2·9	3·5	4·1	4·6	5·2	5·8	6·4	7·0	7·5	8·1	8·7	9·3	9·9	10·4	11·0	11·6	12·2	12·8	13·3	13·9	14·5	15·1	·58
·56	1·7	2·2	2·8	3·4	3·9	4·5	5·0	5·6	6·2	6·7	7·3	7·8	8·4	9·0	9·5	10·1	10·6	11·2	11·8	12·3	12·9	13·4	14·0	14·6	·56
·54	1·6	2·2	2·7	3·2	3·8	4·3	4·9	5·4	5·9	6·5	7·0	7·6	8·1	8·6	9·2	9·7	10·3	10·8	11·3	11·9	12·4	13·0	13·5	14·0	·54
·52	1·6	2·1	2·6	3·1	3·6	4·2	4·7	5·2	5·7	6·2	6·8	7·3	7·8	8·3	8·8	9·4	9·9	10·4	10·9	11·4	12·0	12·5	13·0	13·5	·52
·50	1·5	2·0	2·5	3·0	3·5	4·0	4·5	5·0	5·5	6·0	6·5	7·0	7·5	8·0	8·5	9·0	9·5	10·0	10·5	11·0	11·5	12·0	12·5	13·0	·50
·48	1·4	1·9	2·4	2·9	3·4	3·8	4·3	4·8	5·3	5·8	6·2	6·7	7·2	7·7	8·2	8·6	9·1	9·6	10·1	10·6	11·0	11·5	12·0	12·5	·48
·46	1·4	1·8	2·3	2·8	3·2	3·7	4·1	4·6	5·1	5·5	6·0	6·4	6·9	7·4	7·8	8·3	8·7	9·2	9·7	10·1	10·6	11·0	11·5	12·0	·46
·44	1·3	1·8	2·2	2·6	3·1	3·5	4·0	4·4	4·8	5·3	5·7	6·2	6·6	7·0	7·5	7·9	8·4	8·8	9·2	9·7	10·1	10·6	11·0	11·4	·44
·42	1·3	1·7	2·1	2·5	2·9	3·4	3·8	4·2	4·6	5·0	5·5	5·9	6·3	6·7	7·1	7·6	8·0	8·4	8·8	9·2	9·7	10·1	10·5	10·9	·42
·40	1·2	1·6	2·0	2·4	2·8	3·2	3·6	4·0	4·4	4·8	5·2	5·6	6·0	6·4	6·8	7·2	7·6	8·0	8·4	8·8	9·2	9·6	10·0	10·4	·40
·38	1·1	1·5	1·9	2·3	2·7	3·0	3·4	3·8	4·2	4·6	4·9	5·3	5·7	6·1	6·5	6·8	7·2	7·6	8·0	8·4	8·7	9·1	9·5	9·9	·38
·36	1·1	1·4	1·8	2·2	2·5	2·9	3·2	3·6	4·0	4·3	4·7	5·0	5·4	5·8	6·1	6·5	6·8	7·2	7·6	7·9	8·3	8·6	9·0	9·4	·36
·34	1·0	1·4	1·7	2·0	2·4	2·7	3·1	3·4	3·7	4·1	4·4	4·8	5·1	5·4	5·8	6·1	6·5	6·8	7·1	7·5	7·8	8·2	8·5	8·8	·34
·32	1·0	1·3	1·6	1·9	2·2	2·6	2·9	3·2	3·5	3·8	4·2	4·5	4·8	5·1	5·4	5·8	6·1	6·4	6·7	7·0	7·4	7·7	8·0	8·3	·32
·30	0·9	1·2	1·5	1·8	2·1	2·4	2·7	3·0	3·3	3·6	3·9	4·2	4·5	4·8	5·1	5·4	5·7	6·0	6·3	6·6	6·9	7·2	7·5	7·8	·30
·28	0·8	1·1	1·4	1·7	2·0	2·2	2·5	2·8	3·1	3·4	3·6	3·9	4·2	4·5	4·8	5·0	5·3	5·6	5·9	6·2	6·4	6·7	7·0	7·3	·28
·26	0·8	1·0	1·3	1·6	1·8	2·1	2·3	2·6	2·9	3·1	3·4	3·6	3·9	4·2	4·4	4·7	4·9	5·2	5·5	5·7	6·0	6·2	6·5	6·8	·26
·24	0·7	1·0	1·2	1·4	1·7	1·9	2·2	2·4	2·6	2·9	3·1	3·4	3·6	3·8	4·1	4·3	4·6	4·8	5·0	5·3	5·5	5·8	6·0	6·2	·24
·22	0·7	0·9	1·1	1·3	1·5	1·8	2·0	2·2	2·4	2·6	2·9	3·1	3·3	3·5	3·7	4·0	4·2	4·4	4·6	4·8	5·1	5·3	5·5	5·7	·22
·20	0·6	0·8	1·0	1·2	1·4	1·6	1·8	2·0	2·2	2·4	2·6	2·8	3·0	3·2	3·4	3·6	3·8	4·0	4·2	4·4	4·6	4·8	5·0	5·2	·20
·18	0·5	0·7	0·9	1·1	1·3	1·4	1·6	1·8	2·0	2·2	2·3	2·5	2·7	2·9	3·1	3·2	3·4	3·6	3·8	4·0	4·1	4·3	4·5	4·7	·18
·16	0·5	0·6	0·8	1·0	1·1	1·3	1·4	1·6	1·8	1·9	2·1	2·2	2·4	2·6	2·7	2·9	3·0	3·2	3·4	3·5	3·7	3·8	4·0	4·2	·16
·14	0·4	0·6	0·7	0·8	1·0	1·1	1·3	1·4	1·5	1·7	1·8	2·0	2·1	2·2	2·4	2·5	2·7	2·8	2·9	3·1	3·2	3·4	3·5	3·6	·14
·12	0·4	0·5	0·6	0·7	0·8	1·0	1·1	1·2	1·3	1·4	1·6	1·7	1·8	1·9	2·0	2·2	2·3	2·4	2·5	2·6	2·8	2·9	3·0	3·1	·12
·10	0·3	0·4	0·5	0·6	0·7	0·8	0·9	1·0	1·1	1·2	1·3	1·4	1·5	1·6	1·7	1·8	1·9	2·0	2·1	2·2	2·3	2·4	2·5	2·6	·10
·08	0·2	0·3	0·4	0·5	0·6	0·6	0·7	0·8	0·9	1·0	1·0	1·1	1·2	1·3	1·4	1·4	1·5	1·6	1·7	1·8	1·8	1·9	2·0	2·1	·08
·06	0·2	0·2	0·3	0·4	0·4	0·5	0·5	0·6	0·7	0·7	0·8	0·8	0·9	1·0	1·0	1·1	1·1	1·2	1·3	1·3	1·4	1·4	1·5	1·6	·06
·04	0·1	0·2	0·2	0·2	0·3	0·3	0·4	0·4	0·4	0·5	0·5	0·6	0·6	0·6	0·7	0·7	0·8	0·8	0·8	0·9	0·9	1·0	1·0	1·0	·04
·02	0·1	0·1	0·1	0·1	0·1	0·2	0·2	0·2	0·2	0·2	0·3	0·3	0·3	0·3	0·3	0·4	0·4	0·4	0·4	0·4	0·5	0·5	0·5	0·5	·02

Bushmills Consol Station Dot Sectors

Count of Dots.	True Bearings from Station.									
0	013·8	066·7	103·6	130·2	156·8	193·8	246·7	283·6	310·2	336·8
1	014·2	066·2	103·4	130·0	156·8	193·3	247·2	283·9	310·4	337·0
2	014·7	065 7	103·2	129·8	156·3	192·8	247·6	284·1	310·7	337·3
3	015·2	065·2	102·9	129·6	156·1	192·3	248·1	284·3	310·9	337·5
4	015·7	064·7	102·7	129·4	155·9	191·9	248·6	284·6	311·1	337·8
5	016·2	064·2	102·4	129·2	155·6	191·4	249·0	284·8	311·3	338·0
6	016·8	063·7	102·2	128·9	155·4	191·0	249·4	285·1	311·5	338·3
7	017·3	063·1	101·9	128·7	155·1	190·6	249·9	285·3	311·7	338·5
8	017·9	062·6	101·7	128·5	154·9	190·1	250·3	285·5	311·9	338·7
9	018·5	062·0	101·5	128·3	154·7	189·7	250·7	285·8	312·1	339·0
10	019·0	061·4	101·2	128·1	154·4	189·3	251·1	286·0	312·4	339·2
11	019·6	060·8	101·0	127·9	154·2	188·9	251·6	286·2	312·6	339·5
12	020·3	060·2	100·7	127·7	154·0	188·5	252·0	286·5	312·8	339·7
13	020·9	059·5	100·5	127·4	153·7	188·1	252·4	286·7	313·0	340·0
14	021·6	058·9	100·2	127·2	153·5	187·7	252·8	286·9	313·2	340·2
15	022·2	058·2	100 0	127·0	153·3	187·3	253·2	287·2	313·4	340·5
16	022·9	057·5	099·8	126·8	153·0	186·9	253·6	287·4	313·6	340·7
17	023·7	056·8	099·5	126·6	152·8	186·5	253·9	287·6	313·9	341·0
18	024·4	056·0	099·2	126·4	152·6	186·1	254·3	297·9	314·1	341·2
19	025·2	055·2	099·0	126·2	152·4	185·7	254·7	288·1	314·3	341·4
20	026·1	054·3	098·7	125·9	152·1	185·4	255·1	288·3	314·5	341·7
21	027·0	053·4	098·5	125·7	151·9	185·0	255·4	288·6	314·7	341·9
22	028·0	052·5	098·2	125·5	151·7	184·6	255·8	288·8	314·9	342·2
23	029·0	051·4	098·0	125·3	151·4	184·3	256·2	289·0	315·1	342·5
24	030·2	050·3	097·7	125·1	151·2	183·9	256·5	289·2	315·4	342·7
25	031·5	048·9	097·5	124·9	151·0	183·5	256·9	289·5	315·6	343·0
26	033·0	047·4	097·2	124·7	150·7	183·2	257·3	289·7	315·8	343·2
27	035·0	045·4	097·0	124·4	150·5	182·8	257·6	289·9	316·0	343·5
28	038·6	041·8	096·7	124·2	150·3	182·5	258·0	290·2	316·2	343·7
29			096·5	124·0	150·1	182·1	258·3	290·4	316·4	344·0
30			096·2	123·8	149·8	181·8	258·7	290·6	316·6	344 2
31			095·9	123·6	149·6	181·4	259·0	290·8	316·9	344·5
32			095·7	123·4	149·4	181·1	259·3	291·1	317·1	344·8
33			095·4	123·2	149·2	180·8	259·7	291·3	317·3	345·0
34			095·2	122·9	148·9	180·4	260·0	291·5	317·5	345·3
35			094·9	122·7	148·7	180·1	260·3	291·7	317·7	345·5
36			094·6	122·5	148·5	179·8	260·7	292·0	317·9	345 8
37			094·4	122·3	148·3	179·4	261·0	292·2	318·2	346·1
38			094·1	122·1	148·0	179·1	261·3	292·4	318·4	346·3
39			093·9	121·9	147·8	178·8	261·7	292·6	318·6	346·6
40			093·6	121·6	147·6	178·5	262·0	292·9	318·8	346·9
41			093·3	121·4	147·4	178·1	262·3	293·1	319·0	347·1
42			093·0	121·2	147·1	177·8	262·6	293·3	319·2	347·4
43			092·8	121·0	146·9	177·5	262·9	293·5	319·5	347·7
44			092·5	120·8	146·7	177·2	263·2	293·7	319·7	347·9
45			092·2	120·6	146·5	176·9	263·6	294·0	319·9	348·2
46			092·0	120·3	146·2	176·6	263·9	294·2	320·1	348·5
47			091·7	120·1	146 0	176·3	264·2	294·4	320·3	348·8
48			091·4	119·9	145·8	176·0	264·5	294·6	320·5	349·0
49			091·1	119·7	145·6	175·6	264·8	294·9	320·7	349·3
50			090·9	119·5	145·4	175·3	265·1	295·1	321·0	349·6
51			090·6	119·3	145·1	175·0	265·4	295·3	321·2	349·9
52			090·3	119·0	144·9	174·7	265·7	295·5	321·4	350·1
53			090·0	118·8	144·7	174·4	266·0	295·7	321·6	350·4
54			089·8	118·6	144·5	174·1	266·3	296·0	321·8	350·7
55			089·5	118·4	144·3	173·9	266·6	296·2	322·1	351·0
56			089·2	118·2	144·0	173·6	266·9	296·4	322·3	351·3
57			088·9	117·9	143·8	173·3	267·2	296·6	322·5	351·5
58			088·6	117·7	143·6	173·0	267·5	296·8	322·7	351·8
59			088·3	117·5	143·4	172·7	267·8	297·1	322·9	352·1
60			088·0	117·3	143·2	172·4	268·0	297·3	322·3	352·4

Table 32.

TRUE AMPLITUDES.

Lat.	DECLINATION. 1°	2°	3°	4°	5°	6°	7°	8°	9°	10°	11°	12°	13°	14°	15°	16°	17°	18°	Lat.
°	°	°	°	°	°	°	°	°	°	°	°	°	°	°	°	°	°	°	°
5	1.0	2.0	3.0	4.0	5.0	6.0	7.0	8.0	9.0	10.1	11.1	12.1	13.1	14.1	15.1	16.1	17.1	18.1	5
10	1.0	2.0	3.0	4.1	5.1	6.1	7.0	8.1	9.1	10.1	11.2	12.2	13.2	14.2	15.2	16.2	17.3	18.3	10
14	1.0	2.1	3.1	4.1	5.2	6.2	7.2	8.3	9.3	10.3	11.3	12.4	13.4	14.4	15.5	16.5	17.5	18.6	14
17	1.1	2.1	3.1	4.2	5.2	6.3	7.3	8.4	9.4	10.5	11.5	12.6	13.6	14.7	15.7	16.8	17.8	18.9	17
20	1.1	2.1	3.2	4.3	5.3	6.4	7.5	8.5	9.6	10.6	11.7	12.8	13.8	14.9	16.0	17.1	18.1	19.2	20
22	1.1	2.2	3.2	4.3	5.4	6.5	7.6	8.6	9.7	10.8	11.9	13.0	14.1	15.1	16.2	17.3	18.4	19.5	22
24	1.1	2.2	3.3	4.4	5.5	6.6	7.7	8.8	9.9	11.0	12.1	13.2	14.3	15.4	16.5	17.6	18.7	19.8	24
26	1.1	2.2	3.4	4.5	5.6	6.7	7.8	8.9	10.0	11.2	12.3	13.4	14.5	15.6	16.8	17.9	19.0	20.1	26
28	1.1	2.3	3.4	4.5	5.7	6.8	7.9	9.1	10.2	11.4	12.5	13.6	14.8	15.9	17.1	18.2	19.3	20.5	28
30	1.2	2.3	3.4	4.6	5.8	6.9	8.1	9.3	10.3	11.6	12.7	13.9	15.0	16.2	17.4	18.6	19.7	20.9	30
31	1.2	2.3	3.5	4.7	5.8	7.0	8.2	9.4	10.5	11.7	12.9	14.0	15.2	16.4	17.6	18.8	20.0	21.1	31
32	1.2	2.4	3.5	4.7	5.9	7.1	8.3	9.5	10.6	11.8	13.0	14.2	15.4	16.6	17.8	19.0	20.2	21.4	32
33	1.2	2.4	3.6	4.8	6.0	7.2	8.4	9.6	10.8	12.0	13.2	14.4	15.6	16.8	18.0	19.2	20.4	21.6	33
34	1.2	2.4	3.6	4.8	6.0	7.2	8.4	9.7	10.9	12.1	13.3	14.5	15.7	17.0	18.2	19.4	20.6	21.9	34
35	1.2	2.4	3.7	4.9	6.1	7.3	8.5	9.8	11.0	12.2	13.5	14.7	15.9	17.2	18.4	19.6	20.9	22.2	35
36	1.2	2.5	3.7	4.9	6.2	7.4	8.7	9.9	11.1	12.4	13.6	14.9	16.1	17.4	18.7	19.9	21.2	22.4	36
37	1.2	2.5	3.7	5.0	6.3	7.5	8.8	10.0	11.3	12.6	13.8	15.1	16.4	17.6	18.9	20.2	21.5	22.8	37
38	1.3	2.5	3.8	5.1	6.3	7.6	8.9	10.2	11.4	12.7	14.0	15.3	16.6	17.9	19.2	20.5	21.8	23.1	38
39	1.3	2.6	3.8	5.2	6.4	7.7	9.0	10.3	11.6	12.9	14.2	15.5	16.8	18.1	19.4	20.8	22.1	23.4	39
40	1.3	2.6	3.9	5.2	6.5	7.8	9.1	10.5	11.8	13.1	14.4	15.7	17.1	18.4	19.7	21.1	22.4	23.8	40
41	1.3	2.6	4.0	5.3	6.6	8.0	9.3	10.6	12.0	13.3	14.6	16.0	17.3	18.7	20.0	21.4	22.8	24.2	41
42	1.3	2.7	4.0	5.4	6.7	8.1	9.4	10.8	12.1	13.5	14.8	16.2	17.6	19.0	20.4	21.8	23.2	24.6	42
43	1.4	2.7	4.1	5.5	6.8	8.2	9.6	11.0	12.3	13.7	15.1	16.5	17.9	19.3	20.7	22.1	23.6	25.0	43
44	1.4	2.8	4.2	5.6	7.0	8.3	9.7	11.1	12.6	14.0	15.4	16.8	18.2	19.6	21.1	22.5	24.0	25.4	44
45	1.4	2.8	4.2	5.7	7.1	8.5	9.9	11.3	12.8	14.2	15.6	17.1	18.5	20.0	21.5	22.9	24.4	25.9	45
46	1.4	2.9	4.3	5.8	7.2	8.6	10.1	11.5	13.0	14.5	15.9	17.4	18.9	20.4	21.9	23.4	24.9	26.4	46
47	1.4	2.9	4.4	5.9	7.3	8.8	10.3	11.8	13.3	14.7	16.2	17.7	19.3	20.8	22.3	23.8	25.4	26.9	47
48	1.5	3.0	4.5	6.0	7.5	9.0	10.5	12.0	13.5	15.0	16.6	18.1	19.6	21.2	22.8	24.3	25.9	27.5	48
½	1.5	3.0	4.5	6.0	7.6	9.1	10.6	12.1	13.7	15.2	16.7	18.3	19.8	21.4	23.0	24.6	26.2	27.8	½
49	1.5	3.0	4.6	6.1	7.6	9.2	10.7	12.2	13.8	15.3	16.9	18.5	20.0	21.6	23.2	24.8	26.5	28.1	49
½	1.5	3.1	4.6	6.2	7.7	9.3	10.8	12.4	13.9	15.5	17.1	18.7	20.3	21.9	23.5	25.1	26.8	28.4	½
50	1.5	3.1	4.7	6.2	7.8	9.3	10.9	12.5	14.1	15.7	17.3	18.9	20.5	22.1	23.7	25.4	27.0	28.7	50
½	1.5	3.1	4.7	6.3	7.9	9.5	11.0	12.6	14.2	15.8	17.5	19.1	20.7	22.4	24.0	25.7	27.4	29.1	½
51	1.6	3.2	4.8	6.4	8.0	9.6	11.2	12.8	14.4	16.0	17.6	19.3	20.9	22.6	24.3	26.0	27.7	29.4	51
½	1.6	3.2	4.8	6.4	8.0	9.7	11.3	12.9	14.6	16.2	17.8	19.5	21.2	22.9	24.6	26.3	28.0	29.8	½
52	1.6	3.3	4.9	6.5	8.1	9.7	11.4	13.1	14.7	16.4	18.0	19.7	21.4	23.1	24.9	26.6	28.3	30.1	52
½	1.6	3.3	4.9	6.6	8.2	9.9	11.6	13.2	14.9	16.6	18.3	20.0	21.7	23.4	25.2	26.9	28.7	30.5	½
53	1.6	3.3	5.0	6.7	8.3	10.0	11.7	13.4	15.1	16.8	18.5	20.2	21.9	23.7	25.5	27.3	29.1	30.9	53
½	1.7	3.4	5.0	6.7	8.4	10.1	11.8	13.5	15.2	17.0	18.7	20.5	22.2	24.0	25.8	27.6	29.4	31.3	½
54	1.7	3.4	5.1	6.8	8.5	10.2	12.0	13.7	15.4	17.2	18.9	20.7	22.5	24.3	26.1	28.0	29.8	31.7	54
½	1.7	3.5	5.2	6.9	8.6	10.4	12.1	13.8	15.6	17.4	19.2	21.0	22.8	24.6	26.5	28.3	30.2	32.2	½
55	1.7	3.5	5.2	7.0	8.7	10.5	12.3	14.0	15.8	17.6	19.4	21.2	23.1	24.9	26.8	28.7	30.6	32.6	55
½	1.8	3.6	5.3	7.1	8.9	10.6	12.4	14.2	16.0	17.9	19.7	21.5	23.4	25.3	27.2	29.1	31.1	33.1	½
56	1.8	3.6	5.4	7.2	9.0	10.7	12.6	14.4	16.2	18.1	19.9	21.8	23.7	25.6	27.6	29.5	31.5	33.5	56
½	1.8	3.7	5.5	7.3	9.1	10.9	12.8	14.6	16.5	18.3	20.2	22.1	24.1	26.0	28.0	30.0	32.0	34.0	½
57	1.8	3.7	5.5	7.4	9.2	11.1	12.9	14.8	16.7	18.6	20.5	22.4	24.4	26.4	28.4	30.4	32.5	34.5	57
½	1.9	3.7	5.6	7.5	9.3	11.2	13.1	15.0	16.9	18.9	20.8	22.8	24.8	26.8	28.8	30.9	33.0	35.1	½
58	1.9	3.8	5.7	7.6	9.5	11.4	13.3	15.2	17.2	19.1	21.1	23.1	25.1	27.2	29.2	31.3	33.5	35.7	58
½	1.9	3.8	5.7	7.7	9.6	11.5	13.5	15.4	17.4	19.4	21.4	23.5	25.5	27.6	29.7	31.8	34.0	36.3	½
59	1.9	3.8	5.8	7.8	9.7	11.7	13.7	15.7	17.7	19.7	21.7	23.8	25.9	28.0	30.2	32.3	34.6	36.8	59
½	2.0	3.9	5.9	7.9	9.9	11.9	13.9	15.9	18.0	20.0	22.1	24.2	26.3	28.5	30.7	32.9	35.2	37.5	½
60	2.0	4.0	6.0	8.0	10.0	12.1	14.1	16.2	18.2	20.3	22.4	24.6	26.7	28.9	31.2	33.4	35.8	38.2	60
Lat.	1°	2°	3°	4°	5°	6°	7°	8°	9°	10°	11°	12°	13°	14°	15°	16°	17°	18°	Lat.

Note—Owing to refraction the moment of TRUE Amplitude occurs when the sun's lower limb appears to be about half a diameter above the horizon. In moderate latitudes it is sufficient to judge this moment and observe accordingly. In the case of the higher latitudes, however, observations should either be timed, or the APPARENT Amplitude observed and collected by means of the supplementary table.

Table 32.

TRUE AMPLITUDES.

Lat.	DECLINATION.																		Lat.
	19°	½	20°	½	21°	½	22°	½	23°	½	24°	½	25°	½	26°	½	27°	½	
°	°	°	°	°	°	°	°	°	°	°	°	°	°	°	°	°	°	°	°
5	19.1	19.6	20.1	20.6	21.1	21.6	22.1	22.6	23.1	23.6	24.1	24.6	25.1	25.6	26.1	26.6	27.1	27.6	**5**
10	19.3	19.9	20.3	20.8	21.3	21.8	22.3	22.9	23.4	23.9	24.4	24.9	25.4	25.9	26.4	27.0	27.5	28.0	**10**
14	19.6	20.1	20.6	21.2	21.7	22.2	22.7	23.2	23.8	24.3	24.8	25.3	25.8	26.3	26.9	27.4	27.9	28.4	**14**
17	19.9	20.4	21.0	21.5	22.0	22.5	23.1	23.6	24.1	24.6	25.2	25.7	26.2	26.8	27.3	27.8	28.4	28.9	**17**
20	20.3	20.8	21.3	21.9	22.4	22.9	23.5	24.0	24.6	25.1	25.7	26.2	26.7	27.3	27.8	28.3	28.9	29.4	**20**
22	20.6	21.1	21.7	22.2	22.7	23.3	23.8	24.4	24.9	25.5	26.0	26.6	27.1	27.7	28.2	28.8	29.3	29.9	**22**
24	20.9	21.4	22.0	22.5	23.1	23.7	24.2	24.8	25.3	25.9	26.4	27.0	27.6	28.1	28.7	29.2	29.8	30.4	**24**
26	21.2	21.8	22.4	22.9	23.5	24.1	24.6	25.2	25.8	26.3	26.9	27.5	28.0	28.6	29.2	29.8	30.3	30.9	**26**
28	21.6	22.2	22.8	23.4	24.0	24.5	25.1	25.7	26.3	26.8	27.4	28.0	28.6	29.2	29.8	30.4	30.9	31.5	**28**
30	22.1	22.7	23.3	23.8	24.4	25.0	25.6	26.2	26.8	27.4	28.0	28.6	29.2	29.8	30.4	31.0	31.6	32.2	**30**
31	22.3	22.9	23.5	24.1	24.7	25.3	25.9	26.5	27.1	27.7	28.3	28.9	29.5	30.1	30.8	31.4	32.0	32.6	**31**
32	22.6	23.2	23.8	24.4	25.0	25.6	26.2	26.8	27.4	28.0	28.7	29.3	29.9	30.5	31.1	31.7	32.4	33.0	**32**
33	22.9	23.5	24.1	24.7	25.3	25.9	26.5	27.1	27.8	28.4	29.0	29.6	30.3	30.9	31.5	32.1	32.8	33.4	**33**
34	23.1	23.7	24.4	25.0	25.6	26.2	26.8	27.5	28.1	28.7	29.4	30.0	30.7	31.3	31.9	32.5	33.2	33.8	**34**
35	23.4	24.0	24.7	25.3	25.9	26.6	27.2	27.8	28.5	29.1	29.8	30.4	31.1	31.7	32.4	33.0	33.7	34.3	**35**
36	23.7	24.4	25.0	25.6	26.3	26.9	27.6	28.2	28.9	29.5	30.2	30.8	31.5	32.1	32.8	33.5	34.1	34.8	**36**
37	24.0	24.7	25.3	26.0	26.7	27.3	28.0	28.6	29.3	29.9	30.6	31.3	31.9	32.6	33.3	34.0	34.6	35.3	**37**
38	24.4	25.1	25.7	26.4	27.1	27.7	28.4	29.0	29.7	30.3	31.1	31.8	32.4	33.1	33.8	34.5	35.2	35.9	**38**
39	24.8	25.4	26.1	26.8	27.5	28.1	28.8	29.5	30.2	30.8	31.6	32.3	32.9	33.6	34.3	35.0	35.8	36.5	**39**
40	25.1	25.8	26.5	27.2	27.9	28.6	29.3	30.0	30.7	31.3	32.1	32.8	33.5	34.2	34.9	35.6	36.4	37.1	**40**
41	25.5	26.2	26.9	27.6	28.3	29.0	29.8	30.5	31.2	31.8	32.6	33.3	34.1	34.8	35.5	36.2	37.0	37.7	**41**
42	26.0	26.7	27.4	28.1	28.8	29.5	30.3	31.0	31.7	32.4	33.2	33.9	34.7	35.4	36.2	36.9	37.7	38.4	**42**
43	26.4	27.1	27.8	28.6	29.3	30.1	30.8	31.5	32.3	33.0	33.8	34.6	35.3	36.1	36.8	37.6	38.4	39.1	**43**
44	26.9	27.6	28.4	29.1	29.9	30.6	31.4	32.1	32.9	33.6	34.4	35.2	36.0	36.8	37.5	38.3	39.2	39.9	**44**
45	27.4	28.2	28.9	29.7	30.4	31.2	32.0	32.8	33.5	34.3	35.1	35.9	36.7	37.5	38.3	39.1	39.9	40.8	**45**
46	27.9	28.7	29.5	30.3	31.0	31.8	32.6	33.4	34.2	35.0	35.8	36.7	37.5	38.3	39.1	40.0	40.8	41.7	**46**
47	28.5	29.3	30.1	30.9	31.7	32.5	33.3	34.1	34.9	35.7	36.6	37.5	38.3	39.1	40.0	40.9	41.7	42.6	**47**
48	29.1	29.9	30.7	31.6	32.4	33.2	34.0	34.8	35.7	36.5	37.4	38.3	39.2	40.0	40.9	41.8	42.7	43.6	**48**
½	29.4	30.2	31.1	31.9	32.8	33.6	34.4	35.3	36.1	37.0	37.8	38.7	39.6	40.5	41.4	42.3	43.2	44.2	**½**
49	29.7	30.6	31.4	32.3	33.1	33.9	34.8	35.7	36.5	37.4	38.3	39.2	40.1	41.0	41.9	42.8	43.8	44.7	**49**
½	30.1	30.9	31.8	32.6	33.5	34.4	35.2	36.1	37.0	37.9	38.8	39.7	40.6	41.5	42.4	43.4	44.3	45.3	**½**
50	30.4	31.3	32.1	33.0	33.9	34.8	35.6	36.5	37.4	38.3	39.3	40.2	41.1	42.0	43.0	43.9	44.9	45.9	**50**
½	30.8	31.7	32.5	33.4	34.3	35.2	36.1	37.0	37.9	38.8	39.8	40.7	41.6	42.6	43.6	44.5	45.5	46.5	**½**
51	31.1	32.0	32.9	33.8	34.7	35.6	36.5	37.5	38.4	39.3	40.3	41.2	42.2	43.2	44.2	45.2	46.2	47.2	**51**
½	31.5	32.4	33.3	34.2	35.1	36.1	37.0	37.9	38.9	39.8	40.8	41.8	42.8	43.8	44.8	45.8	46.8	47.9	**½**
52	31.9	32.8	33.7	34.7	35.6	36.5	37.5	38.4	39.4	40.3	41.4	42.4	43.4	44.4	45.4	46.4	47.5	48.6	**52**
½	32.3	33.3	34.2	35.1	36.1	37.0	38.0	38.9	39.9	40.9	41.9	42.9	44.0	45.0	46.0	47.1	48.2	49.3	**½**
53	32.7	33.7	34.6	35.6	36.5	37.5	38.5	39.5	40.5	41.4	42.5	43.6	44.6	45.7	46.7	47.8	49.0	50.1	**53**
½	33.2	34.1	35.1	36.1	37.1	38.0	39.0	40.0	41.1	42.0	43.1	44.2	45.3	46.4	47.5	48.6	49.8	50.9	**½**
54	33.6	34.6	35.6	36.6	37.6	38.6	39.6	40.6	41.7	42.6	43.8	44.9	46.0	47.1	48.2	49.4	50.6	51.8	**54**
½	34.1	35.1	36.1	37.1	38.1	39.1	40.2	41.2	42.3	43.3	44.5	45.7	46.7	47.9	49.0	50.2	51.4	52.7	**½**
55	34.6	35.6	36.6	37.6	38.7	39.7	40.8	41.8	42.9	44.0	45.2	46.3	47.5	48.7	49.8	51.1	52.3	53.6	**55**
½	35.1	36.1	37.1	38.2	39.2	40.3	41.4	42.5	43.6	44.7	45.9	47.1	48.3	49.5	50.7	52.0	53.3	54.6	**½**
56	35.6	36.6	37.7	38.8	39.8	40.9	42.1	43.2	44.3	45.4	46.7	47.9	49.1	50.4	51.6	53.0	54.3	55.7	**56**
½	36.1	37.2	38.3	39.4	40.5	41.6	42.7	43.9	45.0	46.2	47.5	48.7	50.0	51.3	52.6	54.0	55.4	56.8	**½**
57	36.7	37.8	38.9	40.0	41.1	42.3	43.4	44.6	45.8	47.0	48.3	49.6	50.9	52.2	53.6	55.0	56.5	58.0	**57**
½	37.3	38.4	39.5	40.7	41.8	43.0	44.2	45.4	46.6	47.9	49.2	50.5	51.9	53.2	54.7	56.1	57.7	59.3	**½**
58	37.9	39.0	40.2	41.4	42.5	43.8	45.0	46.2	47.5	48.7	50.1	51.5	52.9	54.3	55.8	57.3	58.9	60.7	**58**
½	38.5	39.7	40.9	42.1	43.3	44.6	45.8	47.1	48.4	49.7	51.1	52.5	54.0	55.5	57.0	58.6	60.3	62.1	**½**
59	39.2	40.4	41.6	42.8	44.1	45.4	46.7	48.0	49.3	50.6	52.2	53.6	55.1	56.7	58.3	60.0	61.8	63.7	**59**
½	39.9	41.1	42.4	43.6	44.9	46.2	47.6	48.9	50.3	51.7	53.3	54.8	56.4	58.0	59.7	61.5	63.4	65.5	**½**
60	40.6	41.9	43.2	44.5	45.8	47.1	48.5	49.9	51.4	52.9	54.4	56.0	57.7	59.4	61.2	63.2	65.2	67.4	**60**
Lat.	19°	½	20°	½	21°	½	22°	½	23°	½	24°	½	25°	½	26°	½	27°	½	**Lat.**

Table 32A.

Corr. for APPARENT AMPLITUDE.

ADD to AZIMUTH reckoned from elevated Pole when sun observed with centre on horizon.

Dec.	LAT.					
	0°	20°	40°	50°	55°	60°
°	°	°	°	°	°	°
0	.0	.2	.6	.8	1.0	1.2
15	.0	.2	.6	.9	1.1	1.4
24	.0	.3	.7	1.1	1.4	2.0

Addresses and telephone numbers of
H M Coastguard Rescue Headquarters Stations:

Lerwick, Shetland (76)

Kirkwall, Orkney (3268)

Wick, Caithness (2332)

Stornoway, Isle of Lewis (2013)

Peterhead, Aberdeenshire (3563)

Aberdeen (23312)

Fifeness, Fife (Crail 236)

Seahouses, Northumberland (274)

Tynemouth, Northumberland (North Shields 72691)

Whitby, Yorkshire (2107)

Flamborough Head, Yorkshire (Flamborough 203)

Cromer, Norfolk (2507)

Gorleston, Norfolk (Great Yarmouth 63444)

Thames, Essex (Frinton-on-Sea 5518)

Dover Strait (St. Margaret's Bay 2515)

Shoreham-by-Sea, Sussex (2226)

Needles, Isle of Wight (Freshwater 2265)

Wyke Regis, Dorset (Weymouth 4105)

Brixham, Devon (2156)

Hartland, Devon (235)

Falmouth, Cornwall (314481)

St. Just, Cornwall (657)

Mumbles, Glamorgan (Swansea 66534)

St. Ann's, Pembrokeshire (Dale 218)

Holyhead, Anglesey (2051)

Formby, Liverpool, Lancashire (Formby 72903)

Portpatrick, Wigtownshire (209)

Southend, Campbeltown, Argyll (Southend 232)

Bangor, Co Down, N Ireland (Groomsport 284)

Ramsey, Isle of Man (3255)

INDEX

INDEX

A

B

C

D

Page

E

F

G

H

I